DRUG–DEVICE COMBINATIONS FOR CHRONIC DISEASES

WILEY - SOCIETY FOR BIOMATERIALS SERIES

Bio-Inspired Materials for Biomedical Engineering • Anthony B. Brennan and Chelsea M. Kirschner
Drug–Device Combinations for Chronic Diseases • SuPing Lyu and Ronald Siegel

DRUG–DEVICE COMBINATIONS FOR CHRONIC DISEASES

Edited by

SUPING LYU, Ph.D.
RONALD A. SIEGEL, SC.D.

Wiley-Society for Biomaterials

Published by John Wiley & Sons, Inc., Hoboken, New Jersey
Published simultaneously in Canada

Library of Congress Cataloging-in-Publication Data

Drug-device combinations for chronic diseases / edited by SuPing Lyu, Ronald Siegel.
p. ; cm.
Includes bibliographical references and index.
ISBN 978-1-118-12000-2 (cloth)
I. Lyu, SuPing, editor. II. Siegel, Ronald Alan, editor.
[DNLM: 1. Chronic Disease–drug therapy. 2. Drug Delivery Systems–instrumentation. WT 500]
RS199.5
615'.6–dc23 2015024258

ISBN: 9781118120002

Printed in the United States of America

10 9 8 7 6 5 4 3 2 1

CONTENTS

FOREWORD

Realization of the promise of drug–device combinations has been long in coming. Following the rapid expansion of pharmaceutical and biological research and development in the twentieth century leading to a host of new diagnostics and therapies, methods of their administration improved in tandem. Meanwhile, the effectiveness of drugs and largely mechanically based medical devices improved as the underlying mechanisms of disease were uncovered, increasingly allowing researchers to address the root cause of illness. However, the classes of drugs, devices, and biologics remained largely separate from one another until the 1970s from a commercial and regulatory perspective.

While the benefits of local administration of energy from medical devices in increasingly complex medical devices such as pacemakers, implantable defibrillators, radiation therapy, ablation devices, and diagnostics were recognized earlier, administration of drugs and biologics locally to minimize side effects resulting from systemic exposure took longer to be translated. The convergence of advances in the fields of polymer science, biochemistry, analytical chemistry, and controlled release of small and large molecules from matrices has more recently allowed for early applications of drug–device combinations to emerge as significant advances for patient care and commercial successes. Prominent examples include drug-eluting pacemaker leads, drug infusion pumps, and antirestenotic drug-eluting stents that deliver significant advances in efficacy and ease-of-use for those therapies.

While these and other advances have led to therapies that address acute needs that used to be largely fatal for patients, it has also resulted in an increasing number of patients living in the aftermath with the effects of chronic disease. The resulting shift in obvious clinical need toward chronic disease represents an opportunity for researchers in academia and industry to collaborate in new ways to address these problems. Recent advances in regenerative medicine, including the use of stem cells, improvements in understanding of the underlying mechanism of cell–matrix interactions, genomics, and minimally invasive delivery technologies, promise to disrupt our current treatment modalities. In addition, as we uncover the mechanism and role of the nervous system in modulating the body's response to a diverse range of disease states, the local delivery of energy and drugs by devices via neuromodulation promises to be one of the most exciting tools to address chronic disease.

We believe that we have only scratched the surface of applications in which local applications of pharmaceuticals or biologics, in combination with mechanical action or energy input, can result in significant improvement in patient outcomes. The contributions in this volume represent a cross section of the academic researchers who have devoted their lives to understanding and translating the enabling technologies, as well as the industrial leaders who have commercialized them and made them available to patients globally. Our hope is that the energy and discourse ignited by the discussion in this book, which accounts for some of the successes and challenges presented in research and development of drug–device combinations, will lead to wider understanding of the very real challenges, scientific or otherwise, for advancing this concept to new, unimagined applications.

DAVID KNAPP
Vice President, Corporate Research

TIM GIRTON
Vice President, Interventional Cardiology R&D
Boston Scientific Corporation

PREFACE

Ask not only what devices can do for drugs or what drugs can do for devices; ask also what new things that drugs and devices can do together.

Drugs treat diseases through chemical reactions. Devices treat diseases through physical actions. The differing technical challenges in developing molecules and macroscale devices are obvious. As such, drugs and medical devices have traditionally been developed and used separately. Pharmaceutical companies and device manufacturers operate in different markets and with different goals and business models. Whereas drug companies tend to focus on blockbuster products that are administered repetitively to patients, many medical devices, especially implants, involve long-term use of a single unit. Pharmaceutical companies can expect a relatively long period of profit from a successful drug, as there are few ways to "upgrade" therapy based on a particular molecule, and it is difficult to predict the success of related molecules. Generally speaking, all drugs must go through the same discovery, development, and approval process. Devices, on the other hand, either must undergo a full cycle of research, development, and approval if they are the first of their kind on the market, or they represent incremental changes to predicate products. It is a constant effort for devices to remain at the cutting edge, and many products can only maintain their market share for relatively short periods.

The regulatory paths for the two kinds of therapies have traditionally been separate, with differing sets of hurdles to overcome. Whereas any new drug must be considered on its own, and undergo an exacting and expensive set of phases of study, the path for approval of a device is somewhat less arduous if it is shown to be related to similar products that are already on the market. Original devices, especially those that could have impacts on patients' safety, however, require thorough studies to be performed to win premarket approval. Devices often have numerous components, all of which are subject to extensive studies in order to control and minimize possible mechanical, biocompatibility, electrical, or chemical failure modes following implantation.

Recently, there has been a trend toward drug–device convergence. A number of drug–device combination products have been developed to enable or enhance each other's functions and achieve improved or even new therapies. During 2008–2012, over 1000 new applications of combination products were submitted to U.S. FDA for review, all having drug or biological delivery components. Currently, these types of products represent a market of tens of billions of dollars. However, the definition of drug–device combination products has not been clear. Typically, it refers to products containing both drug and device components that act in concert to achieve functions that otherwise are difficult or impossible to achieve by either component alone. Such synergy is needed to justify the effort in producing combination products.

Drug–device combination products are recent innovation, but drug delivery products can be tracked back to tablet, capsules, and syringes that have long been used and may be considered as early "devices." While these products are still dominant as means for administering drugs, their utility, if not their design, is rather straightforward. Advanced drug delivery for improved efficacy, low toxicity, and convenient uses started in the 1960s. With advances in bioanalytical chemistry and the mathematical and physiological understanding of pharmacokinetics and pharmacodynamics, and the recognition of localized receptors as sites for drug action, it became clear that targeted delivery of drugs could improve therapy and reduce unwanted side effects. Of particular interest was control of the rate and locale of drug release. Rate control could smooth the concentration profile of drug in the blood over time, maintaining drug concentration within its "therapeutic window," wherein the drug is efficacious and nontoxic. On the other hand, release rate could be modulated by need, as in the case of insulin, which should be delivered in concert with intake of carbohydrate. By controlling the location of delivery, the drug could be focused at the site of action and hopefully avoid issues associated with toxicity. Moreover, direct delivery could lessen drug degradation that occurs as it passes through the harsh environment of the gastrointestinal tract and the liver (first pass metabolism).

Based on these considerations, devices designed specifically for drug delivery were developed. Such devices include implantable and externally worn drug pumps, transdermal patches, implantable drug-loaded tubes or rods, injectable drug depots, implantable and resorbable drug-loaded polymer wafers, drug-eluting eye inserts and intrauterine devices, devices for intranasal and inhalation delivery of liquids and dry powders, and a diverse collection of pen injectors, microneedle arrays, and buccal patches. Besides these innovative methods of delivery, there has been a steady improvement in traditional drug delivery devices. For example, syringe needles are now so sharp that they are much less painful, and extended release tablets, capsules, and other oral drug delivery "devices" such as osmotic pumps have stabilized and improved the therapeutic value of drugs. IV catheters can be directed to very specific sites, such as the loci of embolisms, where local administration of streptokinase or tissue plasminogen activator can dissolve the clots.

A more recent development has been the utilization of drugs to improve the function of implanted devices. The "trivial" way to do this is to administer drugs systemically, including antibiotics, blood thinners, and anti-inflammatories, following implantation. However, systemic administration leads to systemic effects, which are often undesirable. By localizing the drug delivery to the site of implantation, these systemic effects can often be reduced or eliminated. By incorporating the drug as a component of the device, not only can such localized delivery be achieved, but also delivery can be controlled in concert with the device's action to achieve synergistic therapeutic outcomes. Steroid-releasing cardiac pacing leads, heparin-coated vascular grafts, drug-eluting stents, antimicrobial pouches, and so on are a few successful examples.

The aim of this book is to summarize general principles surrounding synergistic combination of drugs and devices, to improve the performance of either the drug or the device. Emphasis is placed on the recognition of unmet needs that motivate the development of combination systems, the research and development required to introduce specific products, including recognition of special issues that arise when combining drugs and devices, and in certain cases the special regulatory hurdles that need to be overcome.

An overview of the general issues surrounding the development of drug–device combinations is provided by Avula and Grainger in Chapter 1. This chapter also summarizes progress in particular classes of devices, "case studies" of which are presented in later chapters. In Chapter 2, Peppas et al. provide a historical review of drug delivery devices, with emphasis on general principles and applications. This chapter shows the remarkable progress that has been made in the past 50 years, and demonstrates the ingenuity involved in combining physics, chemistry, engineering, and understanding of anatomy and physiology to create a vast variety of devices for drug delivery. The field has seen a rapid evolution from the relatively crude devices of the 1960s to present systems whose manufacture requires advanced techniques. Many of the latter devices are described in Chapter 3 by Stevenson and Langer.

Chapter 4 by Lyu and Siegel discusses practical aspects of developing and manufacturing drug–device combination products. The chapter starts with a discussion of tests required for combination products that go beyond those needed for simple devices and pharmaceutical products, due to possible interactions between the drug and the device. Selection of materials for combination products is then considered, first in general, and then specifically for drug delivery coatings and catheters. Several physical and chemical interactions between the drug and the device, which play a major role in a products' performance, are then identified. Finally, commonly used technologies for manufacturing combination products are reviewed, including dip coating, spray coating, impregnation, extrusion, molding, powder molding, and reservoir filling.

Chapter 5 by McVenes and Stokes reviews steroid-releasing cardiac pacing leads, which lower the pacing threshold, increasing the safety margin and the battery life of the pacemaker devices. The first such product received FDA approval in 1983. A historical trail describing how engineers and scientists learned that inflammatory reactions result in pacing threshold increase, and how they solved the problem, is presented, including a description of a large animal study to screen drugs for reducing pacing threshold. The chapter also presents the results of studies of steroid release over 7 years. These studies are important since the long-term release of steroid is necessary for certain patient populations.

Chapter 6 by Begovac et al. describes the development of the PROPATEN® Vascular Graft, which is composed of an expanded PTFE vascular graft functionalized with a heparin surface coating. The heparin coating improves blood compatibility of the ePTFE graft, particularly thromboresistance. The chapter starts with a discussion of the recognized needs for thromboresistive vascular grafts, and then presents the steps taken in developing PROPATEN®. The mechanism of action of heparin is discussed, along with the strategy of chemically grafting heparin to a layer-by-layer composite structure formed by cationic and anionic polymers. Product development is discussed in detail, including design requirements, prototyping, manufacturing, quality control, packaging, sterilization, and regulatory standards and pathways. The results of clinical studies are then presented. Challenges as well as potential side effects such as thrombocytopenia (HIT) are identified at the end of the chapter.

Chapter 7 by Hildebrand focuses on pump-based infusion systems and therapies. These systems can be transcutaneous, such as the most currently available insulin delivery pumps, or implantable, such as those used to deliver neurally active drugs such as baclofen and morphine sulfate to the intrathecal space. Pump-based systems can infuse therapeutic agents over days to years. The chapter reviews the basic components, including a pump, a catheter, the therapeutic agent, and accessories such as a digital pump controller, and analyzes the clinical uses of pump-based therapies. Bypass of GI tract and programmable delivery of continuous and bolus doses are shown to provide great advantages over conventional drug administration methods. Potential interactions among pump components, the drugs, and the patient, which impinge on product safety, are also discussed.

Chapter 8 by Chen and Roberson reviews the research and development that was undertaken for the PROMUS Element Plus® drug-eluting stent. The chapter describes the stent, the drug coating, and the delivery technology. Why certain things worked and others did not is discussed in detail from the standpoint of mechanical, medical, and deployment objectives. Of particular interest is the mathematical modeling that was performed to understand drug release kinetics *in vitro* and *in vivo*, distribution into proximal and distal tissues, and pharmacokinetics in the systemic circulation. Pharmacodynamics is also described with emphasis on stent coverage by neointima as a function of time after deployment of the stents. The chapter concludes by describing the results of clinical studies. Safety and efficacy of the PROMUS Element were demonstrated in terms of the 12-month target lesion failure rate.

Chapter 9 by Peckham et al. describes the development of the INFUSE® Bone Graft product, which consists of a pack of recombinant human bone morphogenic protein 2 (hrBMP2) and a sheet of absorbable collagen sponge (ACS) as the drug carrier. A solution of hrBMP2 is reconstituted and loaded into the ACS onsite in the operating room, and the combination is implanted. The chapter reviews the historical and biological background, manufacturing, pharmacological tests, preclinical tests, postclinical studies, and regulatory path. The authors emphasize on how key questions were identified through extensive conversation between the manufacturer and the regulatory agents during the precombination product era, when neither the manufacturer nor the regulatory agent had experience. Insightful discussions are presented to describe how studies were designed and performed to address the scientific and clinical questions.

The research and development pathways for drugs and medical devices are "mature" in the sense that those who wish to develop new products have a clear path. A broad range of manufacturing technologies are available, and the regulatory pathways are well laid out. Drug–device combination, however, is a relatively new area considering almost the entire path from research, development, clinical study, and regulatory approval through postmarket surveillance. The complexity involved is both additive and multiplicative. The establishment of the Office of Combination Products at FDA allows regulators to consider such complexities.

There are several basic questions that presently have no general answers or even methods for study. For example, what is the best information that can be gathered regarding the effect and toxicity of a drug that is released locally? How does one characterize or at least predict local and systemic drug disposition in humans in a way that can be predictive of success or failure? How does the response of tissues to the device affect the drug's pharmacokinetics and pharmacodynamics? Does the drug affect tissue behavior such that its response to the device changes? These questions are specific and fundamental to drug–device combination products. Efforts to addressing these questions and those related to product development path should represent a significant part of academic and industrial efforts to drive maturation of the technologies in the years to come.

The advent of advanced drug–device combination products is relatively recent. As often occurs with new technologies in the biomedical arena, initial fervor is supplanted by a latent period in which unanswered (and sometimes unanticipated) questions need to be explored. Recent examples of this phenomenon include genetically engineered protein drugs, gene therapies, and stem cell-based therapies. Compared to the number of devices and drugs that are on the market, there are relatively few drug–device combinations. However, where drugs and devices can be combined so that one component enables or enhances the function of the other, we expect that there will be continued motivation to advance the science and technological development to further develop such combination products.

SuPing Lyu

Medtronic plc., Minneapolis, MN, USA

Ronald A. Siegel

University of Minnesota, Minneapolis, MN, USA

PART I

BACKGROUND AND CONTEXT

1

ADDRESSING MEDICAL DEVICE CHALLENGES WITH DRUG–DEVICE COMBINATIONS

MAHENDER N. AVULA[1] AND DAVID W. GRAINGER[1,2]

[1]*Department of Bioengineering, University of Utah, Salt Lake City, UT 84112, USA*
[2]*Department of Pharmaceutics and Pharmaceutical Chemistry, University of Utah, Salt Lake City, UT 84112-5820, USA*

1.1 INTRODUCTION

Implanted medical devices (IMDs) comprising synthetic biomaterials have seen exponential growth in their applications and clinical use over the past five decades [1]. The scope and fields of use for IMDs have increased multifold with the advent of new technologies, innovation, and improved understanding of human physiology and its underlying problems. Increasing rates of medical device adoption can be attributed to various factors, including aging median populations worldwide [2], innovations in design and function that increase performance and reliability, rising standards of living among patients in developing nations, and noted improvements in patient quality of life offered by the devices. New IMDs continue to offer improved treatment alternatives for cardiovascular, orthopedic, oncologic, and many other diseases [3]. Given these factors, the global medical device market is expected to continue growing, reaching approximately US$302 billion in 2017 with an annual growth rate of ~6% over the next 6 years (2011–2017) [4]. Tens of millions of people in the United States alone have some kind of IMD in their body. Despite enhanced safety and efficacy, new device design strategies are required to understand and address complex human factors affecting device performance *in vivo*. Innovations in design,

Drug–Device Combinations for Chronic Diseases, First Edition. Edited by SuPing Lyu and Ronald A. Siegel.

biomaterials, surface modifications and biocompatible coatings, and device-based onboard drug delivery mechanisms are among strategies employed to improve clinical IMD performance.

1.1.1 Combination Medical Devices

Drug–device combination medical products are innovative biomedical implants with enhancements to device function provided by the onboard formulation and local pharmacology of selected drugs at the implant site [5]. Combination devices couple a drug loading and releasing mechanism onto an approved prosthetic implant. Together, these seek to provide several improvements to the *in vivo* performance and lifetime of implantable medical devices in various classes and capacities, including cardiovascular, ophthalmic, orthopedic, diabetes, and cancer applications. Drug–device combination products represent relatively new device class among implantable medical devices, one that is drawing increasing attention from both the pharmaceutical and device manufacturing industries and the clinicians to address several long-standing problems associated with IMDs. In 2003, the Food and Drug Administration (FDA) approved a coronary drug-eluting stent (DES) (Cordis CYPHER™, Johnson and Johnson, USA) opening the market to similar officially designated "drug–device combination products" in the United States [6]. Several notable medical devices with locally delivered drugs had earlier precedent, namely, steroid-releasing pacemaker leads, hormone-releasing intrauterine devices, antibiotic-impregnated catheters, aerosolized drug inhalers, drug-infused condoms, and several other precedents. Additionally, several combination products also existed earlier in Europe than elsewhere, for example, antibiotic-releasing bone cements, drug-eluting stents, heparin-coated catheters, and others (approved with the CE mark). FDA's Office of Combination Products (OCP) was established in 2002 to provide a pathway for assigning principal FDA oversite and review policies to drug–biologic–device combinations that could otherwise be confused or compromised by traditional FDA review file assignments [7]. The objective was to provide a streamlined and consistent process for assigning these new products to FDA Centers based on claimed primary modes of action (i.e., device or drug). The OCP defines a "combination device" under 21 CFR 3.2(e) as "A product comprised of two or more regulated components, i.e., drug/device, biologic/device, drug/biologic, or drug/device/biologic, that are physically, chemically, or otherwise combined or mixed and produced as a single entity; or two or more separate products packaged together in a single package or as a unit and comprised of drug and device products, device and biological products, or biological and drug products." Table 1.1 summarizes this classification system. Most combination devices add a drug bioactivity adjunct to an already-approved implanted device to counteract challenges faced by the device in the context of the local host tissue environment. This can include inflammation, fibrosis, coagulation, and infection, improving performance in several conditions. One prominent example is the use of the drug-eluting stent, where local release of micrograms of drug to the vascular bed has reduced the need for surgical intervention by 40–70% over bare metal stents [8–10]. However, combination products are often optimized into an integrated

TABLE 1.1 Diversity of Combination Medical Products Used in Physical or Chemical Combinations, or Copackaged as a Kit, or as Separate Cross-Labeled Products

Combination Product Type	Clinical Examples
Drug and device	Drug-eluting stents, antimicrobial catheters, tibial nail, and sutures
Drug and biologic	Autologous platelet concentrate delivery of gentamycin to an open fracture; demineralized bone matrix delivery of statins to bone defect
Biologic and device	Heparin-coated vascular grafts, insulin infusion pumps, spinal cages with rhBMP-2
Drug and biologic and device	No precedents approved; fictional example: adenoviral NfκB transgene delivery from Taxol-eluting vascular stent

system from separate drug and device products: They were never designed *de novo* to complement each other in structure and function, that is, controlled drug delivery is often an add-on feature to an existing FDA-approved medical device design that is suboptimally adapted to the structural, mechanical, or electronic function of the device [6]. New strategies and new technologies that combine drugs, devices, and biologics *de novo* as coordinated, unified new designs are expected to provide a new generation of combination products, more intelligently incorporating and merging new technologies, changes, and refinements of both existing drug delivery mechanisms and medical device functions, shifts from traditional devices and drugs, while remaining compliant with regulations [6].

Diverse classes of drugs are used in combination devices to enhance medical device and implant performance. Anti-inflammatory, antifibrotic, antiproliferative, antithrombotic, and antibiotic drugs are primary classes of pharmaceutical agents often combined with a controlled delivery mechanism suited to the application. Site- and implant-specific drug interventions before, during, and after medical device implantation can be used to alleviate several adverse host responses, providing a local therapeutic strategy when a device design or systemic drug delivery alone is insufficient. For example, anticoagulants are applied to cardiovascular and intravascular implants to reduce device-based thrombosis, while antifibrotic, anti-inflammatory, and antiproliferative drugs are used for soft tissue implants and endovascular stents susceptible to fibrous tissue in-growth and smooth muscle proliferation. Antibiotics are released from orthopedic implants, shunts, and percutaneous and urinary catheters that exhibit high infection incidence.

Conventional therapeutics are administered in different ways, including nasal, oral, parenteral (intravascular, intramuscular, subcutaneous, and intraperitoneal), topical, transdermal, and other administrative routes [11]. Although systemic administration has its merits, local drug administration can in some cases provide comparable results with significantly lower doses of drugs while limiting the drug efficacy and toxicity to the tissue surrounding the implant site. Drugs are combined with delivery technologies to control rates and local dosing of therapeutics to tissue beds surrounding implanted devices. Typically, drugs are released systematically from the device

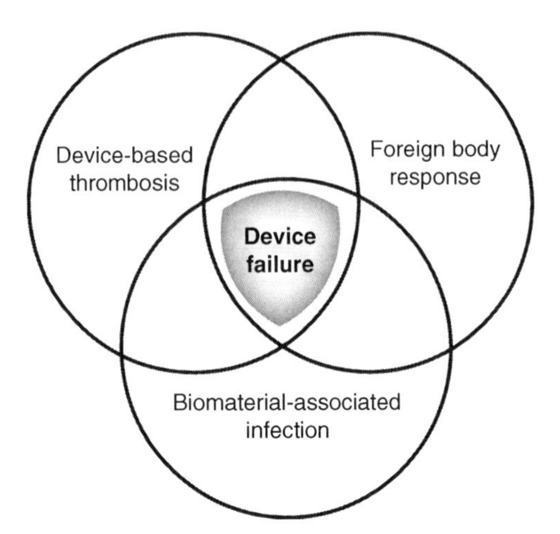

FIGURE 1.1 General host-interfacing challenges facing implanted medical devices.

surface using impregnated resins or rate-controlling polymer films. Occasionally, drugs are eluted from the bulk device as in the case of antibiotic-loaded bone cement. Local drug release limits drug dosing to low quantities, reduces systemic toxicity, increases duration of release, and limits the area of release to the tissue bed surrounding the implant [6]. Local drug release mechanisms offer several advantages over conventional systemic drug administration. An ideal drug delivery system with a combination device should provide continuous and effective drug doses to the site of implantation while also offering possibilities to continue drug release for prolonged periods [12]. Rates and durations of drug delivery depend on several factors such as the implant size, local tissue physiology and morbidity, drug pharmacology and potency in therapy, duration and location of drug release, its kinetics, drug and local clearance, and toxicity.

Due to the widespread development and use of combination products, a comprehensive understanding of drug delivery mechanisms and device functional improvements in the drug's presence is necessary to improve their efficacy and scope of medical applications. Mechanisms involved in drug delivery should be exploited to better match release to the local needs of each specific combination product. The major challenges faced by IMDs in clinical applications are shown in Figure 1.1: (1) nonspecific host response–foreign body reaction; (2) device thrombosis, and (3) biomaterial-associated infections. These all share some interrelated failure mechanisms that may amplify tissue-site adverse reactions and host responses. For example, the link between thrombosis and infection is increasingly identified to be synergistic, as is the relationship between the host foreign body response (FBR) and implant-centered infection. These increasingly complex host response relationships can be difficult to solve using a single device design or biomaterials-based approach alone. Use of local pharmaceutics with the device provides options to exploit device strengths and also drug targeting against multiple challenges in the implant site. The remainder of this chapter serves to describe combination device approaches in the context of the current medical device and implant challenges in host tissue sites.

1.2 THE HOST FOREIGN BODY REACTION

The host's acute and chronic FBR remains an unsolved challenge for many IMDs. As the implantation of almost every medical device creates a wound (e.g., knee arthroplasty and pacemaker), or local disturbance of a tissue bed (e.g., contact lens), a normal host tissue wounding response is spontaneously initiated. This reaction is primarily an abnormal tissue healing response that alters normal wound site healing in the presence of a foreign body (IMD), yielding a chronic unresolved tissue response, often resulting in excessive fibrosis. Extending the functional clinical lifetime of IMDs while reducing their adverse events *in vivo* remains an important goal. Nonetheless, despite many device improvements and design changes, this goal remains elusive. For example, the host's acute and chronic FBR is well known to limit the lifetime of implanted sensors (i.e., glucose real-time monitoring devices) [13–15]. Lack of tissue mechanisms preclude rational implant improvements and other more direct therapeutic approaches. IMDs spontaneously adsorb a diverse array of plasma proteins within the first few seconds of implantation [16]. Neither the types and amounts nor orientations of these proteins on the implant can be controlled *in vivo*, but despite many assertions otherwise, this might not have much significance to the final tissue reaction. Surface properties of the implanted biomaterial certainly govern aspect of protein adsorption, but exactly how this then modulates the host reaction to the implant is less certain. Many biomaterials of distinctly different bulk chemical and surface composition result in very similar endpoints *in vivo* in soft tissue, encased by fibrous overgrowth and an avascular capsule. The IMD as a foreign body destabilizes homeostasis and hemostasis in host tissue and results in a modified "healing response" that adversely affects both the implant's performance and host tissue surrounding it.

The FBR is a consequence of aborted wound healing and the complex interplay between the complement and coagulation cascades with the host immune system. The complement system comprises cascades of blood and cell surface proteins triggered by pathogens and other "foreign" substances, including implanted biomaterials [17]. Blood's potent intrinsic and extrinsic protease cascades are triggered by procoagulant stimulus [18]. In both systems, procoagulant and complement proteins are zymogen proteases activated by the foreign body interacting with the precursor zymogens through proteolytic cleavage [19], and each acting to amplify host cell-signaling and cell-recruiting capacities. FBR results from continuous host exposure to combinations of specific (activating) and nonspecific (activating) proteins on the foreign body and their protease activation. Subsequent chemotaxis and reactions from host immune and inflammatory cells lead to unresolved chronic healing responses, sustained inflammation, recruitment of fibroblasts and fibrotic encapsulation, and foreign body giant cell presence as a terminal response to the implanted device. In this dynamic wound site response, normal wound site acute cell infiltrates comprising neutrophils and other leukocytes, and later monocyte and macrophage invasion stimulate release of

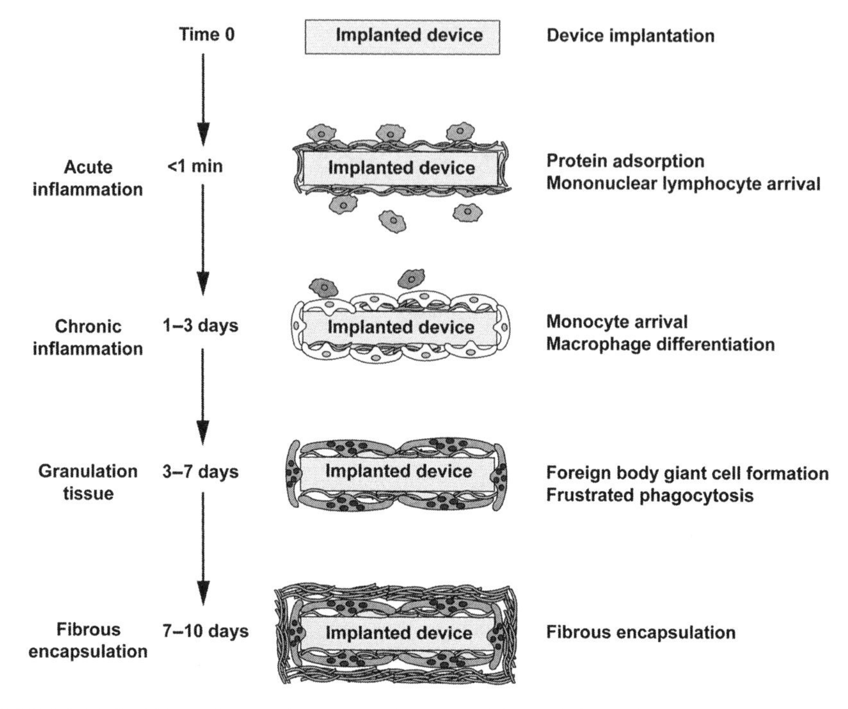

FIGURE 1.2 Illustration of the temporal series of host biological events during the host foreign body response following biomaterial implantation.

inflammatory cytokines such as IL-6, TNF-alpha, IL-4, and IL-13 (i.e., from mast cells) to accelerate recruitment of inflammatory and immune cells to the site of implant [15]. In normal wounds, these abate, but a foreign body provides continuous inflammatory stimulus for sustained, abnormal cell signaling. Fibroblasts then arrive at the implant site and mediate the formation of an avascular fibrous tissue via exuberant collagen production around the implant that can act as a physical barrier blocking access to essential components of the tissue surrounding the implants, an area of local hypoxia and poor perfusion to create an infection niche, and also a physical impediment of prosthetic motion if required (i.e., joint arthroplasty) or adjacent tissue-on-tissue motion (e.g., surgical adhesions) that are highly painful. Chronically, the excess connective tissue remodels into a dense fibrous capsule (fibrosis) that "walls off" the implant, separating the IMD from its physiological surroundings. This foreign body capsule is the hallmark of the FBR, and adversely affects the general performance of IMDs, limiting their reliability and long-term success. Reactions of both the host on the implant and the implant on the host/blood/ tissue need to be understood to enhance IMD performance. Figure 1.2 illustrates the sequence of host-materials events following the implantation of a biomaterial/medical device into host tissue.

While some implants remain unaffected functionally by the FBR, certain types of IMDs are highly compromised. In particular, sensor implants such as continuous glucose monitoring (CGM) sensors [20–22], pacemaker electrical leads [23], and neural deep brain stimulation arrays [24] undergo fibrosis that hinders function. The avascular fibrous tissue surrounding the implant impedes the implant's electrical [25] and chemical contact with the surrounding tissue while also depriving it of essential analytes [26–28] and nutrients, rendering implants less efficient. Pacemaker leads underwent early drug modification, with steroid reservoirs and elution from their porous electrode tips enhancing their impedance and conductance properties with tissue and their functioning lifetime, enhancing battery life and reducing fibrous tissue encapsulation [29,30]. Many CGM sensors are placed subcutaneously where normal sensor fouling, including protein adsorption on or infiltrated into the implanted sensors, as well as inflammatory wound site cellular reactions eventually limit analyte diffusion (mostly glucose and oxygen) into the sensing element, and contribute to the observed continual decreased analyte sensitivity with prolonged implantation [14,21,31]. In addition to ubiquitous sensor fouling and encapsulation, the host's acute inflammatory response to the implanted foreign body produces an immediate, sustained cascade of local tissue cellular reactions that alter the local environment around the implant, substantially modifying local metabolism and homeostasis. This triggers a departure from normal tissue analyte levels and causes the sensors to produce highly altered analyte levels from acute inflammation—an acute reporting phenomenon called "break-in" [32].

As the host foreign body response in soft and hard tissue sites typically produces device-based challenges associated with excess or unresolved inflammation, fibrosis, and infection, combination device strategies seeking to address this issue have used drugs with known pharmacological actions against these specific problems.

1.2.1 Anti-Inflammatory Drug Candidates to Inhibit the Foreign Body Response

Anti-inflammatory steroidal drugs (e.g., dexamethasone) are clinically familiar and used to reduce inflammation and the host FBR in tissues surrounding implant sites [33,34]. Dexamethasone, a glucocorticoid agonist, crosses cell membranes and binds to glucocorticoid receptors controlling different inflammatory pathways with high affinity by inhibiting leukocyte infiltration at sites of inflammation, suppressing humoral immune responses, and reducing edema and scar tissue. Molecular basis for dexamethasone's anti-inflammatory actions are thought to involve the inhibition of cyclooxygenase enzyme [35] that regulates arachidonic acid metabolism responsible for production of inflammatory prostaglandins.

Local controlled release systems containing the steroid, dexamethasone, have been used in intraocular application postsurgery in cataract treatments [36–40]. Local dexamethasone release [41] has also been used to reduce neointimal formation in the arterial wall after balloon angioplasty [42,43] and to prevent restenosis in intravascular drug-eluting stents [44]. Dexamethasone has also been used to improve the performance of pacemaker leads [45]. Dexamethasone release from PLGA microspheres coated onto a cotton suture implant has shown to decrease the acute inflammatory reaction around the implanted suture material [46]. Dexamethasone has also been used in combination with angiogenesis factors such as vascular endothelial growth factor (VEGF) to promote new blood vessel growth while reducing inflammation in the tissue surrounding a hydrogel (PVA) scaffold implant [47]. Sequential or simultaneous release of dexamethasone and VEGF has been shown to improve the performance of implanted biosensors [47–51].

1.2.2 Antiproliferative Drug Candidates to Inhibit the Foreign Body Response

Sirolimus, also called rapamycin, is a potent immunosuppressive drug used in combination with medical devices. As a potent inhibitor of cytokine and growth factor-mediated cell proliferation, sirolimus acts by inhibiting activation of the intracellular protein enzyme, mTOR (mammalian target of rapamycin) [52], a downstream mediator of the PI3K/Akt phosphorylation signaling pathway regulating several key cell functions. Receptor-based inhibition of mTOR results in the blockage of cell cycle proliferation in the late G1 to S phase, causing antiproliferative and antihyperplastic actions [53,54]. Over 70 related "limus" derivatives are known drug candidates. Everolimus, temsirolimus, deforolimus, tacrolimus, and ABT-578 are also used as potent antiproliferative drugs. Paclitaxel is another commonly used antiproliferative drug used with medical devices such as drug-eluting stents. Paclitaxel inhibits cell proliferation, cell motility, shape, and transport between organelles [55]. Both rapamycin and paclitaxel have substantial clinical records as approved therapeutics for a number of indications independent of devices.

1.3 DEVICE-BASED THROMBOSIS

Under normal, steady-state circulation conditions (hemostasis), blood continuously contacts host endothelium with an intrinsic, active anticoagulant and antithrombotic system. Injury to blood vessels exposes subendothelial components, releases procoagulant stimulants, and disrupts hemostasis. Natural host response to this disruption involves blood platelet adhesion, activation, and aggregation in combination with activation of intrinsic and extrinsic coagulation cascades terminating in the formation of a crosslinked fibrin clot. These natural coagulation cascades are depicted in Figure 1.2. The combination of platelet and procoagulant cascade activation rapidly produces a thrombus/clot that stabilizes the injury and prevents further blood loss. Thrombus formation plays an important role in the maintenance of hemostasis. Thrombin-mediated fibrin polymer traps and stabilizes clusters of activated platelets to yield a stable thrombus critical for survival and also contribute powerfully to local wound healing.

Endothelial cells (ECs) lining the walls of the endothelium continuously synthesize and regulate several key molecules necessary for the maintenance of host hemostasis and the intrinsic blood compatibility of vasculature. The EC surface is a dense, brush-like layer of hydrated proteoglycans, called the glycocalyx. Glycocalyx glycoproteins enzyme-grafted with glycosaminoglycans (GAG) side chains [56], including heparan, dextran, and chondroitin sulfate proteoglycans and hyaluronic acid, are negatively charged and highly hydrated, acting as a barrier and a lubricant between the ECs and blood components [57]. ECs also actively produce and release nitric oxide and prostacyclin (PGI_2) that actively prevent platelet adhesion and activation [58,59]. Heparan sulfate proteoglycan synthesized by the ECs inhibits platelet adhesion and activation [60] while also functioning as a catalytic cofactor for binding antithrombin-III and thrombin together to facilitate thrombin inhibition and anticoagulation [61,62]. ECs also produce tissue-type plasminogen activator (t-PA) and urokinase that act to initiate fibrin degradation and aid in clot dissolution [63,64]. This t-PA activity is tightly regulated by the EC-produced plasminogen activator inhibitor type-I [65–67].

Cardiovascular medical devices are placed into contact with patient's blood for varying periods of time, ranging from minutes (e.g., vascular access devices) to many hours (blood pumps, dialysis filters, and central lines), to years (e.g., stents, heart valves, vascular grafts, and pacemaker leads). The blood-contacting surfaces on these devices are critical to their performance, seeking to minimize activation of both platelets and the coagulation cascades. However, no materials chemistry or coatings used on these devices have proven clinically reliable in limiting risks of device-based thrombosis to date. Some blood-contacting biomaterials are grafted with heparin-like coatings, or polymers mimicking the EC glycocalyx [68]. Figure 1.5 shows one example of this device-based surface modification approach using heparin. Other approaches are designed to release anticoagulant and antiplatelet drugs for short durations [69]. No materials yet provide all the passive, active, and functional aspects of ECs in maintaining hemostasis, and, therefore, all induce thrombosis in contact with blood to varying degrees. Device-induced thrombosis is a major cause of failure in blood-contacting biomaterials, mainly cardiovascular implants, which constitute a major class of chronic disease-related IMDs. Implantation of a medical device lacking the properties of a healthy endothelium constitutes the introduction of a foreign object into circulation. Blood–material interactions after implantation spontaneously and immediately trigger a series of complex reactions involving protein and platelet absorption on the biomaterial surface, formation of clots and emboli, and activation of the host's immune system.

1.3.1 Platelet Activation in Device-Based Thrombosis

Platelets are anuclear cytoplasmic fragments present in blood essential for rapid, reliable blood clotting and wound healing [70]. Platelets play an essential role in controlling blood loss and maintaining hemostasis. One common platelet mode of action is the formation of a stable platelet plug when the blood vessel wall is damaged and the endothelial cell layer is disrupted, exposing the underlying basement membrane and extracellular matrix. With every surgical device implantation, blood vessels in the tissue surrounding an implant are injured, exposing collagen IV in the subendothelial layers to blood that results in the activation of circulating platelets. Additionally, platelets also get activated when they undergo shear stress caused by flow disturbances common to implanted devices. Platelet activation is followed by platelet degranulation and then by aggregation and adhesion to each other and to the implanted material. Degranulation serves to release a broad array of potent platelet-derived biochemicals that potentiate local thrombosis by accelerating both local coagulation cascade reactions and platelet activation by release of highly procoagulant stimulants, enzyme substrates, and cofactors. The aggregated platelets are stabilized into a thrombus/clot by the newly formed fibrin polymer. Circulating platelets get activated under three major circumstances: (a) by contacting the basal lamina of the endothelial vessel wall, (b) by contacting with a biomaterial surface, and (c) due to flow disturbances caused in the presence of a biomaterial. Platelet adhesion, activation, and aggregation are combined with simultaneous thrombin-mediated fibrin polymerization that together result in thrombus formation.

1.3.2 Extrinsic and Intrinsic Coagulation Cascades

A biomaterial surface exposed to blood is coated with thousands of plasma proteins within seconds [71]. This adsorption activates some plasma proteins by inducing conformational changes or cleaving small fragments that trigger coagulation and inflammatory responses to the implanted device [72–74]. The coagulation cascade comprises two main branches: the intrinsic pathway (activated by contact with a biomaterial surface) and the extrinsic pathway (induced by EC injury). Both pathways converge at the proteolytic formation of thrombin from its prothrombin zymogen, the penultimate cascade step to converting soluble plasma- and platelet-derived fibrinogen to fibrin polymer. Fibrin polymer is a major protein component of the natural clot. Activation of intrinsic and extrinsic proteolytic reactions following blood contact with biomaterials actively and consistently produces thrombin-mediated fibrin clots unless pharmacological treatments attenuate these natural responses, typically by inhibiting key enzymes. The series of coagulant events triggered by the activation of intrinsic or extrinsic pathways following the implantation of a medical device into blood are shown in Figure 1.3. Adherent platelets—both on the biomaterial and trapped by the clot—activate to release numerous potent thrombotic promoters and catalysts by degranulation. They also recruit more circulating platelets to the device surface. Subsequent device-based thrombosis and thromboemboli formations produce many clinical complications, causing failure in small-diameter grafts, stents, valves, pumps, catheters, and other cardiovascular implants. Furthermore, causal links between device thrombosis and device-centered infection are increasing.

1.4 BIOMATERIALS-ASSOCIATED INFECTION

All implantable devices—from short-term devices, such as contact lens, glucose sensors, urinary catheters, and endotracheal tubes, to long-term surgically implanted devices, such as pacemakers, cardiac valves, endothelial grafts, and orthopedic implants, suffer commonly from varying risks of biomaterials-associated infections (BAIs) or implant-associated infections [41]. BAIs remain a major cause of IMD failure despite years of device innovation, improved quality of care, and surgical techniques [75]. In the United States, approximately 2 million nosocomial infections

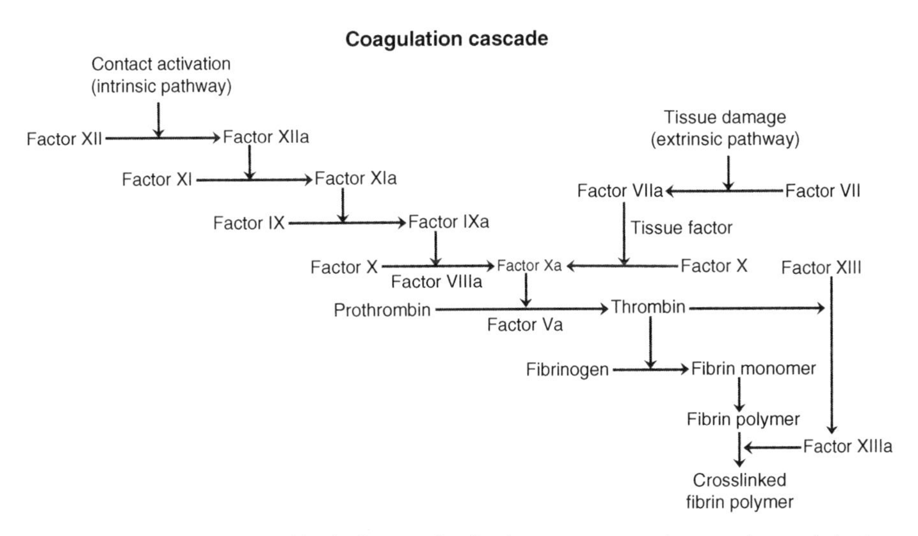

FIGURE 1.3 Extrinsic and intrinsic cascades for the zymogens, active proteins, and clotting factors mediating clot formation after procoagulant stimulus.

costing $11 billion occur annually [76]. A majority of nosocomial infections (60–70%) are biomaterial-associated infections caused from the increasing use of urinary and venous catheters, orthopedic implants, shunts, and other implants [77], and involving significant mortality and economic costs. Infection mitigation is a common problem with IMDs and a primary focus of surgical antibiotic prophylaxis in device placement. BAIs most often result from bacterial contamination of implants intraoperatively during the implantation procedure. They are able to colonize implants using the implant-adherent protein layer and thrombus, proliferating at rates that outpace host wound healing. Bacterial adhesion leading to the formation of mature biofilms on the surface of a biomaterial is shown in Figure 1.4. Bacteria and other pathogens have multiple sources during surgery: no surgical suites, surgical personnel, or patients are sterile, Pathogen seeding of implants and surgical sites is likely, although only small fractions of implants actually colonize and lead to clinically symptomatic infections as BAIs. Nonetheless, BAIs can result in difficult-to-treat systemic infections with costly adverse complications and mortality. BAIs are most prevalent in orthopedic [78,79], dental [80], cardiovascular [81–83], neural, and ophthalmological implants [84,85] and involve a broad spectrum of pathogens, many in polymicrobial implant infections. Rates of infection at the site of implantation postsurgery increase with the severity of the vascular and tissue injury [86]. Upon detection, BAIs often fail systemic administration of antibiotics. Therefore, common treatment most often involves immediate implant removal followed by long-term parenteral administration of antibiotics and then replacement with a second new implant. This often comes with associated morbidity and high treatment costs. Little change in BAI incidence has resulted from changes in surgical practice, device design, or antibiotic usage, prompting re-examination of the entire medical device infection scenario [87]. Since systemic antibiotic therapies have failed to bring down implant

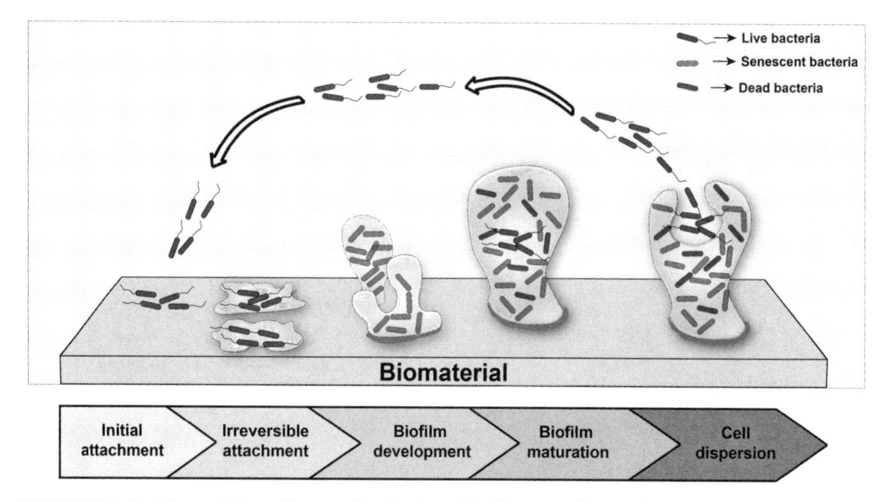

FIGURE 1.4 Bacterial seeding, colonization, biofilm transformation, differentiation, maturation, and further dissemination producing following biomaterial-associated contamination and infection. (See colour plate section.)

infection rates, local release of antiseptics and antibiotics has been sought in combination device form.

1.5 COMBINATION MEDICAL DEVICES

1.5.1 Drug-Eluting Stents

Coronary stent restenosis has been a major challenge since the introduction of percutaneous coronary intervention (PCI) for coronary artery disease [88]. Use of rigid but flexible endovascular scaffolds such as stents prevents the recoil and collapse of the vessel while also mitigating the vessel restenosis experienced after balloon angioplasty [89,90]. Although the development and use of stents in PCIs has demonstrated improvements over balloon angioplasty, vessel restenosis or in-stent restenosis after bare metal stent deployment also poses challenges to successful PCIs, resulting in past patient reinterventions in up to 50% in several patient classes depending on stent placement and patient pathophysiology [91,92]. Systemic administration of drugs to reduce in-stent restenosis is ineffective [93–95] mainly due to poor drug bioavailability, toxicity, and insufficient drug dosing to the implant site. Popularity of drug-eluting stent (DES) is due to proven success in mitigating the effects of tissue hyperplasia-caused vessel occlusion [9,96,97]. DES use has reduced the occurrence of repeated PCIs and surgical revascularization procedures to treat restenosis by 40–70% [8–10]. Emerging classes of DES coated with bioactive agents (DNA, proteins, and viral vectors) and biopharmaceuticals provide improved safety and efficacy in certain cases, but also pose challenges during their fabrication and require specific formulations and delivery mechanisms for reliability and efficacy [98].

In coronary applications, DES devices are typically localized expandable, slotted metal tubes (~4 mm long) coated with a polymer carrying a small dose (micrograms) of pharmacological agent. DES is collapsed around a deployment catheter and installed at a coronary lesion vessel by catheter-initiated intraluminal expansion. This provides structural support to the vessel while releasing drug locally to the vessel wall at the stent implant site [99]. The DES provides the advantage of effective localized drug delivery and therapeutic efficacy at the lesion site while avoiding excessive dose exposures through systemic delivery [100,101]. Other advantages include directional delivery of drug to the vessel wall tissue and only small fractions entering the bloodstream. Additional new stent designs can build drug depots in spatially designated locations on-stent [102], with versatility to carry multiple drugs, releasing with different release kinetics, and also two distinct therapeutic functions: antithrombosis on the blood side and antiproliferatives on the tissue side [103].

Sirolimus-eluting stents originally were the pioneer DES, showing noticeable improvements over early bare metal stent designs in PCI procedures, reducing cell proliferation, migration, and restenosis from the vessel bed at the stenting site. The sirolimus-eluting stent (Cordis CYPHER) was the first DES to receive clinical approval [104,105] in Europe, and the first "official" combination device approved by the FDA in 2003. Sirolimus is now the most extensively studied drug to reduce in-stent neo-intimal hyperplasia following coronary stent deployment [106]. Sirolimus- and paclitaxel-eluting stents (TAXUS®, Boston Scientific, USA) are the two commercially available, first-generation DES. These are coated with a very thin (~μm) nondegradable polymer layer (e.g., polyisobutylene or polymethacrylate copolymers) containing very little drug within the coated polymer (~μg/mm length of stent), released with an early significant burst (up to 50%) within the first 24–36 h postimplantation followed by slower release lasting more than 6 weeks in some cases [6].

After initial enthusiasm with the first-generation DES, controversial debate has ensued over long-term DES safety, with a shift in clinical focus to increased risks of late stent thrombosis [107,108]. Although both CYPHER and TAXUS effectively achieved primary goals of reducing cellular restenosis across almost all lesion and patient subsets over bare metal stents, their safety has been limited by suboptimal polymer biocompatibility, delayed stent endothelialization leading to late stent thrombosis, and local drug toxicity [109–113]. The permanent presence of the noneroding polymer covering the stent struts and wires has been correlated with tissue inflammatory response and local toxicity in preclinical studies [114,115]. Stent thrombosis risk gained primary focus after the dominant restenosis issue had been resolved using drug-eluting stents: DES devices are comparable to bare metal stents in occurrence of stent-associated thrombosis [116]. This has prompted new technologies and designs to overcome the thrombosis problem using new stent designs, stents with multiple drug reservoirs containing both antithrombotics and antiproliferatives, absorbable or biodegradable polymers, nonpolymer drug-loaded surfaces, and changes in the types and doses of currently used antiproliferative drugs placed on-stent.

Recent clinical introduction of the biodegradable polymer-coated DES [117] seeks to overcome stent thrombosis attributed to the permanent polymer layer on first-generation DES. Biodegradable stent coatings have been designed to release loaded drug for an intended amount of time before completely degrading. Clinically familiar poly(L-lactic acid) (PLA), poly(lactic-*co*-glycolic acid) (PLGA), and poly(D,L-lactide) (PDLLA) remain popular choices for biodegradable polymer DES coatings. BioMatrix (Biosensors Inc., USA) is a stainless steel stent containing Biolimus A9 (a derivative of sirolimus) as drug on the abluminal surface facing the vessel wall targeting the mTOR protein, loaded in a PLA coating. The drug is released to the vessel wall over 2–4 weeks and the PLA coating is gradually absorbed between 6 and 9 months. Several other novel DES devices with biodegradable coatings are in development. Cardiomind (Cardiomind Inc., USA), ELIXIR-DES (Elixir Medical Corporation, USA), JACTAX (Boston Scientific Corporation, USA), and NEVO (Cordis, USA) are example stents in development and clinical trials with degradable coatings used to deliver antiproliferative drugs to mitigate neointimal tissue hyperplasia in PCI procedures.

As the model and precedent combination product approved by the FDA, DES is an excellent example of combining a drug with a device to target and address a specific problem unsolved by either component alone or used together but separately. Despite improvements in early prototypes and first-generation stents using new designs, materials, drugs, drug loading methods, drug release kinetics, and release duration and improved understanding of local pharmacology and complications arising several months to years after DES placement, new technology should better address newer DES problems associated with late stent thrombosis, endothelialization, and local drug toxicity. Additionally, expansion of DES use to other challenging luminal lesions, both in vasculature, gut/digestive, and reproductive tissues, will require further innovation of drugs on devices.

1.5.2 Antimicrobial Central Venous Catheters

Central venous catheters are a critical component for fluid delivery and retrieval and parenteral drug and nutritional fluid administration in a variety of clinical settings for critically ill patients. In the United States, physicians insert more than 5 million central venous catheters every year [118]. The two major complications associated with catheters are bacteremia (infection) and thrombosis [119]. Catheters coated with both antimicrobial and antithrombotic agents have been developed and commercialized. Antimicrobial-coated catheter use and efficacy have been studied for more than a decade [120].

Infections associated with catheters are classified as catheter-related bloodstream infections (CRBSI). CRBSI can occur in 3–10% of all patients using central venous catheters [121], affecting over 300,000 patients in the United States annually [122] and causing more than 25,000 patient deaths [123,124]. Systemic administration of antibiotics to treat CRBSI either prophylactically or therapeutically is neither a clinically preferred nor a reliably efficacious route. Local administration of antimicrobial agents from properly designed combination devices seeks to provide small efficacious doses of therapeutics released into local tissue sites without requiring high systemic drug dosing.

Techniques developed to reduce CRBSI incidence include modified catheter designs, use of antimicrobial impregnated catheters, use of cuffed tunneled catheters, local topical treatments, and use of antimicrobial lock solutions [125]. Coating or impregnating the surface of central venous catheters with antimicrobial agents helped to markedly reduce the risk of CRBSI, and their use has now become the standard of care [126,127]. Antimicrobial-coated catheters employ different methods to immobilize the antimicrobial agents onto catheter surfaces–both luminal and external. One method is to simply add the antimicrobial agent to the precursor polymer granules used to fabricate the catheter, similar to adding other constituents such as pigmentation or stabilization compounds prior to injection molding [128]. Another procedure involves electrostatically coating catheter surfaces layer-by-layer with antimicrobial agents and a binding material with opposite electrostatic charge. Hydrophobic alkylated regions of cationic surfactants such as tridodecylmethylammonium chloride (TDMAC) have been adsorbed on catheter surfaces, presenting a cationic surface to anionic drug molecules binding to the surfactant-coated surface [129,130]. Recently, a zwitterionic polymer brush-grafted layer has shown preclinical efficacy as an antimicrobial coating [131,132]. Addition of active drug release capability to this layer would provide enhanced bioactivity.

These strategies facilitate incorporation of different antimicrobials onto catheter surfaces to reduce CRBSI. Multiple antimicrobial agents are preferred (typically combinations of an antiseptic and antibiotic agent) to reduce the development of antimicrobial resistance to any single agent [133]. According to Centers for Disease Control and Prevention (CDC) guidelines, catheters containing combinations of minocycline/rifampin (MR) antibiotics and combinations of chlorhexidine/silver sulfadiazine (CS) antiseptics are the two most effective antimicrobial catheters to treat CRBSI [126,127]. Catheters coated with both antibiotics and antiseptics are FDA and CE approved and commercialized (e.g., CS: ARROWgard®, Arrow international, USA; MR: Cook Specturm® series catheters, Cook Critical Care, USA). Both ARROWgard and Spectrum series catheters have antimicrobial agents impregnated on internal and external surfaces using TDMAC adlayers [134]. Both MR and CS have shown broad-spectrum antimicrobial activity to both Gram-negative and Gram-positive organisms and fungi. Several randomized trials [135,136] conducted with MR and CS showed superior performance from MR-impregnated catheters versus CS-impregnated catheters in preventing CRBSI, especially in patients needing catheter-based access for more than 7–50 days *in situ* [137]. Catheters impregnated with MR have been shown to exhibit higher antiadherence activity and prolonged antimicrobial durability compared to catheters with CS against vancomycin-resistant *Staphylococcus aureus* and multidrug-resistant (MDR) Gram-negative organisms other than *Pseudomonas* [138]. Although MR shows high antimicrobial activity against *Staphylococci* and most of the Gram-negative bacilli [138], they are less effective against *Pseudomonas aeruginosa* (contributing 3–5% of CRBSI) and *Candida* species (contributing about 12% of CRBSI) [138]. *In vitro* studies using catheters coated with a combination of MR and CS have shown to be effective against vancomycin-resistant *S. aureus*, Gram-negative bacilli, *P. aeruginosa*, and *Candida* species [139].

Silver nanoparticle-impregnated catheters (SNPs) (Medex Logicath AgTive®, Smith Medical International Ltd., UK) are CE approved and commercially available in Europe. Catheters coated with silver-based zeolite (SZ) on blood-contacting surfaces (e.g., Lifecath PICC Expert with AgION™ from Vygon International in Europe) use controlled release of silver nanoparticles from the coating to provide antimicrobial properties to the catheter. In a recent study [140] conducted over 14 months involving 246 central venous catheter insertions (122 silver zeolite-impregnated and 124 nonimpregnated catheters), the AgION catheters showed reduced CRBSI compared to uncoated catheters. In silver nanoparticle-impregnated catheters, a recent study has shown that platelets colliding with silver nanoparticles exposed on the coating surface accelerate the process of catheter-related thrombosis while simultaneously exhibiting strong antimicrobial properties [141].

Catheters with antithrombotic coatings are used to reduce the incidence of coagulation. The Carmeda® BioActive Surface (CBAS) on Spire Biomedical® catheter products (Spire Biomedical, Inc., Bedford, MA) and the Trillium® Biosurface developed by BioInteractions Ltd. (UK) are two commercially available antithrombotic-coated catheters. Both catheters use heparin-bonded polymer surfaces as an anticoagulant interface. Heparin is a polysaccharide with anticoagulant properties and has been used as an antithrombotic agent for clinical applications [142]. The CBAS treatment consists of heparin molecules covalently bonded to the catheter surface, exposing active heparin sequences to bind ATIII and thrombin from the bloodstream while shedding other protein components. Schematic representation of thrombin inhibition on a bioactive heparin-coated surface to limit device-based thrombosis is presented in Figure 1.5. The Trillium Biosurface treatment combines a hydrophilic polyethylene oxide layer with negatively charged sulfate polymers to retain hydration at the catheter surface and reduce blood adsorption. In addition, it has heparin covalently bonded to the polyethylene oxide layer for anticoagulation [143]. Although catheters coated with active antithrombotic layers are clinically used, the effects of these coatings on catheter complications are yet to be evaluated in the hemodialysis population where these complications also exist. Importantly, many such technologies have not been shown to produce significant cost-benefit using placebo-controlled blinded prospective studies.

Catheter lock solution (CLS) is another strategy used to reduce CRBSI incidence from central venous catheters. A biocompatible solution containing a combination of antimicrobial and anticoagulant agents constitutes the CLS. The CLS is injected into the lumen of the catheter after a hemodialysis session and retained there to reduce incidence of thrombus and associated biofilm formation. Catheter thrombosis can be limited using heparin solutions or treated by infusing a thrombolytic agent such as urokinase or tissue plasminogen activator (tPA) into the lumen of the catheter [144,145]. In a recent study, athrombogenic Camouflage™-coated (artificial glycocalyx) catheters have exhibited reduced need for urokinase injections for successful catheter tap and blood drawing over uncoated catheters in cancer patients with long-term catheters [68].

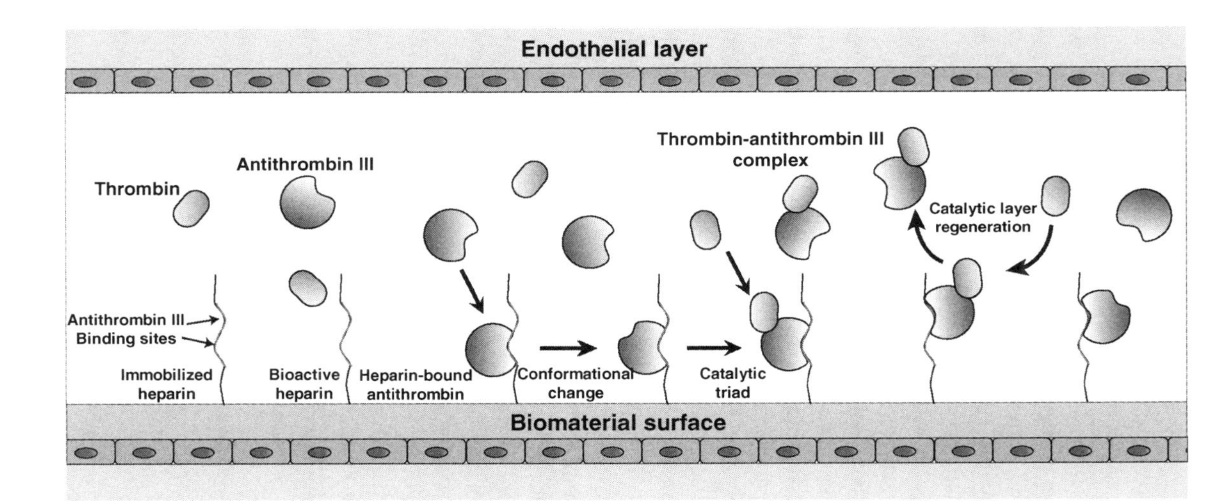

FIGURE 1.5 Bioactive heparin-immobilized surface capable of inhibiting thrombin activation, a key mediator in clot formation. The strategy is used for blood-contacting devices to limit complications from device-based thrombosis. (Reproduced with Permission from W. L. Gore.)

1.5.3 Antimicrobial Urinary Catheters

Urinary catheters allow passage of urine for treatment for patients with urinary retention complications, general surgery recovery, bladder obstruction, paralysis, or loss of sensation in the perineal area [6]. Urinary catheters are generally used to manage urinary incontinence in elderly patients or in patients with long-term spinal cord injuries. More than 30 million urinary catheters are employed in patients annually [146]. Unfortunately, catheter-associated urinary tract infections (CUTI) remain the most common nosocomial infection [147]. Catheter surfaces in contact with the urethral epithelia facilitate bacterial contamination, adhesion, retention, and biofilm formation on both the abluminal and luminal surfaces, eventually leading to infection of the urethra, then the bladder, and ascending into the ureters unless the catheter is exchanged frequently [148]. Microbes in the catheter mediate the breakdown of urea, resulting in an increase in the urine pH [149], inducing formation of mineral crystals on the catheter surface, leading to the formation of urinary infection stones [149] and blockage of the lumen by encrustation, which can produce kidney and bloodstream infections.

Systemic antibiotic therapies, antimicrobial topical ointments, and the use of antimicrobial agents in collection bags are commonly used to treat CUTIs. Silver-impregnated urinary catheters claim 30% reduction in the incidence of CUTI in some studies, although this is not a consensus [150]. Several catheters based on silver and silver oxide coatings are commercially available in the United States (SilvaGard® (I-Flow/Acrymed), KENDALL DOVER® series catheters (Tyco Healthcare), BACTI-GUARD® silver (C.R. Bard)) [75]. A recent UK study involving patients with urethral catheterization for up to 14 days found that silver-coated catheters were ineffective against infection; the incidence of infection is comparable to uncoated PTFE catheters [151].

Ciprofloxacin, gentamicin, norflaxin, nitrofurazone, and combinations of compounds, such as chlorhexidine and protamine sulfate, have been successfully incorporated into catheter coatings [152]. Nitrofurazone-coated catheters (Rochester Medical, MN) are an emerging class of antimicrobial urinary catheters shown to be efficacious against *Escherichia coli* [153] and have exhibited better antimicrobial properties than silver-treated catheters [154] in *in vitro* studies. However, further prospective double-blind powered two-arm clinical studies are required to validate claims for the efficacy of silver- and nitrofurazone-coated catheters in CUTIs.

1.5.4 Orthopedic Drug-Eluting Implants

Bone defects from trauma, disease, surgical intervention, and congenital deficiencies are among the most challenging orthopedic repair problems faced worldwide. Autologous bone grafts are the gold standard to treat bone defects, but are limited, not always appropriate, with harvesting complications, including infection susceptibility. Bone fractures and joint deficiencies are increasingly treated using a variety of implanted biomaterial stabilization devices, including bone cement, hip, knee, shoulder and elbow prosthesis, plates, nails, rods, wires, pins, and screws. Projected market revenues for such orthopedic implants are estimated at $23 billion in 2012 [155]. Bone-implant bonding [156] and long-term stabilization pose significant clinical challenges, including implant infection, bone resorption, and implant loosening [157–159]. Despite the use of advanced stabilization mechanisms and implant instrumentation, some fractures are slow healing or nonunions, requiring revision surgeries at significant expense and patient morbidity. Recent advances in drug delivery are increasingly used with orthopedic implants as combination devices [160]. Increasing reports document effects from delivery of small-molecule osteoinductive agents, drug [161], scaffold [162], gene and cellular delivery [163,164], biologically derived growth factors, antiosteoporotic agents, and osteosynthetic genetic materials such as DNA transgenes and siRNA to bone defects from a variety of implant devices and vehicles [165–172].

BAI remains a major concern in orthopedic implants [173]. Rates of infection are estimated to be 1% for primary hip implants, 4% for knee implants (higher for secondary revisions), and more than 15% for some trauma-associated open fracture implants [155]. Orthopedic implants carry a lifetime risk of infection (acute and hematogenous sources) and are clinically addressed in most cases by revision surgery involving a further substantial risk of infection [155]. A commonly used clinical approach to manage orthopedic implant infection is the use of antibiotics in bone cement, polymethylmethacrylate (PMMA) or PMMA beads. These nondegradable polymer cements have been used to prevent osteomyelitis for four decades [174–176] using either bulk impregnation by the aminoglycoside antibiotics, gentamicin or tobramycin [6,177], or vancomycin (Europe only). The first antibiotic-blended bone cement to be approved in the United States was Simplex P (Stryker Howmedica Osteonics) containing tobramycin [6]. The Palacos™ series of bone cements from Biomet, Inc. (Warsaw, USA) contain gentamycin and have been approved shortly after Simplex P. Recently, Depuy 1 gentamycin-releasing bone cement (Depuy Orthopaedics) has been approved by FDA. *In vivo* studies have demonstrated the efficacy of antibiotic-loaded cements in reducing orthopedic implant infections within a short time after implantation [178–180]. However, despite wide enthusiasm, drawbacks limit clinical applications of antibiotic-loaded bone cements. Pharmacokinetics studies show the inefficiencies of gentamicin release from antibiotic-loaded PMMA bone cements or PMMA beads, with less than 50% of the antibiotic release by 4 weeks [181–184]. The primary concerns with the use of antibiotic-loaded bone cements are possible allergic reactions to the antibiotic used, and the development of drug resistance to the antibiotic at the implant site.

An antimicrobial tibial internal fixation nail coated with a degradable polymer containing gentamicin is marketed in Europe. The polymer coated over the metal nail covers the cannulation, enabling antibiotic delivery to the intramedullary canal and releasing antibiotic for ~2 weeks [185,186]. The FDA recently approved a polyurethane sleeve coated with gentamicin (OrthoGuard AB, Smith & Nephew, UK) that can be used for coating pins and wires used for external fixation devices [187].

1.5.5 Antimicrobial Sutures

According to the CDC, the overall incidence of surgical site infection is estimated to be 2.8% in the United States [188]. Surgical sutures allow microbial adherence and colonization similar to other biomaterials [189] and contribute to surgical site infection incidence. Microbial colonization to suture materials is highly variable, depending on specific microbial species, suture structure, and chemical composition [190]. Braided sutures have been shown to have higher microbial colonization compared to nylon-based monofilament sutures [191]. Triclosan-coated braided polyglactin 910 suture (Vicryl Plus Ethicon, USA) has been developed to mitigate suture-induced surgical site infections. Several *in vitro* and *in vivo* studies [192,193] have shown that the triclosan-coated Vicryl Plus sutures effectively inhibit growth of normal and methicillin-resistant strains of *S. aureus* and *Staphylococcus epidermidis* [192] while showing no difference in physical (strength, breaking force, etc.) and degradation characteristics compared to uncoated polyglactin 910 sutures [193]. Recent clinical trials have shown that use of triclosan-coated Vicryl Plus sutures in a diverse group of 450 patients resulted in a statistically significant reduction in the incidence of surgical site infection [194]. However, some studies advise caution and the need for larger scale studies [195]. Silver-containing sutures are being developed by X-Static. Another antimicrobial suture being developed by Polymedix (PolyCide™) contains the antibiotic polycide that disrupts microbial cell [75]. New antimicrobial strategies should be developed to overcome the limitations of current technologies.

1.5.6 Vascular Grafts with Antithrombotic Coatings

Synthetic vascular grafts have been used to treat vessel occlusion caused by vascular disease for over four decades. Large-diameter grafts have substantially better success rates clinically than those below 5 mm diameter, regardless of biomaterials used. Small-diameter vascular graft failure generally occurs as a consequence of acute thrombus formation on the graft luminal surface, anastomotic intimal hyperplasia, or progression of vascular disease [196]. Although anastomotic hyperplasia and disease progression are important factors for failure, reducing the propensity for acute thrombotic failure by improving graft surface blood compatibility has significant potential for improving clinical performance of small-diameter vascular grafts. Small-diameter expanded polytetrafluoroethylene (ePTFE) vascular grafts containing surface-immobilized heparin are FDA approved for treating vascular occlusion. CBAS-coated vascular grafts (e.g., Gore® PROPATEN® Vascular Graft, Gore® VIABAHN® Endoprosthesis, W.L. Gore, USA) containing immobilized heparin on the graft luminal surface are commercialized. Studies have shown reduction in thrombogenicity for small-diameter ePTFE vascular grafts containing immobilized heparin compared to uncoated ePTFE grafts [197].

1.5.7 Cerebrospinal Shunts

Hydrocephalus is treated using biomaterials-based cerebral shunt implants that drain excess cerebrospinal fluid (CSF) from the cranium to abdomen to relieve intracranial pressure [198]. Infections remain a major clinical complication in using CSF shunt implants and usually require frequent replacement of the shunt system at substantial cost and morbidity (usually in infants and children) [199]. Antibiotic-impregnated CSF shunts demonstrate clinical efficacy in reducing implant infections [200,201]. BACTISEAL® from Depuy and ARES® from Medtronic are two antibiotic-impregnated CSF shunts that contain both clindamycin and rifampicin, released from the shunt surface. Both products demonstrate reduced infection against Gram-positive bacteria for at least 31 days after implantation [202,203]. This area, however, still faces numerous challenges in producing a long-duration product that performs reliably and reduces shunt replacement frequency.

1.6 FUTURE DIRECTIONS

1.6.1 Orthopedic Fixation Plate Sleeves

New biodegradable polymer sleeves formulated with various therapeutics and readily mounted onto orthopedic plates and screw fixation implants intraoperatively prior to implantation provide a patient- and implant-specific customizable therapeutics approach to IMD drug delivery. Sleeves must not interfere with device fixation mechanics and healing (typically on periosteum or in bone) and degrade without adverse incident. Biodegradable sleeves have been prepared using copolymers of glycolide, caprolactone, trimethylene carbonate, and lactide, containing the antimicrobial agents gentamicin sulfate and triclosan (highly potent bactericidal agents against *S. aureus* [204]). These sleeves slip over metallic internal fixation plates (e.g., limited contact dynamic compression plates) and implanted in sheep tibia with induced bone defects. Local release of antimicrobials to mitigate implant-associated bone infection was shown to kill microbes *in vitro* and produce no observed bone irritation or significant FBR in sheep *in vivo* [205]. A sleeve to deliver bone morphogenetic protein-2 (rhBMP-2) within PLGA microparticles through a porous sleeve made of resorbable polypropylene fumarate has also been tested [206]. The porous sleeve is loaded with desired amounts of drug-loaded microspheres prior to implantation, with possibilities to select from a variety of preloaded, preformulated PLGA microsphere/drug combinations. This strategy provides a case-dependent customized solution to surgeons using these implants. However, this intraoperative microsphere loading technique may produce inconsistent results. Multiple variants of sleeves supplied by manufacturers with standardized drug loading and drug delivery mechanisms may result in more standardized results while still allowing surgeons to choose a precise location on the implant to apply it for release.

1.6.2 Customizable Drug-Releasing Adhesive Patches and Intraoperative Custom Coatings

Unless performing drug formulation tasks for device addition off-label, surgeons are currently limited to using drug precoated and preloaded implants as received from a device manufacturer. These types of implants have predetermined amounts of drug, a fixed drug type, and the location of the drug distributed over the implant surface cannot be changed or modified. Such implants are manufactured as "one-size-fits-all" and generally not customizable to any particular patient or condition, or surgeon preference. Increasingly, combinations of multiple drugs are proving more effective than single drugs in a given application. Flexibility for manipulating the drug type, drug loading, and its location over the implant surface can be beneficial to patients receiving certain implant types. New implant coating technologies to address these limitations with the flexibility in design and feasible intraoperative production to be readily customized to patients' needs are desirable. Customizable drug-containing "paints" and patches loaded with desired drugs with a controlled, custom dosing and flexible application locations on a desired implant intraoperatively have been recently proposed [207] to provide a possible solution to such needs. Adhesive drug patches fabricated from resorbable biomaterial laminates or composites in an aseptic environment would be loaded with drugs or drug-loaded degradable microparticles before or during surgery and cut into desired shapes to match the implant, dosing, and intended application. Drug-containing polymer coatings could also be sprayed onto implant surfaces directly using computer-controlled calibrated equipment preprogrammed to match the implant specifications with patient needs and surgeon preferences and applied either pre- or intraoperatively as a validated process. Custom drug-release patches and drug "paintable coatings" would be adhered as thin films to implant sites with surgical glues at desired locations before or during implant surgery.

1.6.3 Shape-Memory Polymeric Biomaterials

Many biomedical implants are polymer based and often require complex surgeries for device implantation and host integration due to their size and shape. Minimally invasive surgeries enable implantations of certain smaller implants with laparoscopes that limit patient risk, procedure cost, and morbidity. Use of biocompatible shape-memory polymers further provides new opportunities for improved implantation of certain medical devices with relative ease and less patient discomfort. Shape-memory processing enables specific material chemistries to "remember" a permanent shape while predeformed into metastable temporary shapes that trigger to the permanent shape with a stimulus (mechanical stress, heat, and light). This property allows modification of the device shape and size to conform to a catheter or a smaller implant readily inserted through smaller incisions using catheters or laproscopes than required for normal surgery. Nitinol is a shape-memory metal commonly used in cardiovascular stent applications due to its ability to be deformed to a small compressed conformation allowing easy insertion in a catheter with minimal implantation trauma and to regain its intended final shape after mechanical balloon-based deployment. Nickel allergy and final metal mechanical properties limit their utility. As an alternative, thermally induced shape-memory polymers can be used in polymer-based suture applications, especially those requiring complex knots, curve shapes, and conformations [208]. Shape-memory polymers have gained increased attention as a proposed biomaterial for minimally invasive surgical devices [208,209]. Medshape (Atlanta, USA) manufactures FDA-approved polymer-based shape-memory implants for suture anchors and soft tissue fasteners. Polymer-based shape-memory implants can also be used for drug delivery to implant locations via impregnation of desired drugs into the material and release upon triggering to final shape after deployment [210].

1.6.4 See-and-Treat Combination Imaging/Drug Delivery Theranostic Agents

Some creative, new medical nanotechnology enables the possibility to combine the imaging, monitoring, and treating of disease condition onto a single platform. Nanoparticles engineered with imaging agents and also containing therapeutic agents permit simultaneous diagnostic and therapeutic functions when circulating *in vivo*. These so-called theranostic agents/devices often incorporate drug conjugates and complexes, dendrimers, liposomes, micelles, core–shell particles, microbubbles, and carbon nanotubes as carriers of either drugs or contrast agents, including optically active small molecules, paramagnetic metals and metal oxides, ultrasonic contrast agents, and radionuclides. This is an emerging area of combination devices that could significantly contribute to improved disease detection and targeted therapy as well as to personalized medicine [211].

Molecular imaging techniques such as magnetic resonance imaging (MRI), radionuclide-based imaging using computed tomography (CT) or positron emission tomography (PET), and high intensity focused ultrasound allow visualization and distinction of tissue, cellular, and subcellular biological processes with the help of contrast agents [211]. These imaging agents, combined into carriers capable of effectively delivering drugs to a biological target, will enable a "see and treat" modality to image the disease condition simultaneously with triggers to delivery therapy from the agent, constituting a theranostic device.

Drug conjugates or complexes with soluble polymers such as poly[*N*-(2-hydroxyl propyl) methacrylamide] (polyHPMA) have been well studied [212]. A contrasting agent visualized by MRI, such as radioactive I-131, conjugated with the doxorubicin-HPMA polymer anticancer prodrug conjugate already synthesized, would enable the complex to be used as a tumor theranostic agent [213]. Dendrimers have been extensively studied and are attractive drug delivery and contrast agent vehicles due to their large number of functional surface chemistry sites on them. Photoactivated drug release using dendrimers with doxorubicin conjugated to a photosensitive compound has been accomplished to target cancer [214]. Other researchers have successfully combined dendrimers with various MRI contrast agents such as high-spin gadolinium and paramagnetic iron oxide [215]. Combination of both of these chemistries onto a common platform constitutes a theranostic agent. Liposomes are another class of carriers recently studied as theranostic agents encapsulating various drugs and conjugated contrast agents [211]. Multiple studies have been performed with liposomal formulations with targeting, therapeutic, and imaging functionalities [216,217].

Several other colloidal and nanoparticulate carriers such as polymersomes, micelles, quantum dots, and carbon nanotubes can be conjugated with drugs and imaging agents for treating a condition simultaneously with detection and diagnosis [211]. While the dual conjugation chemistry is fairly straightforward in many cases, the challenge remains to produce long circulating times to allow these particulate systems to achieve disease site accumulation. The concept of targeting these particles has proven very challenging to date, with very low levels of systemically administered dose (i.e., generally less than 5% of the injected dose) actually reaching the disease site from the bloodstream, with the majority of the dose targeting the liver, spleen, kidney, and lung in most cases. Imaging requires sensitivity, selectivity, and specificity *in vivo* [218]. Therapy requires effective dose delivery without toxic side effects. Building both critical properties onto a single nanoparticle platform is challenging: These two properties are not yet reliably achieved from these nanoparticle systems.

ACKNOWLEDGMENT

The authors gratefully acknowledge support from a University of Utah SEED grant.

REFERENCES

1. A. Simchi, E. Tamjid, F. Pishbin, and A.R. Boccaccini, Recent progress in inorganic and composite coatings with bactericidal capability for orthopaedic applications. *Nanomedicine* 7 (1) (2011) 22–39.
2. World Economic Forum, Global Population Ageing: Peril or Promise? 2012. Available at http://www.icaa.cc/wef/Global-Population-Ageing-Book.pdf. (accessed October 9, 2012).
3. D.B. Kramer, S. Xu, and A.S. Kesselheim, How does medical device regulation perform in the United States and the European Union? a systematic review. *PLoS Med.* 9 (7) (2012) e1001276.
4. Lucintel. Global Medical Device Industry 2012–2017: Trend, Profit, and Forecast Analysis, July 2012.
5. B. Trajkovski, A. Petersen, P. Strube, M. Mehta, and G.N. Duda, Intra-operatively customized implant coating strategies for local and controlled drug delivery to bone. *Adv. Drug Deliv. Rev.* 64 (12) (2012) 1142–1151.
6. P. Wu and D.W. Grainger, Drug/device combinations for local drug therapies and infection prophylaxis. *Biomaterials* 27 (11) (2006) 2450–2467.
7. D.S. Couto, L. Perez-Breva, P. Saraiva, and C.L. Cooney, Lessons from innovation in drug–device combination products. *Adv. Drug Deliv. Rev.* 64 (1) (2012) 69–77.
8. J.W. Moses, M.B. Leon, J.J. Popma, et al., Sirolimus-eluting stents versus standard stents in patients with stenosis in a native coronary artery. *N. Engl. J. Med.* 349 (14) (2003) 1315–1323.
9. G.W. Stone, S.G. Ellis, D.A. Cox, et al., A polymer-based, paclitaxel-eluting stent in patients with coronary artery disease. *N. Engl. J. Med.* 350 (3) (2004) 221–231.
10. G. De Luca, et al., Drug-eluting vs bare-metal stents in primary angioplasty: a pooled patient-level meta-analysis of randomized trials. *Arch. Intern. Med.* 172 (8) (2012) 611–621.
11. H.C. Ansel, N.G. Popovich, and L.V. Allen, *Pharmaceutical Dosage Forms and Drug Delivery Systems*, 6th ed., Williams & Wilkins, Baltimore, 1995.

12. S.J. Liu, S. Wen-Neng Ueng, S.S. Lin, and E.C. Chan, *In vivo* release of vancomycin from biodegradable beads. *J. Biomed. Mater. Res.* 63 (6) (2002) 807–813.
13. M.C. Frost and M.E. Meyerhoff, Implantable chemical sensors for real-time clinical monitoring: progress and challenges. *Curr. Opin. Chem. Biol.* 6 (5) (2002) 633–641.
14. G.S. Wilson and R. Gifford, Biosensors for real-time *in vivo* measurements. *Biosens. Bioelectron.* 20 (12) (2005) 2388–2403.
15. J.M. Anderson, A. Rodriguez, and D.T. Chang, Foreign body reaction to biomaterials. *Semin. Immunol.* 20 (2) (2008) 86–100.
16. D.J. Holt and D.W. Grainger, Host response to biomaterials, in J.O. Hollinger (Ed.), *An Introduction to Biomaterials*, 2nd ed., CRC Press/Taylor & Francis, Boca Raton, FL, 2012, pp. 91–118.
17. A.K. Abbas, A.H. Lichtman, and S. Pillai, *Cellular and Molecular Immunology*, 7th ed., Elsevier/Saunders, Philadelphia, 2012.
18. P.F. Neuenschwander, Coagulation cascade: intrinsic factors, in J.L. Geoffrey and D.S. Steven (Eds.), *Encyclopedia of Respiratory Medicine*, Academic Press, Oxford, 2006, pp. 509–514.
19. C. Janeway, *Immunobiology: The Immune System in Health and Disease*, 5th ed., Garland Pub., New York, 2001.
20. P.H. Kvist, T. Iburg, M. Bielecki, et al., Biocompatibility of electrochemical glucose sensors implanted in the subcutis of pigs. *Diabetes Technol. Ther.* 8 (4) (2006) 463–475.
21. N. Wisniewski, F. Moussy, and W.M. Reichert, Characterization of implantable biosensor membrane biofouling. *Fresenius. J. Anal. Chem.* 366 (6–7) (2000) 611–621.
22. F. Moussy, Implantable glucose sensor: progress and problems. Paper presented at Sensors, 2002. Proceedings of IEEE, 2002.
23. A.E. Epstein, G.N. Kay, V.J. Plumb, S.M. Dailey, and P.G. Anderson, Gross and microscopic pathological changes associated with nonthoracotomy implantable defibrillator leads. *Circulation* 98 (15) (1998) 1517–1524.
24. Y. Zhong and R.V. Bellamkonda, Dexamethasone-coated neural probes elicit attenuated inflammatory response and neuronal loss compared to uncoated neural probes. *Brain Res.* 1148 (2007) 15–27.
25. S. Singarayar, P.M. Kistler, C. De Winter, and H. Mond, A comparative study of the action of dexamethasone sodium phosphate and dexamethasone acetate in steroid-eluting pacemaker leads. *Pacing Clin. Electrophysiol.* 28 (4) (2005) 311–315.
26. A.A. Sharkawy, B. Klitzman, G.A. Truskey, and W.M. Reichert, Engineering the tissue which encapsulates subcutaneous implants: I. Diffusion properties. *J. Biomed. Mater. Res.* 37 (3) (1997) 401–412.
27. A.A. Sharkawy, B. Klitzman, G.A. Truskey, and W.M. Reichert, Engineering the tissue which encapsulates subcutaneous implants: III. Effective tissue response times. *J. Biomed. Mater. Res.* 40 (4) (1998) 598–605.
28. A.A. Sharkawy, B. Klitzman, G.A. Truskey, and W.M. Reichert, Engineering the tissue which encapsulates subcutaneous implants: II. Plasma-tissue exchange properties. *J. Biomed. Mater. Res.* 40 (4) (1998) 586–597.
29. H.G. Mono and K.B. Stokes, The electrode–tissue interface: the revolutionary role of steroid elution. *Pacing Clin. Electrophysiol.* 15 (1) (1992) 95–107.

30. A.S. Radovsky and J.F. Van Vleet Effects of dexamethasone elution on tissue reaction around stimulating electrodes of endocardial pacing leads in dogs. *Am. Heart J.* 117 (6) (1989) 1288–1298.
31. M. Gerritsen, J.A. Jansen, and J.A. Lutterman, Performance of subcutaneously implanted glucose sensors for continuous monitoring. *Neth. J. Med.* 54 (4) (1999) 167–179.
32. S.J. Updike, M.C. Shults, R.K. Rhodes, B.J. Gilligan, J.O. Luebow, and D. von Heimburg, Enzymatic glucose sensors: improved long-term performance *in vitro* and *in vivo*. *ASAIO J.* 40 (2) (1994) 157–163.
33. D.H. Kim and D.C. Martin, Sustained release of dexamethasone from hydrophilic matrices using PLGA nanoparticles for neural drug delivery. *Biomaterials* 27 (15) (2006) 3031–3037.
34. T. Hickey, D. Kreutzer, D.J. Burgess, and F. Moussy, *In vivo* evaluation of a dexamethasone/ PLGA microsphere system designed to suppress the inflammatory tissue response to implantable medical devices. *J. Biomed. Mater. Res.* 61 (2) (2002) 180–187.
35. J.L. Masferrer, B.S. Zweifel, K. Seibert, and P. Needleman, Selective regulation of cellular cyclooxygenase by dexamethasone and endotoxin in mice. *J. Clin. Investig.* 86 (4) (1990) 1375–1379.
36. D.F. Chang and V. Wong, Two clinical trials of an intraocular steroid delivery system for cataract surgery. *Trans. Am. Ophthalmol. Soc.* 97 (1999) 261–274; discussion 274–269.
37. D.F. Chang, I.H. Garcia, J.D. Hunkeler, and T. Minas, Phase II results of an intraocular steroid delivery system for cataract surgery. *Ophthalmology* 106 (6) (1999) 1172–1177.
38. D.T. Tan, S.P. Chee, L. Lim, and A.S. Lim, Randomized clinical trial of a new dexamethasone delivery system (Surodex) for treatment of post-cataract surgery inflammation. *Ophthalmology* 106 (2) (1999) 223–231.
39. M. Kodama, J. Numaga, A. Yoshida, et al., Effects of a new dexamethasone-delivery system (Surodex) on experimental intraocular inflammation models. *Graefes Arch. Clin. Exp. Ophthalmol.* 241 (11) (2003) 927–933.
40. F. Kagaya, T. Usui, K. Kamiya, et al., Intraocular dexamethasone delivery system for corneal transplantation in an animal model. *Cornea* 21 (2) (2002) 200–202.
41. N. Khardori and M. Yassien, Biofilms in device-related infections. *J. Ind. Microbiol.* 15 (3) (1995) 141–147.
42. V. Dev, N. Eigler, M.C. Fishbein, et al., Sustained local drug delivery to the arterial wall via biodegradable microspheres. *Cathet. Cardiovasc. Diagn.* 41 (3) (1997) 324–332.
43. L.A. Guzman, V. Labhasetwar, C. Song, et al., Local intraluminal infusion of biodegradable polymeric nanoparticles: a novel approach for prolonged drug delivery after balloon angioplasty. *Circulation* 94 (6) (1996) 1441–1448.
44. A.M. Lincoff, J.G. Furst, S.G. Ellis, R.J. Tuch, and E.J. Topol, Sustained local delivery of dexamethasone by a novel intravascular eluting stent to prevent restenosis in the porcine coronary injury model. *J. Am. Coll. Cardiol.* 29 (4) (1997) 808–816.
45. G.C. Timmis, J. Helland, D.C. Westveer, J. Stewart, and S. Gordon, The evolution of low threshold leads. *J. Cardiovasc. Electrophysiol.* 1 (4) (1990) 313–334.
46. T. Hickey, D. Kreutzer, D.J. Burgess, and F. Moussy, *In vivo* evaluation of a dexamethasone/ PLGA microsphere system designed to suppress the inflammatory tissue response to implantable medical devices. *J. Biomed. Mater. Res.* 61 (2) (2002) 180–187.

47. S.D. Patil, F. Papadmitrakopoulos, and D.J. Burgess, Concurrent delivery of dexamethasone and VEGF for localized inflammation control and angiogenesis. *J. Control. Release* 117 (1) (2007) 68–79.
48. H.K. Tilakaratne, S.K. Hunter, M.E. Andracki, J.A. Benda, and V.G. Rodgers, Characterizing short-term release and neovascularization potential of multi-protein growth supplement delivered via alginate hollow fiber devices. *Biomaterials* 28 (1) (2007) 89–98.
49. L.W. Norton, E. Tegnell, S.S. Toporek, and W.M. Reichert, *In vitro* characterization of vascular endothelial growth factor and dexamethasone releasing hydrogels for implantable probe coatings. *Biomaterials* 26 (16) (2005) 3285–3297.
50. N.K. Jain, J.L. Vegad, A.K. Katiyar, and R.P. Awadhiya, Effects of anti-inflammatory drugs on increased vascular permeability in acute inflammatory response in the chicken. *Avian Pathol.* 24 (4) (1995) 723–729.
51. J. Sung, P.W. Barone, H. Kong, and M.S. Strano, Sequential delivery of dexamethasone and VEGF to control local tissue response for carbon nanotube fluorescence based microcapillary implantable sensors. *Biomaterials* 30 (4) (2009) 622–631.
52. S. Chan, Targeting the mammalian target of rapamycin (mTOR): a new approach to treating cancer. *Br. J. Cancer* 91 (8) (2004) 1420–1424.
53. A.C. Gingras, B. Raught, and N. Sonenberg, mTOR signaling to translation. *Curr. Top. Microbiol. Immunol.* 279 (2004) 169–197.
54. M. Poon, J.J. Badimon, and V. Fuster, Viewpoint Overcoming restenosis with sirolimus: from alphabet soup to clinical reality. *Lancet* 359 (9306) (2002) 619–622.
55. H.M. Garcia-Garcia, S. Vaina, K. Tsuchida, and P.W. Serruys, Drug-eluting stents. *Arch. Cardiol. Mex.* 76 (3) (2006) 297–319.
56. S. Weinbaum, J.M. Tarbell, and E.R. Damiano, The structure and function of the endothelial glycocalyx layer. *Annu. Rev. Biomed. Eng.* 9 (2007) 121–167.
57. S. Li and J.J. Henry, Nonthrombogenic approaches to cardiovascular bioengineering. *Annu. Rev. Biomed. Eng.* 13 (2011) 451–475.
58. L.C. Best, T.J. Martin, R.G. Russell, and F.E. Preston, Prostacyclin increases cyclic AMP levels and adenylate cyclase activity in platelets. *Nature* 267 (5614) (1977) 850–852.
59. J.C. de Graaf, J.D. Banga, S. Moncada, R.M. Palmer, P.G. de Groot, and J.J. Sixma, Nitric oxide functions as an inhibitor of platelet adhesion under flow conditions. *Circulation* 85 (6) (1992) 2284–2290.
60. C.K. Hashi, Y. Zhu, G.Y. Yang, et al., Antithrombogenic property of bone marrow mesenchymal stem cells in nanofibrous vascular grafts. *Proc. Natl. Acad. Sci. USA* 104 (29) (2007) 11915–11920.
61. M. Bernfield, M. Gotte, P.W. Park, et al., Functions of cell surface heparan sulfate proteoglycans. *Annu. Rev. Biochem.* 68 (1999) 729–777.
62. M. Bernfield, R. Kokenyesi, M. Kato, et al., Biology of the syndecans: a family of transmembrane heparan sulfate proteoglycans. *Annu. Rev. Cell Biol.* 8 (1992) 365–393.
63. H. Matsuno, O. Kozawa, M. Niwa, et al., Differential role of components of the fibrinolytic system in the formation and removal of thrombus induced by endothelial injury. *Thromb. Haemost.* 81 (4) (1999) 601–604.
64. W.P. Fay, N. Garg, and M. Sunkar, Vascular functions of the plasminogen activation system. *Arterioscler. Thromb. Vasc. Biol.* 27 (6) (2007) 1231–1237.

65. J. Schneiderman, M.S. Sawdey, M.R. Keeton, et al., Increased type 1 plasminogen activator inhibitor gene expression in atherosclerotic human arteries. *Proc. Natl. Acad. Sci. USA* 89 (15) (1992) 6998–7002.
66. K. Bajou, A. Noel, R.D. Gerard, et al., Absence of host plasminogen activator inhibitor 1 prevents cancer invasion and vascularization. *Nat. Med.* 4 (8) (1998) 923–928.
67. W.P. Fay, Plasminogen activator inhibitor 1, fibrin, and the vascular response to injury. *Trends Cardiovasc. Med.* 14 (5) (2004) 196–202.
68. F. Hitz, D. Klingbiel, A. Omlin, S. Riniker, A. Zerz, and T. Cerny, Athrombogenic coating of long-term venous catheter for cancer patients: a prospective, randomised, double-blind trial. *Ann. Hematol.* 91 (4) (2012) 613–620.
69. M.C. Frost, M.M. Reynolds, and M.E. Meyerhoff, Polymers incorporating nitric oxide releasing/generating substances for improved biocompatibility of blood-contacting medical devices. *Biomaterials* 26 (14) (2005) 1685–1693.
70. K.D. Mason, M.R. Carpinelli, J.I. Fletcher, et al., Programmed anuclear cell death delimits platelet life span. *Cell* 128 (6) (2007) 1173–1186.
71. A.E. Engberg, J.P. Rosengren-Holmberg, H. Chen, et al., Blood protein–polymer adsorption: implications for understanding complement-mediated hemoincompatibility. *J. Biomed. Mater. Res. A* (2011).
72. J. Andersson, K.N. Ekdahl, R. Larsson, U.R. Nilsson, and B. Nilsson, C3 adsorbed to a polymer surface can form an initiating alternative pathway convertase. *J. Immunol.* 168 (11) (2002) 5786–5791.
73. J. Wettero, T. Bengtsson, and P. Tengvall, C1q-independent activation of neutrophils by immunoglobulin M-coated surfaces. *J. Biomed. Mater. Res.* 57 (4) (2001) 550–558.
74. J. Back, M.H. Lang, G. Elgue, et al., Distinctive regulation of contact activation by antithrombin and C1-inhibitor on activated platelets and material surfaces. *Biomaterials* 30 (34) (2009) 6573–6580.
75. B.D. Brooks, A.E. Brooks, and D.W. Grainger, Antimicrobial medical devices in preclinical development and use, in T.F. Moriarty, S.A.J. Zaat, and H.J. Busscher (Eds.), *Biomaterials Associated Infection*, Springer, 2013, pp. 307–354.
76. J.M. Schierholz and J. Beuth, Implant infections: a haven for opportunistic bacteria. *J. Hosp. Infect.* 49 (2) (2001) 87–93.
77. J.D. Bryers, Medical biofilms. *Biotechnol. Bioeng.* 100 (1) (2008) 1–18.
78. A.F. Widmer, New Developments in diagnosis and treatment of infection in orthopedic implants. *Clin. Infect. Dis.* 33 (Suppl. 2) (2001) S94–S106.
79. W. Zimmerli, A.F. Widmer, M. Blatter, R. Frei, and P.E. Ochsner, Role of rifampin for treatment of orthopedic implant-related staphylococcal infections: a randomized controlled trial. Foreign-Body Infection (FBI) Study Group. *JAMA* 279 (19) (1998) 1537–1541.
80. A. Tanner, M.F.J. Maiden, K. Lee, L.B. Shulman, and H.P. Weber, Dental implant infections. *Clin. Infect. Dis.* 25 (Suppl. 2) (1997) S213–S217.
81. M.D. John, P.L. Hibberd, A.W. Karchmer, L.A. Sleeper, and S.B. Calderwood, *Staphylococcus aureus* prosthetic valve endocarditis: optimal management and risk factors for death. *Clin. Infect. Dis.* 26 (6) (1998) 1302–1309.
82. W. Vongpatanasin, L.D. Hillis, and R.A. Lange, Prosthetic heart valves. *N. Engl. J. Med.* 335 (6) (1996) 407–416.

83. H.J. Gassel, I. Klein, U. Steger, et al., Surgical management of prosthetic vascular graft infection: comparative retrospective analysis of 30 consecutive cases. *Vasa* 31 (1) (2002) 48–55.
84. C. von Eiff, B. Jansen, W. Kohnen, and K. Becker, Infections associated with medical devices: pathogenesis, management and prophylaxis. *Drugs* 65 (2) (2005) 179–214.
85. J.K. Dart, F. Stapleton, and D. Minassian, Contact lenses and other risk factors in microbial keratitis. *Lancet* 338 (8768) (1991) 650–653.
86. S.D. Elek and P.E. Conen, The virulence of *Staphylococcus pyogenes* for man: a study of the problems of wound infection. *Br. J. Exp. Pathol.* 38 (6) (1957) 573–586.
87. H.J. Busscher, H.C. van der Mei, G. Subbiahdoss, et al., Biomaterial-associated infection: locating the finish line in the race for the surface. *Sci. Transl. Med.* 4 (153) (2012) 153rv110.
88. A. Gruntzig, Transluminal dilatation of coronary–artery stenosis. *Lancet* 1 (8058) (1978) 263.
89. D.L. Fischman, M.B. Leon, D.S. Baim, et al., A randomized comparison of coronary-stent placement and balloon angioplasty in the treatment of coronary artery disease. Stent Restenosis Study Investigators. *N. Engl. J. Med.* 331 (8) (1994) 496–501.
90. P.W. Serruys, P. de Jaegere, F. Kiemeneij, et al., A comparison of balloon-expandable-stent implantation with balloon angioplasty in patients with coronary artery disease. Benestent Study Group. *N. Engl. J. Med.* 331 (8) (1994) 489–495.
91. C. Meads, C. Cummins, K. Jolly, A. Stevens, A. Burls, and C. Hyde, Coronary artery stents in the treatment of ischaemic heart disease: a rapid and systematic review. *Health Technol. Assess.* 4 (23) (2000) 1–153.
92. C. Indolfi, A. Mongiardo, A. Curcio, and D. Torella, Molecular mechanisms of in-stent restenosis and approach to therapy with eluting stents. *Trends Cardiovasc. Med.* 13 (4) (2003) 142–148.
93. P.W. Serruys, D.P. Foley, G. Jackson, et al., A randomized placebo-controlled trial of fluvastatin for prevention of restenosis after successful coronary balloon angioplasty: final results of the fluvastatin angiographic restenosis (FLARE) trial. *Eur. Heart J.* 20 (1) (1999) 58–69.
94. P.W. Serruys, D.P. Foley, M. Pieper, J.A. Kleijne, and P.J. de Feyter, The TRAPIST Study. A multicentre randomized placebo controlled clinical trial of trapidil for prevention of restenosis after coronary stenting, measured by 3-D intravascular ultrasound. *Eur. Heart J.* 22 (20) (2001) 1938–1947.
95. D.R. Holmes, Jr., M. Savage, J.M. LaBlanche, et al., Results of Prevention of REStenosis with Tranilast and its Outcomes (PRESTO) trial. *Circulation* 106 (10) (2002) 1243–1250.
96. M.C. Morice, P.W. Serruys, J.E. Sousa, et al., A randomized comparison of a sirolimus-eluting stent with a standard stent for coronary revascularization. *N. Engl. J. Med.* 346 (23) (2002) 1773–1780.
97. G.W. Stone, S.G. Ellis, D.A. Cox, et al., One-year clinical results with the slow-release, polymer-based, paclitaxel-eluting TAXUS stent: the TAXUS-IV trial. *Circulation* 109 (16) (2004) 1942–1947.
98. H. Takahashi, D. Letourneur, and D.W. Grainger, Delivery of large biopharmaceuticals from cardiovascular stents: a review. *Biomacromolecules* 8 (11) (2007) 3281–3293.
99. R. Fattori and T. Piva, Drug-eluting stents in vascular intervention. *Lancet* 361 (9353) (2003) 247–249.

100. F. Liistro and L. Bolognese, Drug-eluting stents. *Heartdrug* 3 (4) (2003) 203–213.
101. D.R. McLean and N.L. Eiger, Stent design: implications for restenosis. *Rev. Cardiovasc. Med.* 3 (Suppl. 5) (2002) S16–S22.
102. A. Finkelstein, D. McClean, S. Kar, et al., Local drug delivery via a coronary stent with programmable release pharmacokinetics. *Circulation* 107 (5) (2003) 777–784.
103. S.-J. Song, K.S. Kim, Y.J. Park, M.H. Jeong, Y.-M. Ko, and D.L. Cho, Preparation of a dual-drug-eluting stent by grafting of ALA with abciximab on a bare metal stent. *J. Mater. Chem.* 19 (43) (2009) 8135–8141.
104. S. Venkatraman and F. Boey, Release profiles in drug-eluting stents: issues and uncertainties. *J. Control. Release* 120 (3) (2007) 149–160.
105. J.E. Sousa, M.A. Costa, A. Abizaid, et al., Lack of neointimal proliferation after implantation of sirolimus-coated stents in human coronary arteries: a quantitative coronary angiography and three-dimensional intravascular ultrasound study. *Circulation* 103 (2) (2001) 192–195.
106. P. Birkenhauer, Z. Yang, and B. Gander, Preventing restenosis in early drug-eluting stent era: recent developments and future perspectives. *J. Pharm. Pharmacol.* 56 (11) (2004) 1339–1356.
107. P.W. Serruys and J. Daemen, Late stent thrombosis. *Circulation* 115 (11) (2007) 1433–1439.
108. E.P. McFadden, E. Stabile, E. Regar, et al., Late thrombosis in drug-eluting coronary stents after discontinuation of antiplatelet therapy. *Lancet* 364 (9444) (2004) 1519–1521.
109. O.F. Bertrand, R. Sipehia, R. Mongrain, et al., Biocompatibility aspects of new stent technology. *J. Am. Coll. Cardiol.* 32 (3) (1998) 562–571.
110. G. Nakazawa, A.V. Finn, E. Ladich, et al., Drug-eluting stent safety: findings from preclinical studies. *Expert Rev. Cardiovasc. Ther.* 6 (10) (2008) 1379–1391.
111. G. Nakazawa, A.V. Finn, M. Joner, et al., Delayed arterial healing and increased late stent thrombosis at culprit sites after drug-eluting stent placement for acute myocardial infarction patients: an autopsy study. *Circulation* 118 (11) (2008) 1138–1145.
112. F. Feres, J.R. Costa, Jr., and A. Abizaid, Very late thrombosis after drug-eluting stents. *Catheter. Cardiovasc. Interv.* 68 (1) (2006) 83–88.
113. N.G. Kounis, G. Hahalis, and T.C. Theoharides, Coronary stents, hypersensitivity reactions, and the Kounis syndrome. *J. Interv. Cardiol.* 20 (5) (2007) 314–323.
114. L.K. Pendyala, J. Li, T. Shinke, et al., Endothelium-dependent vasomotor dysfunction in pig coronary arteries with Paclitaxel-eluting stents is associated with inflammation and oxidative stress. *JACC Cardiovasc. Interv.* 2 (3) (2009) 253–262.
115. T.F. Luscher, J. Steffel, F.R. Eberli, et al., Drug-eluting stent and coronary thrombosis: biological mechanisms and clinical implications. *Circulation* 115 (8) (2007) 1051–1058.
116. L. Mauri, W.H. Hsieh, J.M. Massaro, K.K. Ho, R. D'Agostino, and D.E. Cutlip, Stent thrombosis in randomized clinical trials of drug-eluting stents. *N. Engl. J. Med.* 356 (10) (2007) 1020–1029.
117. E. Grube and L. Buellesfeld, BioMatrix® Biolimus A9®-eluting coronary stent: a next-generation drug-eluting stent for coronary artery disease. *Expert Rev. Med. Devices* 3 (6) (2006) 731–741.
118. I. Raad, Intravascular-catheter-related infections. *Lancet* 351 (9106) (1998) 893–898.

119. S.J. Schwab and G. Beathard, The hemodialysis catheter conundrum: hate living with them, but can't live without them. *Kidney Int.* 56 (1) (1999) 1–17.
120. I. Raad, R. Darouiche, J. Dupuis, et al., Central venous catheters coated with minocycline and rifampin for the prevention of catheter-related colonization and bloodstream infections. A randomized, double-blind trial. The Texas Medical Center Catheter Study Group. *Ann. Intern. Med.* 127 (4) (1997) 267–274.
121. D.G. Maki, D.M. Kluger, and C.J. Crnich, The risk of bloodstream infection in adults with different intravascular devices: a systematic review of 200 published prospective studies. *Mayo Clin. Proc.* 81 (9) (2006) 1159–1171.
122. I. Raad, H. Hanna, and D. Maki, Intravascular catheter-related infections: advances in diagnosis, prevention, and management. *Lancet Infect. Dis.* 7 (10) (2007) 645–657.
123. I. Chatzinikolaou and I.I. Raad, Central venous catheter related infections: the role of antimicrobial catheters, in L.A. Doughty and P. Linden (Eds.), *Immunology and Infectious Disease*, Vol. 3, Springer, 2003, pp. 187–215.
124. M. Viot, Intravenous access: related problems in oncology. *Int. J. Antimicrob. Agents* 16 (2) (2000) 165–168.
125. Y. Jaffer, N.M. Selby, M.W. Taal, R.J. Fluck, and C.W. McIntyre, A meta-analysis of hemodialysis catheter locking solutions in the prevention of catheter-related infection. *Am. J. Kidney Dis.* 51 (2) (2008) 233–241.
126. J.C. Hockenhull, K.M. Dwan, G.W. Smith, et al., The clinical effectiveness of central venous catheters treated with anti-infective agents in preventing catheter-related bloodstream infections: a systematic review. *Crit. Care Med.* 37 (2) (2009) 702–712.
127. P. Ramritu, K. Halton, P. Collignon, et al., A systematic review comparing the relative effectiveness of antimicrobial-coated catheters in intensive care units. *Am. J. Infect. Control* 36 (2) (2008) 104–117.
128. X. Zhang, Anti-infective coatings reduce device-related infections. *Antimicrobial/Anti-Infective Materials: Principles and Applications*, CRC Press, 2000, pp. 149–180.
129. R.S. Greco and R.A. Harvey, The role of antibiotic bonding in the prevention of vascular prosthetic infections. *Ann. Surg.* 195 (2) (1982) 168–171.
130. G.D. Kamal, M.A. Pfaller, L.E. Rempe, and P.J. Jebson, Reduced intravascular catheter infection by antibiotic bonding: a prospective, randomized, controlled trial. *JAMA* 265 (18) (1991) 2364–2368.
131. G. Cheng, G. Li, H. Xue, S. Chen, J.D. Bryers, and S. Jiang, Zwitterionic carboxybetaine polymer surfaces and their resistance to long-term biofilm formation. *Biomaterials* 30 (28) (2009) 5234–5240.
132. R.S. Smith, Z. Zhang, M. Bouchard, et al., Vascular catheters with a nonleaching polysulfobetaine surface modification reduce thrombus formation and microbial attachment. *Sci. Transl. Med.* 4 (153) (2012) 153ra132.
133. J.M. Schierholz, A.F. Rump, G. Pulverer, and J. Beuth, Anti-infective catheters: novel strategies to prevent nosocomial infections in oncology. *Anticancer Res.* 18 (5B) (1998) 3629–3638.
134. M.P. Pai, S.L. Pendland, and L.H. Danziger, Antimicrobial-coated/bonded and -impregnated intravascular catheters. *Ann. Pharmacother.* 35 (10) (2001) 1255–1263.
135. R.O. Darouiche, I.I. Raad, S.O. Heard, et al., A comparison of two antimicrobial-impregnated central venous catheters. Catheter Study Group. *N. Engl. J. Med.* 340 (1) (1999) 1–8.

136. P.E. Marik, G. Abraham, P. Careau, J. Varon, and R.E. Fromm, Jr., The *ex vivo* antimicrobial activity and colonization rate of two antimicrobial-bonded central venous catheters. *Crit. Care Med.* 27 (6) (1999) 1128–1131.
137. R.O. Darouiche, D.H. Berger, N. Khardori, et al., Comparison of antimicrobial impregnation with tunneling of long-term central venous catheters: a randomized controlled trial. *Ann. Surg.* 242 (2) (2005) 193–200.
138. I. Raad, R. Reitzel, Y. Jiang, R.F. Chemaly, T. Dvorak, and R. Hachem, Anti-adherence activity and antimicrobial durability of anti-infective-coated catheters against multidrug-resistant bacteria. *J. Antimicrob. Chemother.* 62 (4) (2008) 746–750.
139. I. Raad, J.A. Mohamed, R.A. Reitzel, et al., Improved antibiotic-impregnated catheters with extended-spectrum activity against resistant bacteria and fungi. *Antimicrob. Agents Chemother.* 56 (2) (2012) 935–941.
140. S. Khare, M.K. Hondalus, J. Nunes, B.R. Bloom, and L. Garry Adams, *Mycobacterium bovis* DeltaleuD auxotroph-induced protective immunity against tissue colonization, burden and distribution in cattle intranasally challenged with Mycobacterium bovis Ravenel S. *Vaccine* 25 (10) (2007) 1743–1755.
141. K.N. Stevens, O. Crespo-Biel, E.E. van den Bosch, et al., The relationship between the antimicrobial effect of catheter coatings containing silver nanoparticles and the coagulation of contacting blood. *Biomaterials* 30 (22) (2009) 3682–3690.
142. J. Hirsh, S.S. Anand, J.L. Halperin, and V. Fuster, Guide to anticoagulant therapy: Heparin: a statement for healthcare professionals from the American Heart Association. *Circulation* 103 (24) (2001) 2994–3018.
143. H.T. Tevaearai, X.M. Mueller, I. Seigneul, et al., Trillium coating of cardiopulmonary bypass circuits improves biocompatibility. *Int. J. Artif. Organs* 22 (9) (1999) 629–634.
144. P. Daeihagh, J. Jordan, J. Chen, and M. Rocco, Efficacy of tissue plasminogen activator administration on patency of hemodialysis access catheters. *Am. J. Kidney Dis.* 36 (1) (2000) 75–79.
145. M.H. Mokrzycki, K. Jean-Jerome, H. Rush, M.P. Zdunek, and S.O. Rosenberg, A randomized trial of minidose warfarin for the prevention of late malfunction in tunneled, cuffed hemodialysis catheters. *Kidney Int.* 59 (5) (2001) 1935–1942.
146. S.T. Reddy, K.K. Chung, C.J. McDaniel, R.O. Darouiche, J. Landman, and A.B. Brennan, Micropatterned surfaces for reducing the risk of catheter-associated urinary tract infection: an *in vitro* study on the effect of sharklet micropatterned surfaces to inhibit bacterial colonization and migration of uropathogenic *Escherichia coli*. *J. Endourol.* 25 (9) (2011) 1547–1552.
147. K.A.M. Wald Hl, Nonpayment for harms resulting from medical care: catheter-associated urinary tract infections. *JAMA* 298 (23) (2007) 2782–2784.
148. J.W. Warren, Catheter-associated urinary tract infections. *Infect. Dis. Clin. North Am.* 11 (3) (1997) 609–622.
149. K.H. Bichler, E. Eipper, K. Naber, V. Braun, R. Zimmermann, and S. Lahme, Urinary infection stones. *Int. J. Antimicrob. Agents* 19 (6) (2002) 488–498.
150. R.M. Donlan and J.W. Costerton, Biofilms: survival mechanisms of clinically relevant microorganisms. *Clin. Microbiol. Rev.* 15 (2) (2002) 167–193.
151. R. Pickard, T. Lam, G. MacLennan, et al., Antimicrobial catheters for reduction of symptomatic urinary tract infection in adults requiring short-term catheterisation in hospital: a multicentre randomised controlled trial. *Lancet* 380 (9857) (2012) 1927–1935.

152. S.M. Jacobsen, D.J. Stickler, H.L.T. Mobley, and M.E. Shirtliff, Complicated catheter-associated urinary tract infections due to *Escherichia coli* and *Proteus mirabilis*. *Clin. Microbiol. Rev.* 21 (1) (2008) 26–59.
153. G. Regev-Shoshani, M. Ko, A. Crowe, and Y. Av-Gay, Comparative efficacy of commercially available and emerging antimicrobial urinary catheters against bacteriuria caused by *E. coli in vitro*. *Urology* 78 (2) (2011) 334–339.
154. T. Bjarnsholt, K. Kirketerp-Moller, S. Kristiansen, et al., Silver against *Pseudomonas aeruginosa* biofilms. *APMIS* 115 (8) (2007) 921–928.
155. A. Trampuz and A.F. Widmer, Infections associated with orthopedic implants. *Curr. Opin. Infect. Dis.* 19 (4) (2006) 349–356.
156. S. Nishiguchi, H. Kato, H. Fujita, et al., Titanium metals form direct bonding to bone after alkali and heat treatments. *Biomaterials* 22 (18) (2001) 2525–2533.
157. S.M. Sporer and W.G. Paprosky, Biologic fixation and bone ingrowth. *Orthop. Clin. North Am.* 36 (1) (2005) 105–111, vii.
158. K. Hirakawa, J.J. Jacobs, R. Urban, and T. Saito, Mechanisms of failure of total hip replacements: lessons learned from retrieval studies. *Clin. Orthop. Relat. Res.* (420) (2004) 10–17.
159. E.W. Morscher, Failures and successes in total hip replacement: why good ideas may not work. *Scand. J. Surg.* 92 (2) (2003) 113–120.
160. D.W. Grainger, Targeted delivery of therapeutics to bone and connective tissues. *Adv. Drug Deliv. Rev.* 64 (12) (2012) 1061–1062.
161. E. Verron, I. Khairoun, J. Guicheux, and J.-M. Bouler, Calcium phosphate biomaterials as bone drug delivery systems: a review. *Drug Discov. Today* 15 (13–14) (2010) 547–552.
162. B. Doll, C. Sfeir, S.R. Winn, J. Huard, and J. Hollinger, Critical aspects of tissue-engineered therapy for bone regeneration. *Crit. Rev. Eukaryot. Gene Expr.* 11 (1–3) (2001) 173–198.
163. J.K. Leach and D.J. Mooney, Bone engineering by controlled delivery of osteoinductive molecules and cells. *Expert Opin. Biol. Ther.* 4 (7) (2004) 1015–1027.
164. S.C. Gamradt and J.R. Lieberman, Genetic modification of stem cells to enhance bone repair. *Ann. Biomed. Eng.*, 32 (1) (2004) 136–147.
165. D. Samartzis, N. Khanna, F.H. Shen, and H.S. An, Update on bone morphogenetic proteins and their application in spine surgery. *J. Am. Coll. Surg.* 200 (2) (2005) 236–248.
166. V. Luginbuehl, L. Meinel, H.P. Merkle, and B. Gander, Localized delivery of growth factors for bone repair. *Eur. J. Pharm. Biopharm.* 58 (2) (2004) 197–208.
167. J.K. Leach and D.J. Mooney, Bone engineering by controlled delivery of osteoinductive molecules and cells. *Expert Opin. Biol. Ther.* 4 (7) (2004) 1015–1027.
168. F. Kandziora, H. Bail, G. Schmidmaier, et al., Bone morphogenetic protein-2 application by a poly(D,L-lactide)-coated interbody cage: *in vivo* results of a new carrier for growth factors. *J. Neurosurg.* 97 (1 Suppl.) (2002) 40–48.
169. Y. Wang and D.W. Grainger, RNA therapeutics targeting osteoclast-mediated excessive bone resorption. *Adv. Drug Deliv. Rev.* 64 (12) (2012) 1341–1357.
170. D.J. Holt and D.W. Grainger, Demineralized bone matrix as a vehicle for delivering endogenous and exogenous therapeutics in bone repair. *Adv. Drug Deliv. Rev.* 64 (12) (2012) 1123–1128.
171. M.D. Kofron and C.T. Laurencin, Bone tissue engineering by gene delivery. *Adv. Drug Deliv. Rev.* 58 (4) (2006) 555–576.

172. C.H. Evans, Gene delivery to bone. *Adv. Drug Deliv. Rev.* 64 (12) (2012) 1331–1340.

173. J.L. Del Pozo and R. Patel, Infection associated with prosthetic joints. *N. Engl. J. Med.* 361 (8) (2009) 787–794.

174. S.B. Trippel, Antibiotic-impregnated cement in total joint arthroplasty. *J. Bone Joint Surg. Am.* 68 (8) (1986) 1297–1302.

175. Y.Y. Huang and T.W. Chung, Microencapsulation of gentamicin in biodegradable PLA and/or PLA/PEG copolymer. *J. Microencapsul.* 18 (4) (2001) 457–465.

176. D. Yu, J. Wong, Y. Matsuda, J.L. Fox, W.I. Higuchi, and M. Otsuka, Self-setting hydroxyapatite cement: a novel skeletal drug-delivery system for antibiotics. *J. Pharm. Sci.* 81 (6) (1992) 529–531.

177. B. Espehaug, L.B. Engesaeter, S.E. Vollset, L.I. Havelin, and N. Langeland, Antibiotic prophylaxis in total hip arthroplasty. Review of 10,905 primary cemented total hip replacements reported to the Norwegian arthroplasty register, 1987 to 1995. *J. Bone Joint Surg. Br.* 79 (4) (1997) 590–595.

178. B. Picknell, L. Mizen, and R. Sutherland, Antibacterial activity of antibiotics in acrylic bone cement. *J. Bone Joint Surg. Br.* 59 (3) (1977) 302–307.

179. R.A. Elson, A.E. Jephcott, D.B. McGechie, and D. Verettas, Bacterial infection and acrylic cement in the rat. *J. Bone Joint Surg. Br.* 59-B (4) (1977) 452–457.

180. M.W. Nijhof, H.P. Stallmann, H.C. Vogely, et al., Prevention of infection with tobramycin-containing bone cement or systemic cefazolin in an animal model. *J. Biomed. Mater. Res.* 52 (4) (2000) 709–715.

181. H. Wahlig, E. Dingeldein, R. Bergmann, and K. Reuss, The release of gentamicin from polymethylmethacrylate beads. An experimental and pharmacokinetic study. *J. Bone Joint Surg. Br.* 60-B (2) (1978) 270–275.

182. S.F. Hoff, R.H. Fitzgerald, Jr., and P.J. Kelly, The depot administration of penicillin G and gentamicin in acrylic bone cement. *J. Bone Joint Surg. Am.* 63 (5) (1981) 798–804.

183. L. Bunetel, A. Segui, M. Cormier, E. Percheron, and F. Langlais, Release of gentamicin from acrylic bone cement. *Clin. Pharmacokinet.* 17 (4) (1989) 291–297.

184. G.H. Walenkamp, T.B. Vree, and T.J. van Rens, Gentamicin-PMMA beads. Pharmacokinetic and nephrotoxicological study. *Clin. Orthop. Relat. Res.* (205) (1986) 171–183.

185. M. Lucke, G. Schmidmaier, S. Sadoni, et al., Gentamicin coating of metallic implants reduces implant-related osteomyelitis in rats. *Bone* 32 (5) (2003) 521–531.

186. G. Schmidmaier, M. Lucke, B. Wildemann, N.P. Haas, and M. Raschke, Prophylaxis and treatment of implant-related infections by antibiotic-coated implants: a review. *Injury* 37 (Suppl. 2) (2006) S105–S112.

187. H. Forster, J.S. Marotta, K. Heseltine, R. Milner, and S. Jani, Bactericidal activity of antimicrobial coated polyurethane sleeves for external fixation pins. *J. Orthop. Res.* 22 (3) (2004) 671–677.

188. P.S. Barie, Surgical site infections: epidemiology and prevention. *Surg. Infect. (Larchmt)* 3 (Suppl. 1) (2002) S9–S21.

189. C.E. Edmiston, G.R. Seabrook, M.P. Goheen, et al., Bacterial adherence to surgical sutures: can antibacterial-coated sutures reduce the risk of microbial contamination? *J. Am. Coll. Surg.* 203 (4) (2006) 481–489.

190. B. Osterberg and B. Blomstedt, Effect of suture materials on bacterial survival in infected wounds. An experimental study. *Acta Chir. Scand.* 145 (7) (1979) 431–434.

191. S. Katz, M. Izhar, and D. Mirelman, Bacterial adherence to surgical sutures: a possible factor in suture induced infection. *Ann. Surg.* 194 (1) (1981) 35–41.

192. S. Rothenburger, D. Spangler, S. Bhende, and D. Burkley, *In vitro* antimicrobial evaluation of Coated VICRYL* Plus Antibacterial Suture (coated polyglactin 910 with triclosan) using zone of inhibition assays. *Surg. Infect. (Larchmt)* 3 (Suppl. 1) (2002) S79–S87.

193. M. Storch, L.C. Perry, J.M. Davidson, and J.J. Ward, A 28-day study of the effect of Coated VICRYL* Plus Antibacterial Suture (coated polyglactin 910 suture with triclosan) on wound healing in guinea pig linear incisional skin wounds. *Surg. Infect. (Larchmt)* 3 (Suppl. 1) (2002) S89–S98.

194. I. Galal and K. El-Hindawy, Impact of using triclosan-antibacterial sutures on incidence of surgical site infection. *Am. J. Surg.* 202 (2) (2011) 133–138.

195. A.E. Deliaert, E. Van den Kerckhove, S. Tuinder, et al., The effect of triclosan-coated sutures in wound healing. A double blind randomised prospective pilot study. *J. Plast. Reconstr. Aesthet. Surg.* 62 (6) (2009) 771–773.

196. G.L. Moneta and J.M. Porter, Arterial substitutes in peripheral vascular surgery: a review. *J. Long Term Eff. Med. Implants.* 5 (1) (1995) 47–67.

197. J.M. Heyligers, H.J. Verhagen, J.I. Rotmans, et al., Heparin immobilization reduces thrombogenicity of small-caliber expanded polytetrafluoroethylene grafts. *J. Vasc. Surg.* 43 (3) (2006) 587–591.

198. J.M. Drake, J.R. Kestle, and S. Tuli, CSF shunts 50 years on: past, present and future. *Childs Nerv. Syst.* 16 (10–11) (2000) 800–804.

199. I.K. Pople, R. Bayston, and R.D. Hayward, Infection of cerebrospinal fluid shunts in infants: a study of etiological factors. *J. Neurosurg.* 77 (1) (1992) 29–36.

200. S.H. Farber, S.L. Parker, O. Adogwa, D. Rigamonti, and M.J. McGirt, Cost analysis of antibiotic-impregnated catheters in the treatment of hydrocephalus in adult patients. *World Neurosurg.* 74 (4–5) (2010) 528–531.

201. R. Eymann, S. Chehab, M. Strowitzki, W.I. Steudel, and M. Kiefer, Clinical and economic consequences of antibiotic-impregnated cerebrospinal fluid shunt catheters. *J. Neurosurg. Pediatr.* 1 (6) (2008) 444–450.

202. G.K. Wong, M. Ip, W.S. Poon, C.W. Mak, and R.Y. Ng, Antibiotics-impregnated ventricular catheter versus systemic antibiotics for prevention of nosocomial CSF and non-CSF infections: a prospective randomised clinical trial. *J. Neurol. Neurosurg. Psychiatry* 81 (10) (2010) 1064–1067.

203. A. Pattavilakom, D. Kotasnas, T.M. Korman, C. Xenos, and A. Danks, Duration of *in vivo* antimicrobial activity of antibiotic-impregnated cerebrospinal fluid catheters. *Neurosurgery* 58 (5) (2006) 930–935; discussion 930–935.

204. J. Webster, Handwashing in a neonatal intensive care nursery: product acceptability and effectiveness of chlorhexidine gluconate 4% and triclosan 1%. *J. Hosp. Infect.* 21 (2) (1992) 137–141.

205. S.C. von Plocki, D. Armbruster, K. Klein, et al., Biodegradable sleeves for metal implants to prevent implant-associated infection: an experimental *in vivo* study in sheep. *Vet. Surg.* 41 (3) (2012) 410–421.

206. A.M. Henslee, P.P. Spicer, D.M. Yoon, et al., Biodegradable composite scaffolds incorporating an intramedullary rod and delivering bone morphogenetic protein-2 for stabilization and bone regeneration in segmental long bone defects. *Acta Biomater.* 7 (10) (2011) 3627–3637.

207. B. Trajkovski, A. Petersen, P. Strube, M. Mehta, and G.N. Duda, Intra-operatively customized implant coating strategies for local and controlled drug delivery to bone. *Adv. Drug Deliv. Rev.* 64 (12) (2012) 1142–1151.

208. A. Lendlein and R. Langer, Biodegradable, elastic shape-memory polymers for potential biomedical applications. *Science* 296 (5573) (2002) 1673–1676.

209. A. Shmulewitz, R. Langer, and J. Patton, Convergence in biomedical technology. *Nat. Biotechnol.* 24 (3) (2006) 277–277.

210. C.M. Yakacki, R. Shandas, C. Lanning, B. Rech, A. Eckstein, and K. Gall, Unconstrained recovery characterization of shape-memory polymer networks for cardiovascular applications. *Biomaterials* 28 (14) (2007) 2255–2263.

211. S.M. Janib, A.S. Moses, and J.A. MacKay, Imaging and drug delivery using theranostic nanoparticles. *Adv. Drug Deliv. Rev.* 62 (11) (2010) 1052–1063.

212. J. Kopecek and P. Kopeckova, HPMA copolymers: origins, early developments, present, and future. *Adv. Drug Deliv. Rev.* 62 (2) (2010) 122–149.

213. Z.R. Lu, Molecular imaging of HPMA copolymers: visualizing drug delivery in cell, mouse and man. *Adv. Drug Deliv. Rev.* 62 (2) (2010) 246–257.

214. S.K. Choi, T. Thomas, M.H. Li, A. Kotlyar, A. Desai, and J.R. Baker, Jr., Light-controlled release of caged doxorubicin from folate receptor-targeting PAMAM dendrimer nanoconjugate. *Chem. Commun.* 46 (15) (2010) 2632–2634.

215. S.D. Konda, M. Aref, S. Wang, M. Brechbiel, and E.C. Wiener, Specific targeting of folate-dendrimer MRI contrast agents to the high affinity folate receptor expressed in ovarian tumor xenografts. *MAGMA* 12 (2–3) (2001) 104–113.

216. S. Erdogan and V.P. Torchilin, Gadolinium-loaded polychelating polymer-containing tumor-targeted liposomes. *Methods Mol. Biol.* 605 (2010) 321–334.

217. W. Al-Jamal, K.T. Al-Jamal, B. Tian, A. Cakebread, J.M. Halket, and K. Kostarelos, Tumor targeting of functionalized quantum dot liposome hybrids by intravenous administration. *Mol. Pharm.* 6 (2) (2009) 520–530.

218. I.K. Kwon, S.C. Lee, B. Han, and K. Park, Analysis on the current status of targeted drug delivery to tumors. *J. Control. Release* (2012)164 (2) (2009) 108–114.

2

HISTORICAL SURVEY OF DRUG DELIVERY DEVICES

NICHOLAS A. PEPPAS,[1,2,3] MARY E. CALDORERA-MOORE,[1,2] AND STEPHANIE D. STEICHEN[1]

[1]*Department of Biomedical Engineering, University of Texas at Austin, Austin, TX 78712, USA*
[2]*Department of Chemical Engineering, University of Texas at Austin, Austin, TX 78712, USA*
[3]*Division of Pharmaceutics, University of Texas at Austin, Austin, TX 78712, USA*

2.1 INTRODUCTION

In the broadest sense, drug delivery devices are systems that have the ability to administer therapeutic agents in a controlled manner over a given period of time. Since their infancy, approximately 35 years ago, controlled drug release devices have evolved into sophisticated systems that can provide real-time control over drug administration noninvasively. Drug delivery systems (DDS) have a growing number of applications and are of increasing importance in health care. These advanced systems have ushered in a new age of growth and exploration within the pharmaceutical industry. To truly appreciate the advancements that have been made in the way patients receive therapeutic agents, we need to first investigate the early pioneering controlled release devices. In this chapter, the authors will first explore the historical road of drug delivery device development. The different types of drug delivery systems will then be presented, along with the status of devices in clinical use. Finally, current research trends in the field of drug delivery devices will be investigated.

Drug–Device Combinations for Chronic Diseases, First Edition. Edited by SuPing Lyu and Ronald A. Siegel.

2.2 THE ORIGINS OF DRUG DELIVERY DEVICES

In the early years (1960–1980), the term "drug delivery" was rarely used in the pharmaceutical field. Pharmaceutical companies rarely placed first priority on novel release systems prepared by anything more than the standard tableting and coating methods used in the field. Unfortunately, to make things even more difficult, drug delivery research did not attract major funding from federal organizations. It was not until effective and efficient controlled drug release systems were developed that the field of drug release received any attention from the pharmaceutical industry. Thus, the field of controlled release systems as we know now could not have been established if it was not for the introduction of mathematical modeling and molecular design principles in pharmaceutical formulations.

There were two main research sparks that ignited the field of drug delivery and the corresponding drug delivery systems. The first was the seminal fundamental work of Professor Takeru Higuchi. Higuchi's equation was the first mathematical model to provide researchers the ability to predict the release of drugs as a function of time [1]. Higuchi's classic equation became the standard of design of drug delivery systems in the period 1960–1985 and continues to be widely used in the design of many ethical and generic products, especially of the oral/transmucosal delivery type. The second was the ground-breaking research of Dr. Judah Folkman, of Harvard Medical School, who developed the first *in vivo* controlled drug delivery system in drug-loaded shunts.

2.2.1 The Higuchi Model: Mathematical Modeling of Drug Transport and Diffusion

Dr. Takeru Higuchi originally derived the now legendary Higuchi equation to model the release of therapeutic agents from a thin ointment film into skin [2]. The following conditions were taken into consideration:

(1) Transport of the drug through the ointment is rate limited, while transport through the skin is fast.
(2) The skin is a "perfect sink," that is, the drug concentration within the skin is negligible.
(3) The size of the drug carrier vehicle (particles) is significantly smaller than the thickness of the film.
(4) Initially the drug is homogeneously dispersed throughout the film.
(5) The initial drug concentration in the film is significantly higher than the solubility of the drug in the ointment/film base.
(6) The dissolution of drug molecules within the ointment base is rapid compared with the diffusion of dissolved drug molecules within the ointment base.
(7) The diffusion coefficient of the drug within the ointment base is constant and is not dependent on the time or location within the film.

(8) The surface of the film exposed to the skin is large in comparison to its thickness and, therefore, the mathematical model of drug diffusion can be limited to one dimension (i.e., edge effects are negligible).
(9) The ointment base does not swell or dissolve during release of therapeutic agents.

Under these conditions, when a thin film of a polymer containing dissolved drugs is exposed to the perfect sink conditions of the skin, the drugs will begin to diffuse out of the devices. At the beginning, this only occurs close to the surface of the ointment film. Since the concentration of drug is in excess and the drug dissolution is rapid, the drug molecules that permeate out of the system are quickly restored by the partial dissolution of nondissolved molecules at the surface. Therefore, the concentration of dissolved therapeutic molecules within the film remains constant as long as the film is saturated with the nondissolved drug, that is, the drugs exist in excess. When all drug particles located in the region adjacent to the surface are dissolved, the concentration of dissolved molecules in this region decreases below saturation concentration. Subsequently, due to concentration gradient, the dissolved drug located further away from the film's surface diffuses through the ointment base into the skin. The concentration of dissolved therapeutics in this new region remains constant as long as the nondissolved drug remains in excess in this region.

After a given period of time (t), the drug's "concentration–distance profile," as demonstrated in Figure 2.1, is achieved in the ointment film. On the y-axis, the drug concentration is plotted against distance on the x-axis. As shown in the cross-sectional schematic of the film, at time t, regions of the film have been depleted of drug and at a specific distance from the surface a sharp front is observed at which the drug concentration steeply increases from "saturation concentration," c_s, to "initial

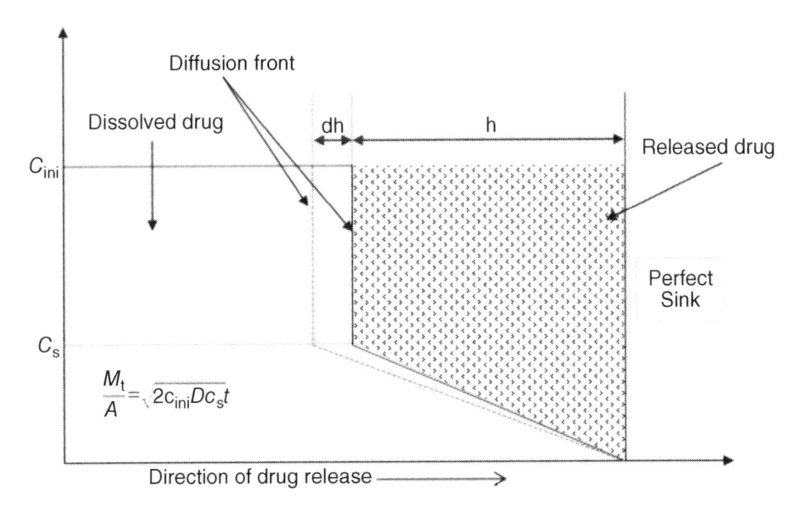

FIGURE 2.1 A schematic representation of the Higuchi model of drug transport from a drug-laden ointment to a perfect sink. (Adopted from Ref. [1].)

concentration," c_{ini}. This front is formed at distance h from the film's surface and is sometimes referred to as the "diffusion front." To calculate the amount of drug released from the ointment film at this time point t, the drug concentration–distance profile within the area of the ointment depleted of excess drug must be known.

To model the drug concentration gradient in the ointment region located between the "diffusion front" and the perfect sink, Higuchi used a "pseudo-steady-state" approximation, which is valid only for systems initially containing a large excess of drug. Since the ointment base does not change physically or chemically, that is, it does not swell or dissolve during the release of drug, pseudo-steady-state conditions are achieved for drug diffusion: a saturated drug solution on the one side, perfect sink on the other side, and a specified distance in between.

Applying Fick's second law of diffusion, the drug concentration–distance profile between the surface of the film and the "diffusion front" is linear at time t (shown by solid line in Figure 2.1). The diffusion front moves against the direction of drug release ($h + \mathrm{d}h$) as time increases ($t + \Delta t$) (shown by the dotted line in Figure 2.1). The amount of drug released from the film at time t can be represented by the patterned trapezoid region. The "area" of this region corresponds to the cumulative amount of therapeutic released, M_t, divided by the surface area of the film exposed to the skin, A. Using this simple geometry, it is clear that the cumulative amount of drug released from the ointment film at time t, M_t, can be calculated as

$$\frac{M_t}{A} = h\left(c_{ini} - \frac{c_s}{2}\right) \tag{2.1}$$

However, if h is unknown, the equation can be expressed as

$$\frac{\mathrm{d}M}{A} = c_{ini}\mathrm{d}h - \frac{c_s}{2}\mathrm{d}h \tag{2.2}$$

where h is a function of time and concentration. Fick's first law of diffusion can be used to quantify the amount of drug released from the ointment film within time interval $\mathrm{d}t$ (considering a saturated drug solution at distance h from the surface and perfect sink conditions):

$$\frac{\mathrm{d}M}{\mathrm{d}t} = AD\frac{c_s}{h} \tag{2.3}$$

Combining Eqs. (2.2) and (2.3) allowed Higuchi to solve for h:

$$h = 2\sqrt{\frac{Dtc_s}{2c_{ini} - c_s}} \tag{2.4}$$

Substituting Eq. (2.4) into Eq. (2.1) and simplifying gives

$$\frac{M_t}{A} = \sqrt{(2c_{ini} - c_s)Dtc_s} \tag{2.5}$$

For a high initial concentration of drug ($c_{ini} >> c_s$), this equation can be further simplified to obtain the classical Higuchi equation:

$$\frac{M_t}{A} = \sqrt{2c_{ini}Dc_st} \tag{2.6}$$

As observed, the Higuchi equation describes "square root of time" release kinetics, that is, the release rate decreases proportionally to the square root of time. More specifics on the derivation, applications, and adaptations of the Higuchi model can be found in a comprehensive review by Siepmann and Peppas [1].

In an attempt to provide a simple analysis of controlled release data from polymeric devices of different geometries, Ritger and Peppas proposed a new empirical, exponential expression that relates the fractional release of drug M_t/M_∞ to the release time, t [3]:

$$\frac{M_t}{M_\infty} = kt^n \tag{2.7}$$

where k is a constant that incorporates the characteristics of both the macromolecular network system and the drug, and n is the diffusional exponent, which is indicative of the transport mechanism. This equation can be used to analyze drug release from sheets, cylinders, spheres, disks, and polydisperse microspheres under perfect sink conditions.

Figure 2.2 shows the effect geometry plays on drug diffusion over time for slab, cylinder, and sphere systems. The drug release in Eq. (2.7) is quantitated as the fractional release, M_t/M_∞, or the amount of drug released at time t, compared to the total drug release at infinite time. Furthermore, this empirical relationship can describe various drug release kinetics, namely, zero- and first-order release behaviors (shown in Figure 2.3). In systems where the drug release is independent of the mass of remaining drug in the device, the release kinetics is considered *zero order*. For devices whose drug release is proportional to the amount of drug remaining within the system, release follows *first-order* release kinetics [4–6].

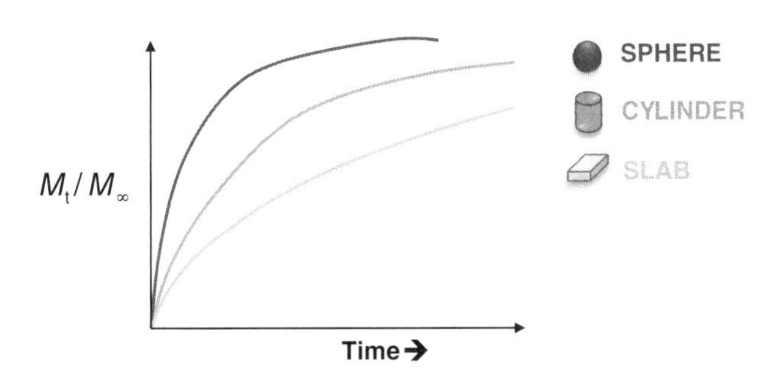

FIGURE 2.2 The effects of geometry on the release kinetics of a therapeutic from spherical, cylindrical, or slab systems. (Adopted from Ref. [3].)

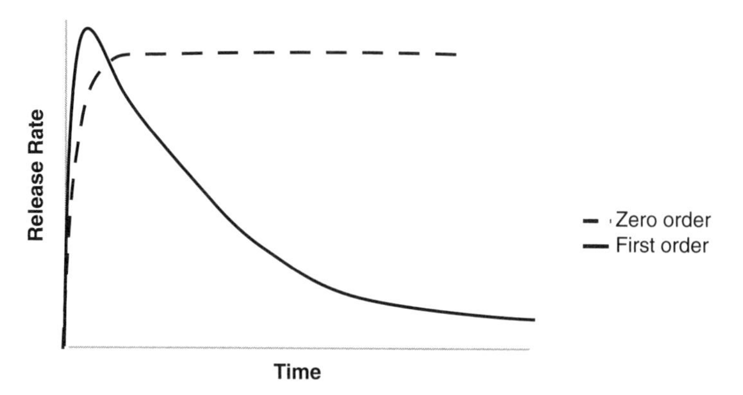

FIGURE 2.3 Generalized release rates of zero- and first-order drug delivery systems.

2.2.2 Development of the First In Vivo Zero-Order Drug Delivery Devices

The second "spark" in the early drug delivery history was the ground-breaking research of Dr. Judah Folkman, at Harvard Medical School. In the mid-1960s, Folkman and his associates were using Silastic® (silicone rubber) arterio-venous shunt for circulation of rabbit blood when he discovered that when he exposed the silicone tubing to anesthetic gases on the outside, the rabbits would fall asleep. From this, he hypothesized that sealed segments of silicone tubing containing a drug could be implanted, and if the tubing did not change in composition or dimension, the implant would be a constant rate drug delivery device. He also demonstrated that the release rate decreased as the tubing thickness increased. This was the first *in vivo* example of zero-order controlled drug delivery. Folkman and Long's first publication of this pioneering research [7] ushered in a new age in "drug delivery."

In 1966, the Population Council developed a contraceptive subcutaneous implant called the Norplant®, which was a direct extension of Folkman and Long's work. The Norplant was comprised of six silicone rubber tubes filled with a contraceptive steroid, levo-Norgestrel. Around the same period (1970–1980s) the World Health Organization (WHO) developed a doughnut-shaped contraceptive device for vaginal insertion made of silicone rubber and that used the silicone rubber as the rate controlling membrane. The membrane was composed of the contraceptive steroid blended with silicone rubber and coated with a drug-free poly(dimethyl siloxane) (PDMS).

2.2.3 The Commercialization of Drug Delivery: The ALZA Corporation

The small flame that was created by these two independent research sparks was kindled into a blazing research field by the ALZA Corporation, founded in 1967, by Dr. Alejandro Zaffaroni. ALZA assembled a dream team of expert scientists and engineers to work together to design and create some of the first-order controlled drug delivery systems on the market. ALZA Corporation's early contributions to the field include the development of reservoir (membrane) and osmotic drug delivery devices [8]. These systems progressed into commercial systems for ocular therapy, contraceptives, and transdermal delivery.

Some of the first-order controlled drug delivery devices of ALZA were an ophthalmic insert called the Ocusert®, which released the antiglaucoma drug, pilocarpine, at a constant rate directly to the eye, and an intrauterine device (IUD) called Progestersert®, which released progesterone at a constant rate in the uterine cavity. Both of these systems utilized a poly(ethylene-*co*-vinyl acetate) membrane (popularly known as an EVAc membrane or film) to control the rate of drug release.

ALZA was also working on the development of transdermal delivery patches capable of administering a therapeutic via a rate-controlled adhesive patch. In 1971, Alejandro Zaffaroni was issued one of the first patents for such systems, which he referred to as a "bandage for administering drugs" [9]. These systems were classified as membrane-controlled (or reservoir) devices capable of achieving zero-order release kinetics. The most common polymers utilized for transdermal delivery systems were either EVAc or a porous polypropylene (PP). The first commercially available transdermal patches were loaded with scopolamine for the treatment of motion sickness. The most popular current transdermal patches include Nicoderm® developed by GlaxoSmithKline to treat nicotine addiction and Ortho Evra® made by Ortho-McNeil, which serves as a contraceptive.

Along with the membrane-controlled reservoir devices, ALZA also introduced a family of osmotic pump capsules named Oros® or Osmose®. These capsules were designed for controlled release of drugs in the gastrointestinal (GI) tract. The oral capsule provided zero-order release within the enteric-colonic tract and some were designed to have timed controlled release capabilities so the therapeutic agent could be released within a specific region of the GI tract.

ALZA was also the chief contributor for the development of a variety of oral and implanted (Duros®) osmotic delivery devices in the 1990s. The majority of these capsules were composed of cellulose acetate, sometimes blended with a small amount of a low molecular weight poly(ethylene glycol) (PEG) to facilitate the start of water diffusion.

As previously mentioned, transdermal patches and oral osmotic capsules are diffusion-controlled, zero-order release devices that employ a release-controlling membrane to achieve the desired release characteristics. In the patch, the membrane controls drug diffusion out of the patch to maintain a constant rate, while in the osmotic capsule the membrane controls a constant rate of water diffusion into the capsule, forcing an equal volume of the drug solution out of the capsule through a fixed-size hole.

2.2.4 Early-Controlled Drug Delivery Research in the Academic Field

Independent research in academia also aided in the expansion of the field of controlled drug release. In the mid-1970s, Folkman and his postdoctoral researcher Robert Langer demonstrated that protein-based drugs could be slowly released from macroscopic hydrophobic polymer matrices. Even though this device never reached the clinical setting, its contributions were significant as it was the first example of successful delivery of sensitive biomolecules in an active form from a hydrophobic, nondegradable polymer matrix [10].

Langer and Folkman continue to make significant contributions to the field of controlled drug delivery devices. Langer was the first to develop porous polymer systems for peptide delivery. Using these systems, the group demonstrated that large molecular weight proteins could be released from polymers over the course of several years. This seminal work opened up the field of drug delivery to all academics and to industry including larger pharmaceutical companies, as well as smaller, specialized companies and start-ups. Langer and Brem [11] also developed the first successful biodegradable systems for treatment of brain tumors [12,8]. More recently, Langer et al. developed the first microscale drug delivery device to reach clinical trials [13,14]. This system will be discussed in more detail in Section 9.7.

In 1981, Langer and Peppas offered the first systematic, mechanism-based classification of controlled release systems [12]. In this analysis, the systems were classified as diffusion-controlled (matrix and reservoir systems), chemically activated (biodegradable and pendant chain systems), solvent-activated (swelling controlled and osmotic systems), and pulsatile delivery systems. The last group included pH or temperature sensitive, electrically, magnetically, or ultrasonically activated systems.

2.3 CLASSIFICATION OF DRUG DELIVERY DEVICES

Drug delivery systems are often classified according to delivery mechanism. It is convenient to classify drug delivery systems into four broad categories: (1) diffusion controlled, (2) swelling controlled or solvent activated, (3) chemically controlled, and (4) pulsatile delivery [12].

Diffusion-controlled devices include monolithic (or matrix) devices and membrane-controlled (or reservoir) devices. A monolithic device is a nonswelling and nonbiodegradable matrix with a dispersed drug that diffuses out through the matrix. A membrane-controlled device is a device consisting of a drug containing core from which the drug diffuses through a surrounding rate-controlling polymeric membrane.

Swelling-controlled drug delivery systems can also be designated at solvent-activated systems. An osmotic system utilizes an osmotic pressure gradient to draw fluid into a rigid container through a semipermeable membrane, subsequently forcing drug housed in a flexible, nonpermeable interior membrane, out of the same container (Figure 2.4). Swelling-controlled systems are typically composed of hydrophilic polymer in which the drug is dispersed. The system initially exists in a glass/dry state in which the therapeutic is immobilized within the polymeric matrix. Upon introduction to a swelling medium, typically a biological or physiological fluid, the system transitions into a rubbery/wet state mobilizing the therapeutic and allowing it to diffuse out of the now-present pores in the polymer matrix [15].

Chemically controlled systems are inert until acted upon by some chemical mechanisms that release the therapeutic and initiate drug delivery. Reservoir and

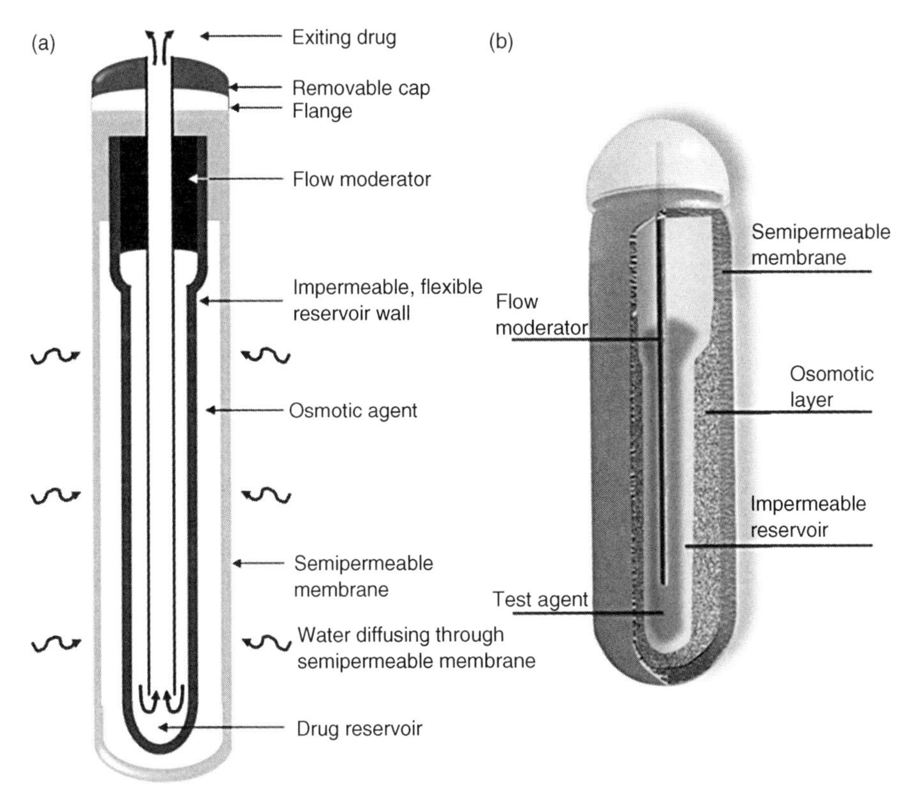

FIGURE 2.4 (a) Cross-sectional schematic of an osmotic pump. Adopted from schematics in an Alzet® catalog. (b) Alzet Osmotic Pump for drug delivery. (Obtained from Alzet catalog.). (See colour plate section.)

monolithic systems, like those described above, can be considered chemically controlled if the system contains any biodegradable elements. The biodegradation can either release a therapeutic too large for diffusion out of the intact polymeric matrix or exert some temporal control over the release rate. Also included are systems in which drugs are covalently bonded as pendants to biodegradable polymer backbones. These drugs are only released upon degradation of those polymeric linkers.

Pulsatile drug delivery systems encompass the fourth type of drug delivery system. These systems were initially developed to provide more physiologically relevant release patterns of therapeutics. The idea was that a polymeric reservoir would release a specific amount of drug upon an external trigger such as temperature change or electric stimulus, or an internal trigger such as the presence of a particular protein or a change in pH [16–18]. Pulsatile drug delivery systems can also be included under the wider umbrella of responsive systems, sometimes referred to as "smart" or "intelligent." These systems respond in some way to stimuli such as temperature, pH, solvent, electricity, light, mechanical stimulus, magnetic stimulus, ultrasound, or self-induced stimuli.

2.4 AN EARLY EXAMPLE OF DESIGNED DRUG DELIVERY SYSTEMS: OSMOTIC DRUG DELIVERY SYSTEMS

Osmotic drug delivery systems have found numerous applications in medicine. Such systems were once primarily used in animal applications, but have since been used increasingly in humans because of their reliability and functionality independent of local physiological function [19]. While there are some osmotic systems intended for implantation in the body, the vast majority of osmotic systems are designed to be taken orally. These systems offer several advantages over other controlled oral delivery systems such as matrix- and reservoir-type capsules. Although both osmotic systems and reservoir-containing systems employ a drug reservoir and a membrane, reservoir systems rely on passive drug diffusion across a semipermeable membrane, while osmotic systems utilize osmotic pressure generated across a membrane to force a therapeutic agent out through a small orifice [20]. As a result, osmotic systems are not sensitive to local pH, the presence of food, or other factors that may affect reservoir and matrix systems [21]. They will continue to release their agent at their designated rate, meaning there is no disruption in therapeutic efficacy [20].

Osmotic systems may be used for oral delivery of numerous types of drugs, although different types of osmotic systems are required for drugs that are either water soluble or water insoluble. An elementary osmotic pump is suitable for delivering water-soluble drugs. It is composed of a drug-containing core that may or may not include an additional osmotic agent. This core is surrounded by a rate-controlling membrane that has a small orifice to allow for the release of its contents as shown in Figure 2.4. A rigid semipermeable membrane encapsulates the pump, allowing for the diffusion of water into the system. These devices will release 60–80 percent of their contents at a constant rate, and typically require a lag time of up to 1 h before the system is hydrated enough such that zero-order release begins [20].

An advantage of osmotic drug delivery systems is that they are suitable for drugs of all solubility [21]. If a drug is not water-soluble, then a push–pull osmotic system must be used for successful delivery. A push–pull osmotic system is a capsule consisting of two chambers separated by a movable partition. The entire system is encapsulated in a semipermeable membrane. The first chamber contains the therapeutic agent and may also contain an additional osmotic agent. It also has a small orifice leading to the surrounding environment. The second chamber contains an osmotic agent and can expand by moving the partition. As water is taken into the second chamber due to an osmotic pressure gradient with the surrounding environment, it expands to occupy the volume originally taken by the first chamber. This expansion forces the drug contained within the first chamber into the surroundings. Both elementary and push–pull osmotic systems are illustrated in Figure 2.5. There are many creative examples and variants of these two basic concepts in service today for drug delivery applications [20].

The rate and duration of drug release from both elementary osmotic pumps and push–pull osmotic pumps can be engineered based on modifying the parameters related to the osmotic pressure, delivery orifice, and membrane characteristics. Osmotic pressure is important for the release rate of a drug from osmotic systems,

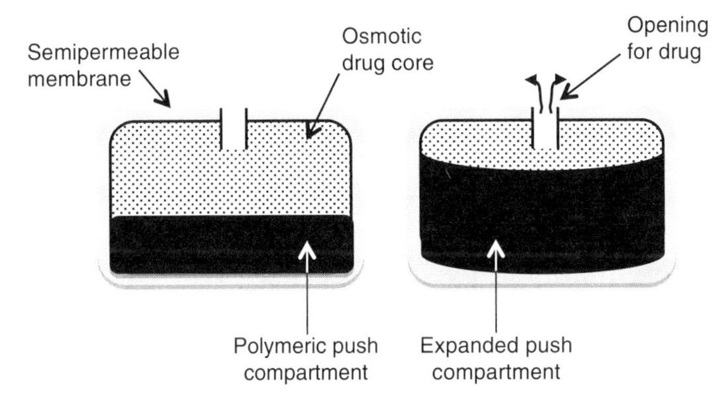

FIGURE 2.5 Schematic of an elementary push–pull osmotic pump device.

and it must therefore be controlled. Constant osmotic pressure can be achieved if a saturated solution of the drug is achieved and maintained, which in turn can be utilized to reach zero-order release. Since the desired drug may not generate large enough osmotic pressures on its own, an additional osmotic agent may be added to the device. This agent can be any one of a variety of compounds, such as many water soluble salts, water soluble amino acids, and organic polymers used to generate needed osmotic pressures. A membrane must be selected such that it is permeable to water but not to the solute. Commonly used materials include cellulose esters, especially cellulose acetate. Other materials used include cellulose triacetate, cellulose propionate, and cellulose acetate butyrate. Cellulose acetate is a common selection because it is insoluble and has favorable water permeability. Additionally, its solubility may be reduced by increasing acetylation [20].

2.4.1.1 Alza OROS Osmotic Systems The first and perhaps the most notable oral osmotic system was developed by the Alza Corporation, now a part of Johnson & Johnson, who currently have a number of commercially available products utilizing OROS technology. Push–pull osmotic pumps were developed by Alza in products such as Ditropan XL (oxybutynin chloride), Procardia XL (nifedipine), Sudafed 24hr (pseudoephedrine), and Glucotrol XL (glipizide) [22,20]. Alza Corporation also developed the L-OROS for delivery of lipophilic liquid formulas, as well as the OROS-CT for targeted delivery to the colon. The OROS-CT may have as many as six push–pull osmotic units within a hard gelatin capsule. The gelatin capsule is designed to dissolve once it comes in contact with gastrointestinal fluids revealing an enteric coating. This enteric coating prevents the fluids of the stomach from entering the system. Once in the small intestine, the enteric coating dissolves to allow water into the system. The drug is then released into the colon [20].

2.4.1.2 Implantable Osmotic Systems Osmotic systems are also employed as implantable devices such as the DUROS systems developed by the Durect

Corporation. These are small devices measuring only 4 mm across and 45 mm in length. They are designed for the zero-order release of potent drugs over extended periods of time. For example, Viadur® (leuprolide acetate) is a commercially available product capable of continuous delivery of leuprolide acetate, a super active LHRH, at a rate of 125 ug/day for over 1 year [20].

2.5 ANOTHER EARLY DELIVERY SYSTEM BASED ON MEMBRANE-CONTROLLED RELEASE: OCULAR DELIVERY SYSTEMS

Ocular drug delivery encompasses many forms of delivery techniques including gels, injectables, inserts, ointments, orals, solutions, and suspensions with solutions, ointments, and suspensions being by far the most common [23]. Solutions are incredibly common, because they are patient friendly and simple to use. However, from a therapeutic standpoint, solutions are often not ideal because of the amount of drug lost during use and the subsequent short-term bioavailability. These factors result in the need for frequent reapplication [24]. It is often desirable to lengthen the duration of drug availability. Gels and ointments are an alternative to low viscosity solutions or suspensions and serve to enhance residence times. However, these systems often result in uncomfortable stickiness, irritating sensation, or blurred vision [25]. To overcome the problematic nature of simplistic ocular delivery techniques, several more sophisticated systems that address these challenges have been developed or are currently under development [24].

2.5.1.1 Ocular Inserts Ocular inserts encompass a wide range of ocular insertable devices intended to provide more desirable release behavior than simple solutions, suspensions, or ointments. Most insertable devices are insoluble and do not erode, therefore, must eventually be removed from the eye. Others are soluble and erode while a drug is being released and therefore do not require eventual removal [26]. One study describes the effective use of an erodible insert comprised of gel forming poly(ethylene oxide) (PEO) to achieve traditional zero-order release. These inserts were used to deliver ofloxacin (OFX) as the inserts eroded, with the erosion rate controlled by varying amounts of Eudragit L100 [27]. Typical applications of ocular inserts include antibacterials, antivirals, antifungals, antifilarials, antiallergenics, anti-inflammatories, fibrinolytics, antiglaucoma, immunosuppressants, and growth factors.

Of these applications, antiglaucoma applications have been the most successful. Perhaps the most notable of these drugs is the Ocusert pilocarpine ocular therapeutic system introduced by Alza Corporation. This system is a membrane-type reservoir device and was developed when the advantages of a zero-order release of pilocarpine for treatment of glaucoma was recognized. The Ocusert was extremely successful at maintaining intraocular pressure in patients for up to 7 days by encasing pilocarpine alginate in ethylene-vinyl acetate [26].

2.6 EARLY SUCCESSES IN THE USE OF MEMBRANE AND HYBRID SYSTEMS: TRANSDERMAL PATCHES

Transdermal delivery refers to the delivery of drugs across the skin for therapeutic effect elsewhere in the body. It is a promising alternative to oral delivery and intravenous injections. Transdermal delivery avoids degradation of the therapeutic agent molecules in the acidic environment of the gastrointestinal tract and a removal of the drug by the liver associated with oral delivery methods. This delivery method also avoids the painful effects that can occur with injections. Delivery through the skin also offers the ability to continuously control the delivery rate over an extended period of time. The primary challenge of transdermal delivery is getting the therapeutic agent across the transdermal corneum barrier. For transdermal delivery to be effective, new methods for getting molecules across the skin barrier need to be developed. Because the skin is in fact a highly impermeable and resilient protective covering of the body, drug delivery across this barrier is more difficult than may be supposed. The skin prevents bodily fluid loss and temperature fluctuation. Furthermore, the skin also protects against microbes, toxic substances, and electromagnetic radiation. It is believed that an acid mantle, which possesses a pH of between 4.8 and 6, controls microorganism growth on the skin. At the skin's surface is an impermeable layer of dead skin cells known as the stratum corneum. Beneath the stratum corneum are the living epidermal cells that form a layer known as the viable epidermis. Deeper below the surface is the dermis, which is a highly vascularized layer. It is the stratum corneum that is principally responsible for the poor transport of drugs across the skin. The dead skin cells of the stratum corneum are closely packed with a lipid matrix that fills in the gaps. Drug absorption across this layer is passive, slow, and obeys Fickian diffusion [28]. The rate of drug transport across the stratum corneum is usually the rate-determining step of transdermally applied compounds [29]. In some cases, enhancers are used to speed up absorption for the purposes of drug uptake. Some examples of compounds that improve skin permeability include dimethyl sulfoxide, dimethyl formamide, and *N*-methyl-2-pyrrolidone [28].

In practice, few drugs are actually suitable for transdermal delivery, because a suitable drug must be highly hydrophobic, oil soluble, have low crystallinity, and be effective in small doses. Some examples of successful drugs used in transdermal applications are fentanyl, clonidine, nicotine, nitroglycerine, scopalamine, testosterone, and 17-β-estradiol. Lidocaine and pilocaine have also been used, but these are merely employed to kill skin tissue prior to minor surgery [28]. Transdermal drug delivery includes both the use of semisolid formulas such as ointments and gels as well as patches as shown in Figure 2.6. Transdermal patches come in many forms and exist as both monolithic and membrane-controlled reservoir devices [29].

2.6.1.1 Transdermal Delivery of Fentanyl One application of the use of a patch for transdermal delivery is for chronic cancer pain management using fentanyl in a membrane-controlled reservoir-containing patch [30]. Fentanyl is a synthetic opioid, and demonstrates short-term analgesic behavior when applied intravenously or

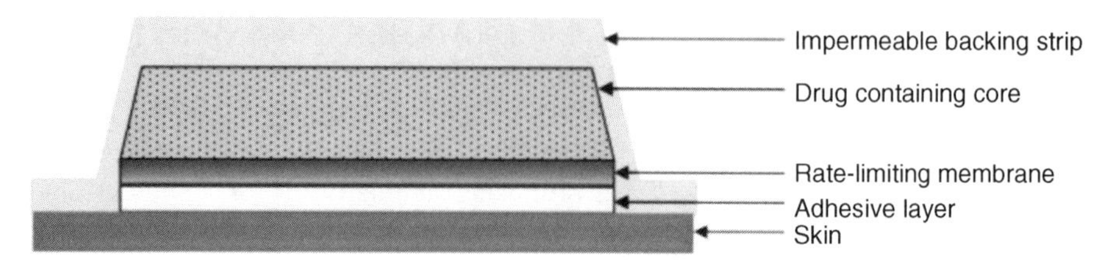

FIGURE 2.6 Side-view of a transdermal delivery patch including impermeable backing strip, drug-containing core, and rate-limiting membrane.

subcutaneously. Although many compounds are not suitable for transdermal application, fentanyl will pass through the skin due to its low molecular weight and lipid solubility. Also, its high potency means that only small doses are required. Systems utilizing fentanyl employ an ethylene-*co*-vinyl acetate membrane for controlled release and are designed for zero-order release for up to 3 days before reapplication is required [31,32]. Although fentanyl may also be injected, patches result in similar plasma concentrations 8–12 h after application. They are best suited for patients who have difficulty in swallowing, have a need to rotate pain management medications because of side effects, or have variable consciousness [30].

2.7 EARLY EFFORTS TO TARGET DRUG DELIVERY: MUCOADHESIVE DRUG DELIVERY SYSTEMS

Mucoadhesive drug delivery systems were created with the intent of targeting drugs to certain mucus-containing regions of the body for extended periods of time. Mucoadhesion is a form of bioadhesion, which involves an interfacial adhesion between an introduced material and biological tissue. Mucoadhesion refers specifically to the attachment of a polymer to the mucin layer of mucosal tissue, which is present in several areas of the body. As a result, mucoadhesive principles may be applied to achieve extended residence time of drug delivery systems for oral, vaginal, rectal, nasal, and ocular applications using polymer drug carriers to deliver therapeutic agents such as proteins, organic molecules, antibiotics, vaccines, and DNA [33,34].

There are several advantages of using mucoadhesion, making it a widely investigated form of drug delivery. With the ability to adhere to mucosal surfaces, one can achieve targeted delivery to a specific site or organ. Additionally, by adhering drug-containing polymer particles to a mucosal surface, the particles will not be easily carried away by the body. This results in enhanced residence time of the drug in the body. Furthermore, the intimate contact between the drug-containing polymer and the mucosal layer of the targeted region or organ results in greater availability of the drug to the targeted site, as well as direct absorption through the nearby cells. Four mechanisms by which therapeutic agents may be transported into the body through the outer layer of cells are illustrated in Figure 2.7. One of these mechanisms—

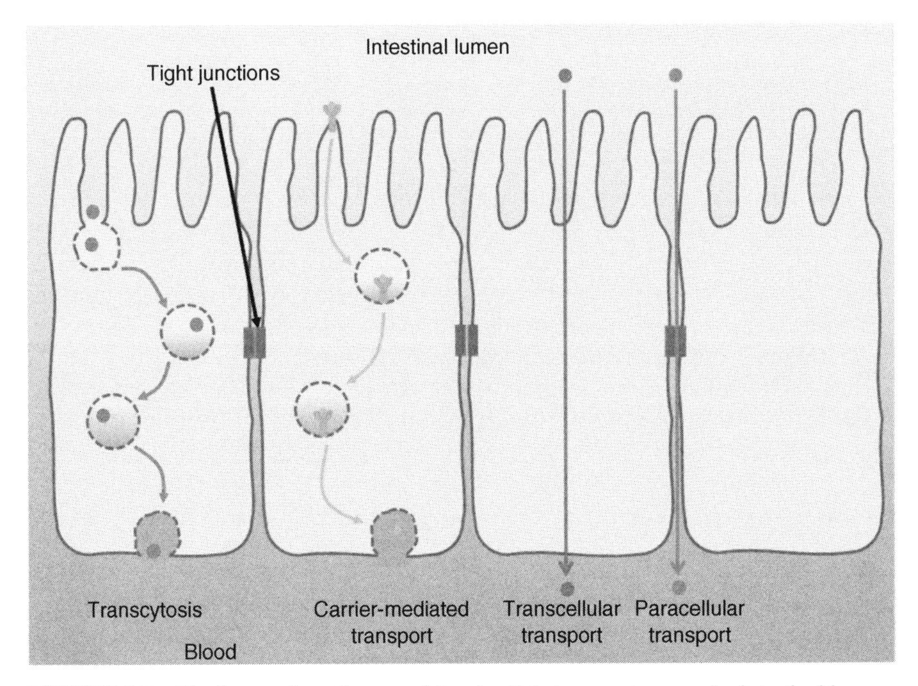

FIGURE 2.7 The four main pathways of drug/particle transport across the intestinal lumen: transcytosis, carrier-mediated transport, transcellular transport, and paracellular transport. (See colour plate section.)

transcellular transport, paracellular transport, carrier mediated transport, and transcytosis—must occur for successful therapeutic benefit [35].

Several factors related to the polymer used for mucoadhesive drug delivery will impact the adhesive effectiveness. A polymer's molecular weight and cross-linking density are both critical, and each has a critical or optimal value for which adhesive strength will be greatest. Actual mucoadhesion occurs as a result of close contact between an incorporated polymer and the mucin layer. A polymer within the drug delivery system must make this close contact with the mucus layer in order to be an effective mucoadhesive system. An important factor in mucoadhesion is the flexibility and mobility of the polymer chains. They must reach the mucus and create an entangled network of polymer chains and mucus. This interaction will then allow for hydrogen bonding and electrostatic and hydrophobic interaction, giving rise to the bond [25]. If a polymer's free chains are too short because of high cross linking, then its mucoadhesion strength will be reduced [34]. An example of a commonly used polymer for mucoadhesion is poly(acrylic acid), which has been shown to be effective for adhesion in multiple areas including the gastrointestinal tract. Poly(acrylic acid) is especially effective within the stomach due to its low pH [36,37].

2.7.1.1 Nasal Mucoadhesive Drug Delivery Nasal mucoadhesive drug delivery is a widely explored means of drug delivery utilizing mucoadhesion principles for several reasons. First, the nasal epithelium has a large surface area, which can be used for drug deposition and absorption. Furthermore, the nasal epithelium is relatively thin, porous, and richly vascularized. This is important because it will result in substantial absorption, as well as rapid introduction, of absorbed substances into circulation. Nasal delivery also has the advantage of releasing drugs directly into circulation, unlike basic oral delivery, which must contend with the first pass metabolic effect. Drug delivery to the nose also has the advantage that patients may easily self-medicate. As a result, patient compliance is higher. Finally, the risk of overdose with nasal delivery is relatively low. In spite of these advantages, the potential for nasal drug delivery is limited for a number of reasons, which will not be discussed here. However, by utilizing mucoadhesion principles, a number of these limitations can be surmounted allowing for successful nasal delivery for several applications [38].

In specific applications, mucoadhesive nasal delivery has been demonstrated to be effective for delivery of small organic molecules, antibiotics, vaccines, DNA, and proteins. An application involving a small organic molecule is the nasal administration of apomorphine. Apomorphine is used by patients who suffer from Parkinson's disease and experience variable severity of their symptoms. Nasal administration is particularly appropriate given the rapid uptake and therefore rapid therapeutic benefit that may be achieved. The polymers shown to be effective for nasal delivery of apomorphine include poly(acrylic acid), also known as Carbopol, and carboxymethylcellulose. It was demonstrated that the use of such mucoadhesive polymers increased the residence time and sustained bioavailability of apomorphine, resulting in comparable bioavailability to subcutaneous injections [38].

For the use of antibiotics, nasal mucoadhesive delivery has been shown to be effective for a variety of antibiotic–polymer combinations. Although there are several concerns related to nasal administration of antibiotics, such as the potential for superinfection of the nasal cavity, these issues may be ameliorated by careful selection of antibiotics. One possible application of nasal delivery for antibiotics that has been demonstrated in animals is the use of hyaluronic acid/chitosan microspheres to deliver gentamicin [38].

Additionally, vaccinations are an attractive application for nasal mucoadhesive delivery, because it would facilitate the vaccination of large numbers of people without the need for needles. Furthermore, nasal administration of vaccines is a natural choice given that most pathogens gain access to the body by mucosal contact. One nasal-delivered vaccination includes the use of PEG-coated poly(lactic acid) nanospheres, chitosan-coated poly(lactic–glycolic acid) nanospheres and chitosan nanospheres to encapsulate tetanus toxoid. Also, chitosan has been utilized to immunize against influenza, pertussis, and diphtheria with positive results [34]. The favorable results when using chitosan are largely attributed to the increased residence time on the nasal mucosa as a result of mucoadhesion, as well as the ability to open the epithelial tight junctions on the nasal epithelium. Chitosan is a positively charged linear polysaccharide that adheres strongly to the nasal epithelial cells and the mucus that overlies them. For such applications, a commonly used form of chitosan resulting in strong biointeraction is chitosan glutamate salt with a molecular weight of approximately 250 kDa and greater than 80% deacetylation [38].

2.8 DRUG DELIVERY IN COMBINATION PRODUCTS: DRUG-ELUTING STENTS

For patients who have undergone some form of percutaneous coronary intervention such as angioplasty, stenting following the procedure has been shown to greatly improve the outcome. Therefore, intravascular stenting has become standard practice for patients with coronary artery disease. Although bare metal stents have been used in patients with coronary artery disease, they tend to exhibit a problem known as in-stent restenosis. A bare metal stent has the tendency to trigger an increase in growth factors that promote the in-growth of the smooth muscle cells into and beyond the stent itself, occluding the artery. This mechanism is known as neointimal proliferation. In order to prevent in-stent restenosis due to neointimal proliferation, some patients receive drug-eluting stents because of their demonstrated therapeutic advantage over bare metal stents. These stents prevent the in-growth of cells by incorporating antiproliferative agents such as sirolimus or paclitaxel in such a manner that they are slowly released from the stent. Other drugs studied for drug-eluting stents include actinomycin, everolimus, ABT-578, and 7-hexanoyltaxol [39,40].

2.8.1.1 Cypher The Cordis Corporation's drug-eluting coronary stent known as the Cypher was approved in April of 2003 by the Food and Drug Administration [41]. The Cypher is an example of a monolithic device designed for gradual drug release. The frame of this stent is constructed of 316L stainless steel. In order to incorporate a drug, the steel substrate is first coated with parylene C. Above this initial coating is a matrix of poly(ethylene-vinyl acetate) and poly(butyl methacrylate) throughout which sirolimus is dispersed. The stent is designed such that 80% will be released within the first month. Sirolimus is a macrolide antibiotic that serves primarily to inhibit proliferation of smooth muscle cells into the stent. Sirolimus has the added benefit that it also acts as a strong anti-inflammatory agent [39]. Because of these functions, the Cypher effectively prevents restenosis.

2.8.1.2 Taxus Another commercially available drug-eluting stent is the Taxus stent manufactured by the Boston Scientific Corporation. The Taxus stent gained approval in March of 2004, shortly after the Cypher appeared on the market [41]. The Taxus stent is another example of a monolithic device, although in some cases it can be coated with a rate controlling membrane. It also contains a stainless steel frame that is coated with a matrix of Translute® polymer engineered for controlled release and a dispersion of paclitaxel. Translute polymer was selected because it is highly compatible with the vasculature and is biostable. It is a hydrophobic triblock elastomeric copolymer. Paclitaxel is a common chemotherapeutic agent that also blocks many of the pathways associated with restenosis due to neointimal proliferation. Unlike sirolimus, it is not significantly anti-inflammatory [39]. Recent advances in drug-eluting stents can be found in the comprehensive review by Puranik et al. [40].

2.9 THE ADVENT OF MICROFABRICATED DRUG DELIVERY DEVICES

As coined by Dr. Allan Hoffman in his 2008 *Journal of Controlled Release* review of the origins and evolution of "controlled" drug delivery systems, approximately the first 10 years of the controlled release devices field was the "MACRO era." [9] Most systems produced during this time period were macroscopic in scale. Advancements in micro- and nanofabrication process has now allowed for devices to be miniaturized and lead to the beginning of a "MICRO era" of drug delivery devices.

2.9.1 Microfabricated Drug Delivery Devices

Advancements in micro- and nanofabrication technology have enabled the creation of higher performing therapeutic delivery systems for a wide variety of biological applications. A number of successful implantable delivery systems composed of silicone silicon, glass, silicone elastomers (PDMS) have been fabricated using standard microfabrication techniques [42–47]. Controlled release drug delivery is still an important challenge when treating pathophysiological conditions. Microfabrication of biomedical microelectromechanical systems (bioMEMs) devices for controlled release is a potential avenue to develop new treatment methods [48].

Patient compliance is an essential issue in treatment regimens for chronic diseases. It can be quite challenging to get patients to follow a specific drug regiment, especially when it requires giving themselves frequent painful injections. In most delivery schemes, the initial drug concentration immediately after administration is above the toxicity threshold and it will then gradually diminish over time to an insufficient therapeutic level. This is a very ineffective and potentially dangerous way of delivering drugs. High dosages of nontargeted drugs are often administered to achieve an effective blood concentration for treatment, which could be damaging to the entire body [49]. Furthermore, the duration of the therapeutic efficacy is dependent on the frequency of administration and the half-life of the drug.

Even with the advancements in pharmacokinetics leading to the development of better, more sophisticated protein- and DNA-based drugs, the same delivery challenges that exist with conventional therapies still exist. Most of these new drugs have a very narrow therapeutic window and variations in systemic concentration can lead to toxicity when too high and lost therapeutic benefit when too low. While only small amounts of therapeutic are needed to achieve the desired therapeutic effects, the aforementioned challenges render conventional drug delivery methods inefficient.

Conventional drug administration methods are also limited in their ability to provide long-term treatment, a narrow therapeutic window, complex dosing schedule, combination therapy, and personalized dosing [50]. To overcome these limitations, developments of combination drug and medical device systems, which have the ability to protect active ingredients, have precise control of drug release kinetics (time of dose and the amount administered), and deliver multiple doses. New medical device and drug combination systems must be easily controlled and able to adjust the release of therapeutic agents based upon external stimuli. This could be accomplished by integrating sensors and feedback sensors into the device. This will help eliminate the need for frequent injection or even surgery for implantable drug release systems [50]. Microfabrication techniques give researchers the power to create therapeutic delivery systems that have a combination of structural, mechanical, and electronic feedback sensor features that can overcome the limitations of conventional delivery therapy. To tackle the unmet needs related to dosing, a variety of microelectronic-controlled drug delivery devices have been developed, including iontophoresis patches, transdermal microneedle patches, extracorporeal pumps, and implanted pumps and microchips. In this section, these systems will be highlighted.

2.9.2 Microelectronic Iontophoresis Patch

Iontophoresis patches employ a small electric current to drive molecules through the skin. As mentioned earlier, transdermal delivery of therapeutic agents is beneficial because it allows for real-time, on-demand control over drug delivery in a noninvasive, nonsurgical method. Through the addition of a low electrical current, the amount of drug transported across the skin can be significantly increased [51]. Iontophoresis patches controlled by microelectronics have the added benefit of being lightweight and easy to wear without needing to be connected to an external power source. The majority of iontophoresis patches currently contain a liquid form of the intended drug within a reservoir. A minor limitation of iontophoresis patch is the total number of doses contained in one patch—usually only a couple of days' supply. The ease of application of a new patch overcomes this limitation. However, as presented earlier, the main limitation of transdermal delivery from iontophoresis patches is the transport of drugs across the skin barrier.

The application of a low electric current increases the penetration of drugs through the skin; but transport is still limited to small-molecular-weight hydrophilic drugs (less than 7 kDa). Because of this limitation, ionophoresis patches are used only for administration of fentanyl to treat acute pain after surgery, water for the treatment of hyperhidrosis, and pilocarpine to induce sweating as part of a cystic fibrosis diagnostic test [52].

2.9.3 Transdermal Microneedle Drug Delivery Systems

Another alternative for transdermal delivery of drugs are microneedle patches. Microneedles are one approach to create microscale "conduits" to transport molecules across the stratum corneum painlessly [53]. For microneedles to be effective, they need to be long enough to penetrate the skin, with a large enough diameter to transport drugs containing macromolecules, while being short enough to avoid stimulating nerves (and therefore pain) [54].

Arrays of microscale needles are a minimally invasive transdermal delivery of therapeutic agents. Numerous types of microneedles have been developed for specific applications. Microfabrication techniques have been developed for microneedles composed of silicon, metal, and biodegradable polymers [55]. Needles have been designed with solid or hollow bores, tapered or beveled tips, and feature sizes from 1 to 1000 μm. Solid needles can increase skin permeability by orders of magnitude *in vitro* for both macromolecules and particles up to 100 μm in size. Using more sophisticated microfabrication techniques like coherent porous silicon etching technology produced needle arrays with different pitches and diameters [56]. Drugs can also be directly dry-coated onto the microneedles for direct drug application.

Arrays of solid silicon microneedles have been fabricated with individual needles measuring 80 μm in diameter at the base, 150 μm in height, and a radius of curvature less than 1 μm [53]. Hollow needles have also been fabricated with similar dimensions but have a hollow hole in the range of 5–70 μm in diameter, depending on the application the needle is required for. Biocompatibility and durability of the microneedles have been studied. Solid microneedles required approximately 10 N force to penetrate the skin. Only a small percent of the microneedles are damaged [53,54]. The microneedle array could also be removed from the skin without additional damage and could also be reinserted numerous times.

The design of the microneedles has been altered to provide better control over the release kinetics of therapeutic agents. Silicon microhypodermic needles have been developed in combination with heat-controlled bubble pumps [54]. Hollow metal microneedles have also been fabricated by electrodepositing metal into epoxy micromolds [54]. Polysilicon microneedles have formed using reusable molds [54]. Micromolding fabrication using polysilicon is the more cost-effective approach to creating entire arrays of needles. Polysilicon microneedles would also be a better design for single use disposable platforms.

Combination of polysilicon microneedles and a pressurized reservoir to create a drug delivery pump have already been incorporated into a wearable drug infusion system to deliver insulin [55]. Work is also being done to further modify the needle dimension and design to incorporate multiple channels and ports, optimized microhypodermic needle and microprobes for better systemic delivery.

2.9.4 Extracorporeal Pumps as Controlled Release Systems

Real-time controlled release of therapeutic agents can also be accomplished using extracorporeal pumps. These pumps employ indwelling catheters to overcome the limitations of transport across the skin barrier. Because of this, drugs ranging from small molecular weight drugs to macromolecules can be administered over a range of doses. These indwelling catheters do accomplish delivery of drugs but they are also invasive and prone to infections. Also, despite the use of microelectronics, extracorporeal pumps remain bulky in comparison to transdermal patches. Extracorporeal pumps are currently in clinical use for administration of insulin to diabetic patients and for patient-controlled analgesia, which employs a microprocessor-controlled infusion pump to deliver a preprogrammed dose of drug [52].

2.9.5 Implanted Pump

The advancements in microfabrication have allowed for the fabrication of microscale pumps. These pumps can be incorporated into implantable devices for administration of therapeutic agents. Real-time controlled release of liquid drug formulations can be preprogrammed into the device or adjusted using a wireless controller. As their name implies, these devices require implantation within the body and, therefore, involve invasive surgery. Another downside to implantable devices is that another invasive procedure is required to refill the device when the drug reservoir is depleted. Implanted microscale pumps are currently used for intrathecal delivery of baclofen to treat severe spasticity and morphine to treat intractable pain [57].

2.9.6 Implanted Microchips

The concept of the programmable implantable microchip-based drug delivery device was first demonstrated preclinically in 1999 by Langer et al. [13]. These devices are able to provide real-time wireless management of drug release. Minor surgery is still required for implantation of the microchip. Since the implantable microchip does not employ a pump for transport of the drugs, microchip devices are an improvement over implantable pumps because the solid-state forms of drugs can be loaded within the chip, thus increasing drug stability. Macromolecules can be administered but due to size limitations of the chip, there is a limit of doses per device. Recently, the first clinical study in human was conducted to evaluate the effectiveness of delivery of human parathyroid hormone fragment from an implanted microchip.

2.9.7 Reservoir-Type Microchip Drug Delivery Systems

Microfabrication technology has produced a new sophisticated class of controlled release systems for therapeutic delivery based on programmable microdevices. The small size, ability to integrate microelectronics, and their ability to store and release drugs on demand, makes them very attractive systems for controlled drug release [45]. With advancements in microfabrication technology and biosensors, implantable responsive drug release systems are becoming more realistic for medical applications.

Implantable drug delivery devices for administration of a precise amount of therapeutic agents at a specific time would be an important tool for treatment of numerous diseases that require repeat administration of drugs. There are some major limitations to implantable drug delivery devices. The device will require surgery for implantation and, therefore, needs to have the ability to release drugs over a long period of time and, consequently, a large amount of encapsulated/loaded drugs. Even with most drugs requiring only microgram daily doses, sufficient storage capacity for a year's worth of therapeutic is a lofty design challenge. Additionally, if a year's worth of therapeutic is to be loaded, it must be done very safely, as its premature release would be potentially harmful or even lethal to the patient.

The ideal implant system would protect the drug from the body until it is needed, allow continuous or time-specific delivery of therapeutic agents, and be controlled externally without surgery. These requirements can be achieved with a microscale array of individually sealed reservoirs that can be opened on command to release the contained therapeutic agent into the environment. Individual drug containing reservoir microchip design for drug delivery applications has one important advantage over other designs. They have the ability to totally control drug delivery amount and timing: continuous or pulsatile delivery. This design is also very flexible for numerous different applications, because its release characteristics can be governed independently by the release mechanism, drug formulation, or reservoir geometry.

The first microreservoir microchip developed for drug delivery applications was fabricated by Langer et al. [13]. The device was fabricated by the sequential processing of a silicon wafer using microelectronic processing techniques including UV photolithography, chemical vapor deposition, electron beam evaporation, and reactive ion etching [13]. The fabricated prototype was $17\,mm \times 17\,mm \times 310\,\mu m$ square silicon device containing an array of 34 square pyramidal reservoirs etched completely through wafers [13]. Reservoirs of volume 25 nl were sealed at one end by a thin membrane of gold to serve as an anode in an electrochemical reaction. Electrodes were placed on the device to serve as a cathode. The reservoirs were filled through the open end by microsyringe pumps or inkjet printing in conjugation with a computer-controlled alignment apparatus. The open end was then covered with a thin adhesive plastic and sealed with waterproof epoxy [13].

When the microfabricated reservoir system was submerged in an electrolyte solution, ions form a soluble complex with the anode material in its ionic form [45]. An applied electric potential oxidizes the anode membrane, forming a soluble complex with the electrolyte ions. The complex dissolves in the electrolyte, the membrane disappears, and the solution within the reservoir is released [45]. The release time for each individual reservoir is determined by the time at which the reservoir's anode membrane is removed [45]

This microchip reservoir system laid the foundation of current microchip reservoir systems. Current research on microchip drug delivery systems is focused on integrating active components including battery clocks, reference electrodes, and biosensors to achieve a single package for simpler implantation [50]. Development of a passive, polymer microchip that contains no electronics, power sources, or microprocessors is another important possibility for creating better drug delivery devices. Polymer microchips that are biodegradable are also a beneficial product because surgery would not be required to retrieve the depleted device. MicroCHIPS have developed a self-contained drug delivery chip shown in Figure 2.8. The microchip, hardware, power supply, electrical components, and wireless communication system are embedded and sealed inside the device. The microchip is composed of a silicon/glass-bonded substrate containing 100 individual 300 nl reservoirs. Individual membranes cover each reservoir, and are composed of platinum and titanium layers. The membranes covering a reservoir are removed by local resistive heating from an applied current [50]. The MicroCHIPS drug delivery systems have been shown to deliver a controlled pulsatile release of polypeptide leuprolide from specific reservoirs in a canine model for 6 months. The MicroCHIPS device is the first completely self-contained microchip implant that provides constant programmed delivery of therapeutic drugs.

A biodegradable polymer version multireservoir drug delivery systems consist of reservoirs covered with resorbable membranes [58]. The chemical composition or the physical properties of the membranes will vary with their degradation time and,

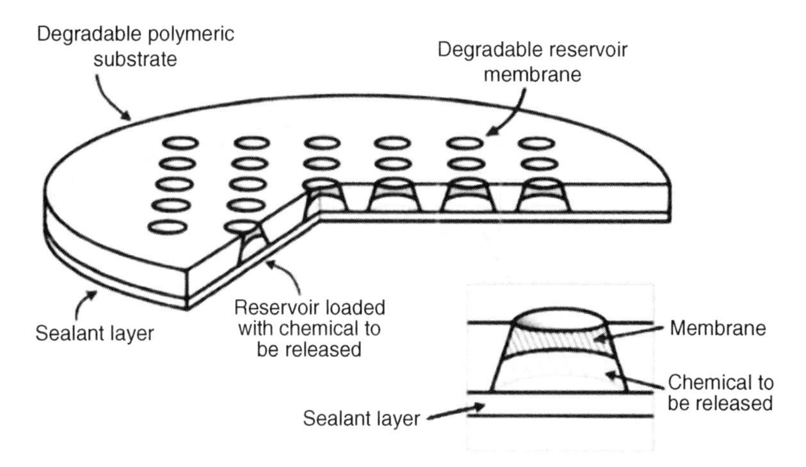

FIGURE 2.8 A polymeric biodegradable self-contained drug delivery chip developed by MicroCHIPS. All components are embedded and sealed within the device. This chip is capable of delivering drugs in a pulsatile release pattern [50].

consequently, the release time of the therapeutic agent within the reservoir. These polymer devices and reservoirs are made out of compressed-molded polylactic acid (PLA). The ratio of PLA and PGA and the molecular weight of the polymers are varied between membranes to control release [58].

The concept of a programmable implantable microchip-based drug delivery device was first evaluated preclinically in 1999 by Dr. Langer and colleagues; and was recently assessed in the first human clinical study for delivery of human parathyroid hormone fragment 1–34 (hPTH[1-34]). In this study, the microchip-based delivery system accomplished pharmacokinetics similar to multiple injections and did not produce any significant adverse side effects.

2.10 CONCLUSIONS AND DIRECTIONS

Over the last 50 years, technological advancements have revolutionized the practice of medicine and drastically improved quality of life. For instance, 2 of the top 20 greatest achievements of the twentieth century, as designated by the National Academy of Engineering, are related to the field of medicine: imaging and health technologies [59]. As we proceed into the twenty-first century, drug delivery technology is becoming progressively controlled at the molecular level. Molecular understanding is driving the next generation of commercial diagnostic and therapeutic devices. The application of micro/nanodevices for biological and medical applications is leading to fundamental insights into the behavior and function of tissues, intra- and intercellular communication, forces and flows and the effects on individual cells, and the structure, function, and behavior of proteins, DNA, and other biological molecules. Therefore, micro- and nanotechnologies are expected to lead to unparalleled progress of medical science and technology and to countless technological innovations in medicine.

Application of nanoscale science and engineering to create novel materials and devices for therapeutics will profoundly impact the practice of medicine. In biological and medical applications, controlling interactions at the level of natural building blocks, from proteins to cells, facilitates the novel exploration, manipulation, and application of living systems and biological phenomena. Nanoscale science and engineering has an unlimited potential to positively affect the diagnosis of medical conditions and the corresponding therapeutic methods for improved prognosis and rational treatment. The unlimited potential is clear, especially for application in medical science and technology leading to technological innovations in medicine. In the near future, it is expected that new materials, advanced designs, and recognitive intelligent systems will take a leading position in the field and will dramatically influence the nature and practice of medicine.

REFERENCES

1. J. Siepmann and N.A. Peppas, Higuchi equation: derivation, applications, use and misuse. *Int. J. Pharm.* 418 (1) (2011) 6–12.
2. T. Higuchi, Rate of release of medicaments from ointment bases containing drugs in suspension. *J. Pharm. Sci.* 50 (10) (1961) 874–875.
3. P.L. Ritger and N.A. Peppas, A simple equation for description of solute release II: Fickian and anomalous release from swellable devices. *J. Control. Release* 5 (1) (1987) 37–42.
4. C.S. Brazel and N.A. Peppas, Modeling of drug release from swellable polymers. *Eur. J. Pharm. Biopharm.* 49 (1) (2000) 47–58.
5. J. Siepmann, H. Kranz, et al., Calculation of the required size and shape of hydroxypropyl methylcellulose matrices to achieve desired drug release profiles. *Int. J. Pharm.* 201 (2) (2000) 151–164.
6. J. Siepmann and N.A. Peppas, Modeling of drug release from delivery systems based on hydroxypropyl methylcellulose (HPMC). *Adv. Drug Deliv. Rev.* 48 (2–3) (2001) 139–157.
7. J. Folkman and D.M. Long, The use of silicone rubber as a carrier for prolonged drug therapy. *J. Surg. Res.* 4 (3) (1964) 139–142.
8. N.A. Peppas, Historical perspective on advanced drug delivery: how engineering design and mathematical modeling helped the field mature. *Adv. Drug Deliv. Rev.* 65 (1) (2012) 5–9.
9. A.S. Hoffman, The origins and evolution of "controlled" drug delivery systems. *J. Control. Release* 132 (3) (2008) 153–163.
10. R. Langer and J. Folkman, Polymers for the sustained release of proteins and other macromolecules. *Nature* 263 (5580) (1976) 797–800.
11. H. Brem and R. Langer, Polymer-based drug delivery to the brain. *Sci. Med.* 3 (1996) 52–61.
12. R.S. Langer and N.A. Peppas, Present and future applications of biomaterials in controlled drug delivery systems. *Biomaterials* 2 (4) (1981) 201–214.
13. J.T. Santini, M.J. Cima, et al., A controlled-release microchip. *Nature* 397 (6717) (1999) 335–338.

14. R. Farra, N.F. Sheppard, et al., First-in-human testing of a wirelessly controlled drug delivery microchip. *Sci. Transl. Med.* 4 (122) (2012) 122ra21.
15. P. Colombo, R. Bettini, et al., Swellable matrices for controlled drug delivery: gel-layer behaviour, mechanisms and optimal performance. *Pharm. Sci. Technolo. Today* 3 (6) (2000) 198–204.
16. N.A. Peppas, P. Bures, et al., Hydrogels in pharmaceutical formulations. *Eur. J. Pharm. Biopharm.* 50 (1) (2000) 27–46.
17. N.A. Peppas, Devices based on intelligent biopolymers for oral protein delivery. *Int. J. Pharm.* 277 (1–2) (2004) 11–17.
18. W.B. Liechty, M. Caldorera-Moore, et al., Advanced molecular design of biopolymers for transmucosal and intracellular delivery of chemotherapeutic agents and biological therapeutics. *J. Control. Release* 155 (2) (2011) 119–127.
19. R.K. Verma, S. Arora, et al., Osmotic pumps in drug delivery. *Crit. Rev. Ther. Drug Carrier Syst.* 21 (6) (2004).
20. R.K. Verma, D.M. Krishna, et al., Formulation aspects in the development of osmotically controlled oral drug delivery systems. *J. Control. Release* 79 (1) (2002) 7–27.
21. A. Thombre, L. Appel, et al., Osmotic drug delivery using swellable-core technology. *J. Control. Release* 94 (1) (2004) 75–89.
22. A. Naik, Y.N. Kalia, et al., Transdermal drug delivery: overcoming the skin's barrier function. *Pharm. Sci. Technolo. Today* 3 (9) (2000) 318–326.
23. J.C. Lang, Ocular drug delivery conventional ocular formulations. *Adv. Drug Deliv. Rev.* 16 (1) (1995) 39–43.
24. J.R. Robinson and G.M. Mlynek, Bioadhesive and phase-change polymers for ocular drug delivery. *Adv. Drug Deliv. Rev.* 16 (1) (1995) 45–50.
25. A. Ludwig, The use of mucoadhesive polymers in ocular drug delivery. *Adv. Drug Deliv. Rev.* 57 (11) (2005) 1595–1639.
26. M.F. Saettone and L. Salminen, Ocular inserts for topical delivery. *Adv. Drug Deliv. Rev.* 16 (1) (1995) 95–106.
27. G. Di Colo, S. Burgalassi, et al., Gel-forming erodible inserts for ocular controlled delivery of ofloxacin. *Int. J. Pharm.* 215 (1–2) (2001) 101–111.
28. C. Ramachandran and D. Fleisher, Transdermal delivery of drugs for the treatment of bone diseases. *Adv. Drug Deliv. Rev.* 42 (3) (2000) 197–223.
29. Y.N. Kalia and R.H. Guy, Modeling transdermal drug release. *Adv. Drug Deliv. Rev.* 48 (2–3) (2001) 159–172.
30. S. Menahem and P. Shvartzman, High-dose fentanyl patch for cancer pain. *J. Am. Board Fam. Pract.* 17 (5) (2004) 388–390.
31. W. Jeal and P. Benfield, Transdermal fentanyl. *Drugs* 53 (1) (1997) 109–138.
32. H. Kwon, H. Song, et al., Transdermal therapeutic system of narcotic analgesics using nonporous membrane (I): effect of the ethanol permeability on vinylacetate content of EVA membrane. *Pollimo* 23 (3) (1999) 421–426.
33. K.P.R. Chowdary and Y. Srinivasa Rao, Mucoadhesive microspheres for controlled drug delivery. *Biol. Pharm. Bull.* 27 (11) (2004) 1717–1724.
34. M.I. Ugwoke, R.U. Agu, et al., Nasal mucoadhesive drug delivery: background, applications, trends and future perspectives. *Adv. Drug Deliv. Rev.* 57 (11) (2005) 1640–1665.

35. J. Blanchette, N. Kavimandan, et al., Principles of transmucosal delivery of therapeutic agents. *Biomed. Pharmacother.* 58 (3) (2004) 142–151.
36. L. Brannon-Peppas and N.A. Peppas, Equilibrium swelling behavior of pH-sensitive hydrogels. *Chem. Eng. Sci.* 46 (3) (1991) 715–722.
37. L. Serra, J. Domenech, et al., Drug transport mechanisms and release kinetics from molecularly designed poly(acrylic acid-*g*-ethylene glycol) hydrogels. *Biomaterials* 27 (31) (2006) 5440–5451.
38. L. Illum, "Nasal drug delivery" possibilities, problems and solutions. *J. Control. Release* 87 (1–3) (2003) 187–198.
39. M.J. Eisenberg and K.J. Konnyu, Review of randomized clinical trials of drug-eluting stents for the prevention of in-stent restenosis. *Am. J. Cardiol.* 98 (3) (2006) 375–382.
40. A.S., Puranik, E.R. Dawson, et al., Recent advances in drug eluting stents. *Int. J. Pharm.* 441 (1–2) (2013) 665–679.
41. S.V. Rao, R.E. Shaw, et al., Patterns and outcomes of drug-eluting coronary stent use in clinical practice. *Am. Heart J.* 152 (2) (2006) 321–326.
42. J.L. Wilbur, A. Kumar, et al., Microcontact printing of self-assembled monolayers: applications in microfabrication. *Nanotechnology* 7 (4) (1996) 452.
43. T.A. Desai, W.H. Chu, et al., Microfabricated immunoisolating biocapsules. *Biotechnol. Bioeng.* 57 (1) (1998) 118–120.
44. M.L. Reed, C. Wu, et al., Micromechanical devices for intravascular drug delivery. *J. Pharm. Sci.* 87 (11) (1998) 1387–1394.
45. J.T. Santini, Jr, A.C. Richards, et al., Microchips as controlled drug-delivery devices. *Angew. Chem., Int. Ed.* 39 (14) (2000) 2396–2407.
46. A. Ahmed, C. Bonner, et al., Bioadhesive microdevices with multiple reservoirs: a new platform for oral drug delivery. *J. Control. Release* 81 (3) (2002) 291–306.
47. Y. Lu and S. Chen, Micro and nano-fabrication of biodegradable polymers for drug delivery. *Adv. Drug Deliv. Rev.* 56 (11) (2004) 1621–1633.
48. M. Caldorera-Moore and N.A. Peppas, Micro- and nanotechnologies for intelligent and responsive biomaterial-based medical systems. *Adv. Drug Deliv. Rev.* 61 (15) (2009) 1391–1401.
49. S.L. Tao and T.A. Desai, Microfabricated drug delivery systems: from particles to pores. *Adv. Drug Deliv. Rev.* 55 (3) (2003) 315–328.
50. M. Staples, K. Daniel, et al., Application of micro-and nano-electromechanical devices to drug delivery. *Pharm. Res.* 23 (5) (2006) 847–863.
51. L. Brown and R. Langer, Transdermal delivery of drugs. *Annu. Rev. Med.* 39 (1) (1988) 221–229.
52. X.D. Guo and M.R. Prausnitz, Microelectronic control of drug delivery. *Expert Rev. Med. Devices* 9 (4) (2012) 323–326.
53. S. Henry, D.V. McAllister, et al., Microfabricated microneedles: a novel approach to transdermal drug delivery. *J. Pharm. Sci.* 87 (8) (1998) 922–925.
54. D.V. McAllister, M.G. Allen, et al., Microfabricated microneedles for gene and drug delivery. *Annu. Rev. Biomed. Eng.* 2 (1) (2000) 289–313.
55. D.V. McAllister, P.M. Wang, et al., Microfabricated needles for transdermal delivery of macromolecules and nanoparticles: fabrication methods and transport studies. *Proc. Natl. Acad. Sci. USA* 100 (24) (2003) 13755–13760.

56. S. Rajaraman and H. Henderson, A unique fabrication approach for microneedles using coherent porous silicon technology. *Sens. Actuators B* 105 (2) (2005) 443–448.
57. A. Koulousakis, J. Kuchta, et al., Intrathecal opioids for intractable pain syndromes. *Acta Neurochir. Suppl.* 97 (1) (2007) 43–48.
58. L. Leoni, A. Boiarski, et al., Characterization of nanoporous membranes for immunoisolation: diffusion properties and tissue effects. *Biomed. Microdevices* 4 (2) (2002) 131–139.
59. National Academy of Engineering, Grand challenges for engineering. (2008) Available at http://www.engineeringchallenges.org/Object.File/Master/11/574/Grand%20Challenges%20final%20book.pdf (retrieved May 20, 2013).
60. M.R. Prausnitz, S. Mitragotri, et al., Current status and future potential of transdermal drug delivery. *Nat. Rev. Drug Discov.* 3(2) (2004) 115–124.

3

DEVELOPMENT OF COMBINATION PRODUCT DRUG DELIVERY SYSTEMS

Cynthia L. Stevenson[1] and Robert Langer[2]

[1]*Pharmaceutical Consultant, Mountain View, CA 94040, USA*

[2]*Department of Chemical Engineering, Massachusetts Institute of Technology, Cambridge, MA 02139, USA*

3.1 INTRODUCTION

Combination product drug delivery systems have become smaller and incorporate more mechanisms that allow more precise control over the drug delivery rate, enable the patient or physician to actively start/modify/stop drug release in an interactive format, and provide ways to reach difficult-to-treat locations. These targeted delivery systems allow the drug to be delivered preferentially to a relatively inaccessible site, a specific tissue type, or simply to limit side effects due to systemic exposure. Many therapeutically active agents, especially biomolecules, have short plasma half-lives and require frequent dosing to maintain blood levels in the therapeutic window. A more controlled therapeutic blood level can result in fewer adverse reactions because a high C_{max} after injection is not required to keep the drug in the therapeutic window for the required time period. Drug delivery systems can improve therapeutic response by providing more consistent blood levels than immediate release or sustained release depot parenterals. Therefore, most delivery systems provide a less frequent, more efficacious dosing regimen with increased patient comfort, safety, and compliance.

This chapter covers combination product drug delivery systems, as defined and regulated by the Food and Drug Administration (FDA): A product comprised of two or more regulated components (drug and device, biologic and device, etc.) that are

Drug–Device Combinations for Chronic Diseases, First Edition. Edited by SuPing Lyu and Ronald A. Siegel.

physically, chemically, or otherwise combined to produce a single entity. The chapter highlights three delivery routes (oral, transdermal, and implantable), with an emphasis on microtechnology and implantable systems. In order to understand the evolution of these implantable combination products, an understanding of oral and transdermal systems is required to provide context on how microtechnology innovations have been incorporated. Other FDA-approved combination products (prefilled syringes, pen injectors, transdermal patches, inhalers, and nasal sprays) are beyond the scope of this chapter.

Combination product drug delivery systems routinely use microtechnology in their design and manufacture, which results in delivery aspects in the microscale, forming either a complete device or a critical component of a larger delivery system. Key features of each technology are highlighted, such as microtechnology, working principles, microfabrication methods, dimensional constraints, and performance criteria. These systems provide a well-controlled environment for the drug formulation, allowing increased targeted or systemic drug delivery, drug stability, and prolonged delivery times with various delivery schemes, including zero-order, pulsatile, and on-demand dosing.

3.2 ORAL DELIVERY

Alza Corporation (now Johnson & Johnson) developed a highly successful reservoir-based oral osmotic system (OROS®). OROS technology can be found in many approved products including Procardia® (approved in 1989), Ditropan® (approved in 1998), Glucotrol® (approved in 1999), and Concerta® (approved in 2000) [1,2]. An OROS capsule is coated with a rigid semipermeable membrane containing a precision laser-drilled orifice (0.5–1.4 mm in diameter) for drug release. After ingestion, water diffuses through the semipermeable membrane into the tablet reservoir, and drug is released through the orifice as the osmotically active polymer excipients expand. The membrane controls the rate at which water enters the tablet core, which in turn controls the rate of drug release [3]. OROS tablet technology differentiated from single compartment to multicompartment systems capable of delivering solids and solutions (push–pull, trilayer technologies). The scientific principles in osmotic delivery and controlled release through a constricted orifice have been translated into many smaller implantable systems, as described later.

A second oral delivery example is the application of microfabrication and MEMS (microelectromechanical systems) technologies, resulting in more complicated multi-dose delivery systems. Silicon microparticles (50 μm × 50 μm × 2 μm) with 25 μm × 25 μm × 1 μm deep wells were designed for targeted oral drug delivery [4]. Polymer-based systems measuring ~150 μm × 150 μm × 14 μm followed [5–7]. These systems provided a protected drug reservoir, a functionalized surface for cytoadhesion, and asymmetric drug delivery to the intestinal wall instead of the lumen [7–9]. Devices were also made from polymethylmethacrylate (PMMA), poly-lactic/glycolic acid (PLGA), and poly(ethylene glycol) dimethacrylate (PEGDMA) [6–9]. These systems remain in preclinical testing, and have not undergone manufacturing scale-up or development of product specifications to prove reproducible drug delivery. However, multireservoir microfabrication technology appears in many other drug delivery systems.

3.3 DERMAL DELIVERY

Microneedles can be used in place of larger needles on parenteral syringes or integrated into a transdermal patches, and are considered combination products. Microneedles were developed to increase skin permeability without causing pain because the microneedles penetrate the epidermal layer, but do not extend to the nerve endings found deeper in the skin. These microneedle systems can be differentiated into biodegradable or durable systems.

3.3.1 Biodegradable Microneedles

Biodegradable microneedles formed from drug/polymer mixtures result in a higher drug payload, but commonly result in poor drug stability due to moisture ingress and subsequent drug degradation [10]. Dissolving microneedle patches for influenza vaccination have been tested in mice and pigs [11,12]. The polyvinylpyrrolidone (PVP) microneedles measured 650 μm high and encapsulated 3 μg of inactivated influenza virus per patch [11]. The patch was applied with the thumb, penetrated ~200 μm and deposited the vaccine into the epidermis. This technology would be practical for vaccination in remote areas of the world, if cost of goods (COGs) can be reduced and secondary packaging provided a moisture barrier.

3.3.2 Durable Microneedles

Many durable silicon-based microneedle designs are fabricated by deep reactive ion etching (DRIE) technology and include solid or hollow microneedles with tapered or beveled tips with drug coated onto the array [13–15]. These coated arrays are not reservoir based, but they do allow an initial bolus dose and can be coupled with a reservoir system.

Reservoir-based microneedles are hollow and deliver drug from an external reservoir or from drug stored inside the hollow space of the needle. They can be made using a variety of geometries, including barbed tips [16], microfilters at the needle base [17], bubble pumps for delivery fluids [18], blunt cylinders [19], tapered cylinders [20], volcano shapes [21,22], citadel structures [23,24], and sawtooth structures [25,26]. These microneedles can be integrated into transdermal patches and a variety of injectable systems.

For example, volcano style needles were used to deliver methyl nicotinate in mice [21,22]. A MEMS syringe was configured with an array of silicon needles and a PDMS reservoir attached at the back [21,27]. The entire reservoir assembly was ~10 mm, holding an array of eight needles, which were 200 μm long. The dose was delivered by pressing the device against the skin for a few seconds.

Citadel side-opened needles were licensed by Debiotech for the Microject/Nanoject platform. Hollow microneedles with side openings provide two advantages: an ultrasharp tip for efficient skin penetration and alleviation of tissue coring/channel blockage during insertion [23,24]. The microneedles were fabricated in a $3 \times 3\ mm^2$ chip array, are ~210 µm long, and can inject 500 µl in 5 s [23] (http://www.debiotech.com/). Debiotech also provides needles from 300–1000 µm in length in 25 microneedle arrays, and the drug reservoir can be sealed with a gold film, which is ruptured upon application to the skin (http://www.debiotech.com/) [28]. The microneedles were also incorporated into a MEMS micropump for delivery of insulin [29].

Sawtooth structures are used in the MicroPyramid MicronJet Needle offered by NanoPass Technologies. The MicronJet Needle structures are 150–350 µm in length and ~250 µm wide at the base. The MicronJet needle can be coupled with a standard syringe, and has been used in clinical trials for anesthesia, insulin, and vaccines [25,26].

Tapered cylindrical hollow microneedles have been made from silicon, metal, and polymer [22]. The silicon needles had a constant bore diameter of 60 µm and an increased wall thickness at the base. The metal needles have a constant wall thickness of 10 µm and a bore that widens at the base, with widths ranging from 35 to 350 µm and lengths ranging from 150 to 1000 µm. Polyethylene terephthalate needles measuring 50 µm in length with a 75 µm tip diameter, in an array of 16 microneedles, were tested in rats [30]. The V-Go™ disposable insulin delivery device from Valeritas (a BioValve subsidiary) uses tapered cylindrical, hollow microneedles and was licensed from Georgia Tech [20,30]. V-Go utilizes the hPatch™ platform designed to deliver drugs into the subcutaneous space and the Micro-Trans™ microneedle array. V-Go received 510(k) clearance for the continuous subcutaneous delivery of insulin (http://www.valeritas.com/). The device measures 2.4 in. × 1.3 in. × 0.5 in. and is applied to the skin daily.

Hollow microneedles have also been used to deliver therapeutics to the eye [31]. Solution and suspension formulations of sulforhodamine (15–35 µl) were delivered to the suprachoroidal space through 800–1000 µl needles. Small particles (20–100 nm) can flow through the scleral tissue, and larger particles (500–1000 nm) can be delivered with longer microneedles. The use of hollow microneedles, alone or coupled with an implantable system, highlights an area of innovative targeted applications.

3.4 IMPLANTABLE DELIVERY

Implantable drug delivery systems have continued to evolve into innovative combination products of ever-diminishing dimensions. These implantable combination products can deliver drugs to targeted sites or for systemic therapy. Implantable drug delivery systems can be categorized as passive or active [14,32–35].

A passive system can be defined as a system where drug release is predetermined by the materials, fabrication methods, or drug formulation and cannot be controlled after implantation.

An active system is a system in which drug release is controlled after implantation, using mechanical, electrical, magnetic, laser, or other means. Passive systems utilize diffusion, osmotic potential, or concentration gradients as their driving forces, while active systems include mechanical pumping, electrolysis, and other actuation methods. These micro and nano implants provide new solutions to unmet medical needs and have resulted in some of the most elegant drug delivery concepts.

3.4.1 Passive Delivery Systems

3.4.1.1 Membrane-Controlled Diffusion Systems Diffusion-controlled systems rely on diffusion of drug out of or through a polymer layer that may be nonporous or microporous. The rate-determining step may be diffusion through the membrane structure or transport of drug through the static aqueous diffusion layer. For membrane-controlled diffusion, the diffusion coefficient is dictated by the size of the drug molecule and the pore size or space between the polymer chains. This technology has been used in oral, transdermal, and implantable systems.

For example, several ocular implants utilized a scaled-down version of a diffusion-controlled implant.

The Medidur™ platform is used in two approved products and three more in development. Vitrasert® (approved in 1996) and Retisert® (approved in 2005) are both reservoir-based ocular implants, where drug is diffused through polyvinylalcohol (PVA) [3]. Both of these products were developed by pSivida based on their Medidur platform and are distributed by Bausch & Lomb (Figure 3.1) [36] (http://www.psivida.com/products.html).

The Vitrasert implant is approved for the treatment of CMV retinitis in patients with AIDS. A 4.5 mg ganciclovir tablet reservoir (3.5 mm) is coated with PVA and

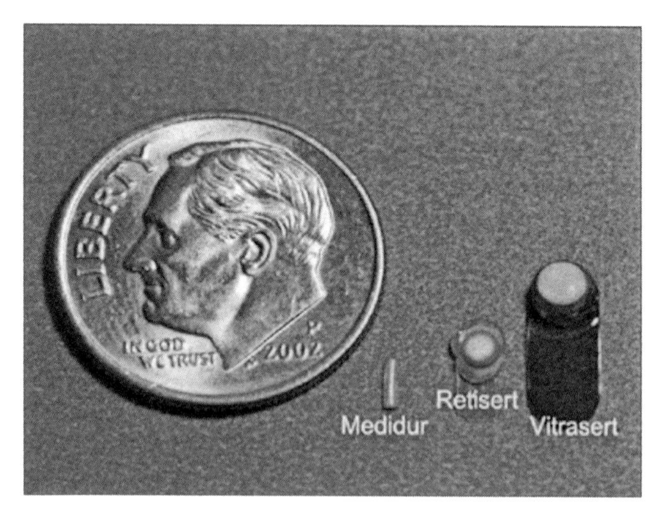

FIGURE 3.1 Medidur delivery platform showing Vitrasert, Retisert, and Medidur. (Reproduced with permission from pSivida, Ltd.) (See colour plate section.)

ethylene vinyl acetate (EVA) polymers and attached to a tab, which is sutured to the inside wall of the eye. The PVA controls the drug release and the EVA controls the surface area of the device through which ganciclovir can permeate. The device releases the drug into the vitreous of the eye for a period of ~5–8 months [3,36–38].

The Retisert implant is approved for uveitis and is the second-generation Medidur technology. The implant consists of 0.59 mg fluocinolone acetonide in a 1.5 mm tablet reservoir. The tablet is encased in a silicone elastomer cup containing a release orifice and a PVA membrane positioned between the tablet and the orifice [3]. The silicone elastomer cup assembly is attached to a PVA suture tab with silicone adhesive, measures 3 mm × 2 mm × 5 mm, and is surgically implanted through 3–4 mm incision. Retisert releases fluocinolone acetonide into the vitreous of the eye at a nominal initial rate of 0.6 μg/day, decreasing over the first month to a steady state between 0.3 and 0.4 μg/day over ~30 months [36,39,40].

The Iluvien™ implant utilizes the third generation of Medidur technology and demonstrates the technological capability of fabricating smaller devices. Alimera has completed phase 3 clinical trials with Iluvien for the treatment of diabetic macular edema (DME) and received approval in Europe. Iluvien is inserted into the vitreous of the eye with a 25 G needle in an outpatient procedure and is not sutured to the eye, in contrast to Vitrasert and Retisert. The implant is a cylinder 3.5 mm long and 0.37 mm in diameter and consists of a polyimide tube reservoir containing 190 μg fluocinolone in a PVA matrix (http://www.alimerasciences.com/Products/iluvien.aspx) [36,41,42]. The tube is capped with rate-controlling membranes. Implants with two release rates were considered during development: a higher dose implant with an initial *in vitro* release rate of 0.45 μg/day and a lower dose implant with an initial *in vitro* release rate of 0.23 μg/day (http://www.alimerasciences.com/Products/iluvien.aspx). The release rates of both implants steadily decreased *in vitro* and were nearly the same at ~0.15 μg/day by 18 months [42]. Clinical results indicate improved best corrected visual acuity (BVCA) after 2 or 3 years of treatment [43] (http://investor.alimerasciences.com/releasedetail.cfm?ReleaseID=574002).

Durasert™ is the next follow-on implant from pSivida. Durasert is a bioerodible reservoir implant for the delivery of latanoprost for high intraocular pressure and is in phase 2 clinical trials (http://www.psivida.com/products.html). Finally, pSivida is also developing Tethadur™ for biomolecules and the delivery of more than one therapeutic agent (http://www.psivida.com/products.html). Tethadur utilizes nanostructured pores to control the delivery of varying size molecules, is targeted to last 6 months, and reported to be fully bioerodible. The addition of nanopores to diffusion-based platforms provides another mechanism for controlling release rate and is discussed later. The successful evolution of this ocular platform is an example of a well-developed delivery system, adaptable to a variety of drugs, acceptable to patients and physicians, and manufactured with acceptable COGs.

Iveena is developing the capsule drug ring (CDR™), reminiscent of the larger and less sophisticated pilocarpine Ocusert™ system (developed by Alza and approved in 1974 for glaucoma). Intraocular lenses inserted during cataract surgery leave unfilled space in the lens capsule. The CDR (13 mm outside diameter and 9.2 mm inside diameter) was designed to be implanted in the peripheral lens capsule during or after

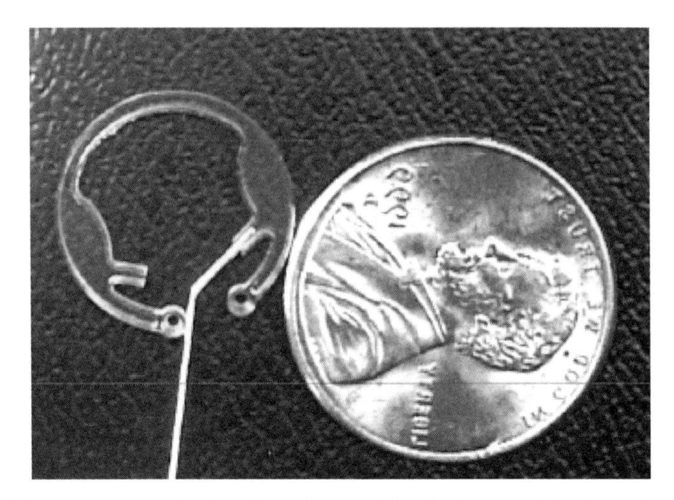

FIGURE 3.2 Iveena capsule drug ring (CDR) with a 27 G cannula in the valve access port (Reproduced with permission from Figure 3 in Ref. [44]. Copyright 2010, Elsevier.)

cataract surgery (Figure 3.2) [44]. The implant is made of a polymethyl methacrylate (PMMA) shell, a semipermeable membrane to control the delivery rate, and silicon valves for refilling the drug reservoir. The refillable drug reservoir holds 50 μl, and a linear delivery rate of 0.06–0.08 mg/day Avastin® (bevacizumab) was achieved [44].

*3.4.1.2 **Matrix-Controlled Diffusion Systems*** A subclass of diffusion-controlled systems takes advantage of a combination of drug release from a polymer matrix via porosity (tortuous path) and polymer erosion. Pulsatile release from a passive system for ocular use was proposed by the Ocular Drug Delivery Group at UC Irvine. The micromachined rod-like device is designed with alternating drug-loaded and drug-free polymer zones, enabling drug levels to vary in a preprogrammed manner as each layer is sequentially exposed to the body [36,45].

The I-vation™ implantable system, from Surmodics, was developed for the treatment of diabetic macular edema [36,46–48]. The intravitreal implant is a titanium helical coil coated with 925 μg triamcinolone acetonide formulated with poly(methyl methacrylate) and ethylene vinyl acetate (Figure 3.3). The sustained release implant is 5 mm long and 2 mm in diameter and is placed in the posterior chamber of the eye. The implant is inserted through a 25 G trocar into the sclera and the cap anchors it to the eye.

Intravascular stents are used to hold open blood vessels and are constructed as an expandable wire mesh. Many stents are simply coated with a drug/polymer matrix to reduce restenosis. This coating technology does not deliver drug from a reservoir, but from the polymer matrix. However, several new stent technologies have incorporated reservoirs.

Debiotech developed a polymer-free nanocoating technology based on biocompatible nanostructured ceramic coatings with a variety of porosities, and this coating can be manipulated to form drug reservoirs [27,49]. The Debiostent loading

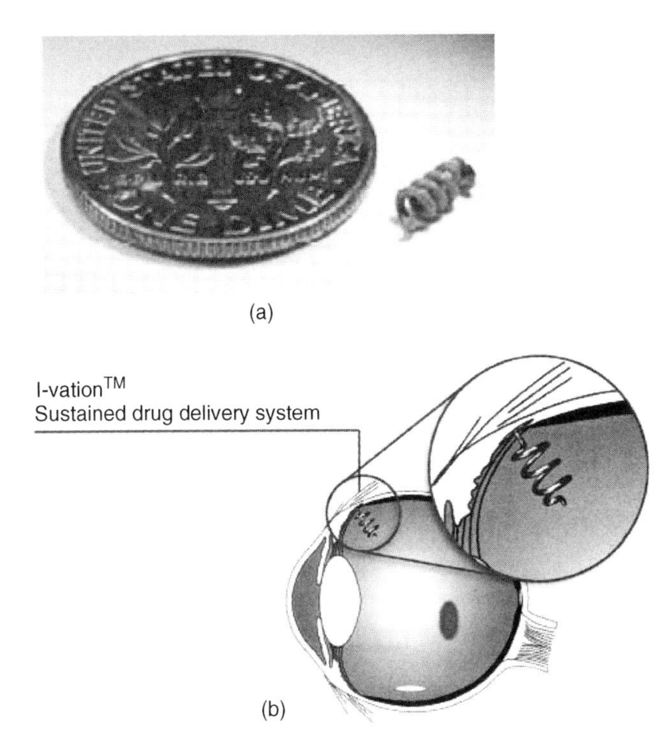

FIGURE 3.3 I-vation ocular implant. (Reproduced with permission from Surmodics, Inc.)

capacity is defined by the size of the drug reservoirs in a lower coating layer, where reservoirs represent up to 30% of the total coating volume, and drug loading can reach up to 10 μg/mm^2. An upper coating layer seals the reservoirs and controls the drug elution kinetics. The location and number of drug reservoirs can be controlled to focus the delivery of drug from specific areas of the stent (i.e., only the adluminal regions in contact with the vessel) for maximum efficacy [27].

Conor Medsystems (now Johnson & Johnson) developed cobalt–chromium stents with hundreds of laser-cut reservoirs in the stent struts [50,51]. The reservoirs are loaded with drugs in a bioresorbable PLGA matrix. Stents containing various drugs, for example, Nevo™ (sirolimus), CoStar™ (paclitaxel), Corio (pimecrolimus), and Symbio™ (paclitaxel and pimecrolimus in adjacent reservoirs), have been tested in clinical trials [50,51].

The JACTAX™ stent by LabCoat Ltd. (now Boston Scientific) formulated the drug in polymer reservoirs or microdots placed on the surface of the stent strut [52,53]. A bare metal stent is coated on the abluminal side with 2750 discrete microdots. The microdots consist of a 50/50 mixture of polylactic acid polymer and paclitaxel.

Medtronic and Intesect ENT have also applied the same concept of formulating steroid in a PLGA matrix for chronic sinusitis. The Propel® implant is a dissolvable implant inserted after sinus surgery. The biodegradable spring-like implant expands the ethmoid sinus and delivers 370 μg mometasone furoate over 30 days [54].

3.4.1.3 Microchannel Systems Micro- or nanochannel implants use a membrane-like structure, where the release rate is controlled by diffusion along a constrained channel. Specifically, the size of the pore and the size of the drug substance are correlated to ensure molecular constraint. Modifying the properties of the microfluidic devices, such as surface effects, charge interaction, concentration polarization, and streaming current phenomena, are used to obtain a zero-order release profile [55–59].

The Hydron® Implant Technology (acquired from Valera Pharmaceuticals by Endo Pharmaceuticals) is a cylindrical, nonbioerodible implant 3 mm in diameter and 3.5 cm in length. The implant is made of a hydrogel polymer blend (2-hydroxyethyl methacrylate, 2-hydroxypropyl methacrylate, trimethylolpropane trimethacrylate, benzoin methyl ether, Perkadox-16, Triton X-100) called MedLaunch,™ and is spun-cast into small tube reservoirs [3]. The implant is supplied prehydrated and contains micropores that allow drug diffusion in a zero-order fashion for a year or longer. Vantas® for prostate cancer (approved 2004) and Supprelin® LA for precocious puberty (approved 2007) are both implanted with a trocar, contain 50 mg histrelin acetate, and last for 1 year with a release rate of 41 μg/day. The histrelin is formulated with stearic acid to form a solid drug core placed in the reservoir [3]. Endo Pharmaceuticals also has an 84 mg octreotide implant that lasts for 6 months in phase 3 clinical trials for acromegaly [60].

iMEDD Inc. developed a small cylindrical titanium implant for continuous release of α-interferon for 3–6 months for the treatment of hepatitis C [56]. The implant is designed to be inserted under the skin and to maintain drug plasma level above 50 pg/ml in order to maintain dosing in a tight therapeutic window.

The NanoGATE™ implant is 4–5 mm in diameter, 20–35 mm in length, with a 75–300 μl reservoir that can contain a suspension or solid-state drug formulation (Figure 3.4) [55,61]. The titanium implant is capped at both ends, and a 2 mm × 3 mm nanopore membrane is affixed over a small bore opening in a cylindrical methacrylate inset carrier. The carrier is fitted with two silicone O-rings and inserted into the titanium encasement. Upon aligning the membrane with the titanium grate opening, the device is filled with formulation [55]. The nanopore membrane controls diffusion of the drug from the reservoir and is made of silicon films with parallel rectangular channel arrays of 10–100 nm in diameter [61]. The nanopore membrane provides a non-Fickian zero-order release, unrelated to drug concentration. The pore size is designed to approximate the diameter of the drug substance, so that the flux of molecules occurs "single file" [56,57,59]. *In vitro* release rate studies with bovine serum albumin, a 13 nm nominal pore size and 4 μm membrane thickness revealed a release rate of 900 μg/day. It should be noted that both the NanoGATE and Hydron technologies are diffusion-controlled instead of osmotically driven and provide more space within the implant for drug formulation.

Taris Biomedicals is in phase 2 trials with a lidocainc intravesical system (LiRIS®) for interstitial cystitis [62,63]. The implant has a shape retention memory, and is placed in the bladder with catheterization and flexible cystoscopy. The implant is 5 cm × 5 cm, and is comprised of a solid dry powder drug core contained in a semipermeable polymeric silicone tube with a micromachined release orifice for drug elution (Figure 3.5). A biodegradable version has also been developed for

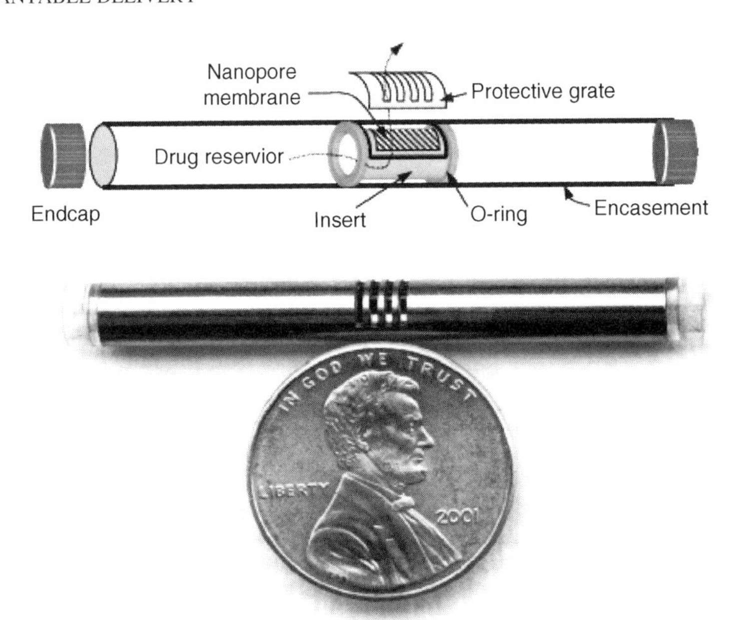

FIGURE 3.4 NanoGATE microchannel implant. (Reproduced with permission from Figure 4 in Ref. [55]. Copyright 2005, Elsevier.)

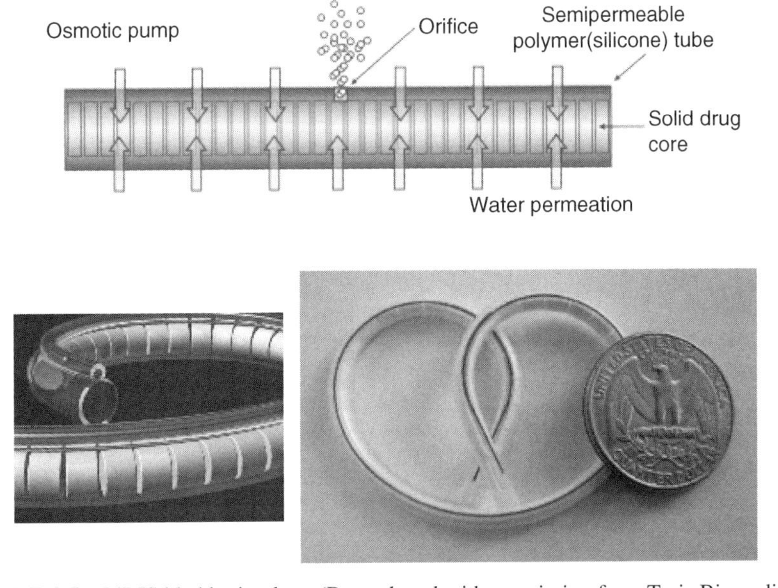

FIGURE 3.5 LiRIS bladder implant. (Reproduced with permission from Taris Biomedical.)

controlled release of ciprofloxacin. Poly(glycerol-*co*-sebacic acid) (PGS) was cast into a tube to provide the semipermeable membrane [64].

Debiotech offers the DebioStar™ technology, a silicon nanoporous membrane used in an implant over a reservoir to deliver drug from weeks to months [27]. The pores can be controlled between 1 and 250 nm, where the number of pores per membrane can be as high as 1 billion pores/cm^2 and the total thickness of the membrane can range from 50 nm to ~200 μm. Delivery rate control has been demonstrated with somatostatin by altering pore diameter and surface properties.

Finally, molecular constraint has also been used in a silicon-based nanochannel delivery system (nDS) implanted in proximity to unresectable metastatic melanoma lesions. Local, targeted delivery of α-INF to the tumor site may alleviate the need for multiple injections and side effects observed with systemic administration [56].

3.4.1.4 Osmotic Systems Osmotic pumps are comprised of a drug reservoir, a piston, a semipermeable membrane, and an osmotic engine. In operation, water is drawn through the semipermeable membrane in response to an osmotic gradient between an osmotic engine and moisture in the surrounding interstitial fluid. Expansion of the osmotic engine drives a piston forward, expelling the drug formulation through an orifice.

The Viadur® leuprolide acetate implant was developed by Alza and utilized the DUROS® platform. Viadur (approved in 2000) for the treatment of prostate cancer was discontinued in 2007 due to generic competition. The DUROS titanium implant is a cylinder 4 mm in diameter and 45 mm in length, and provides osmotically driven, zero-order drug delivery (Figure 3.6) [65]. The implant is placed under the skin on the inside of the upper arm with the aid of a trocar, and explanted at the end of the 1-year delivery duration. Viadur delivered leuprolide (370 mg/ml leuprolide in dimethyl sulfoxide) continuously over 1 year at ~120 μg/day (0.4 μl/day) from a 150 μl drug reservoir [66]. *In vitro* and *in vivo* release rate data demonstrated zero-order delivery for 1 year, and were consistent with decreased serum testosterone and steady leuprolide levels for 12 months [65,67].

Durect Corp. licensed the DUROS platform from Alza for pain indications. The Chronogesic® reservoir contained ~155 μl sufentanil in benzyl alcohol and provided zero-order release at 5 μg/h [68]. Durect initiated phase 3 trials with sufentanil for moderate to severe chronic pain. However, the trial was suspended in 2003 due to premature shutdown of devices [69]. Drug delivery systems containing a drug payload large enough to cause adverse events or death must have safety designs to remediate drug dose dumping and patient death. Furthermore, drugs susceptible to drug abuse must have safety attributes to retard drug abuse (i.e., device lock outs or formulation strategies that circumvent a drug high).

Intarcia Therapeutics also licensed the DUROS implant for the delivery of exenatide (type 2 diabetes) and ω-interferon (Hepatitis C) (www.intarcia.com). Exenatide was formulated for a 6 month (45 μg/day), 9 month (30 μg/day), and 12 month (10 μg/day) duration and showed good stability at 25 and 37 °C for 12 months [70]. Furthermore, exenatide was dosed at 20–80 μg/day for 6 months

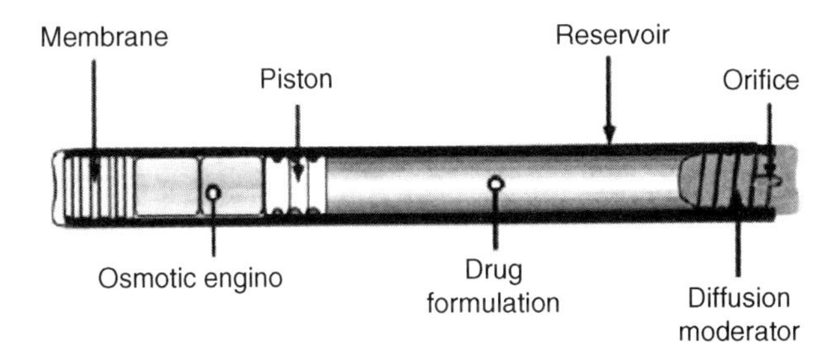

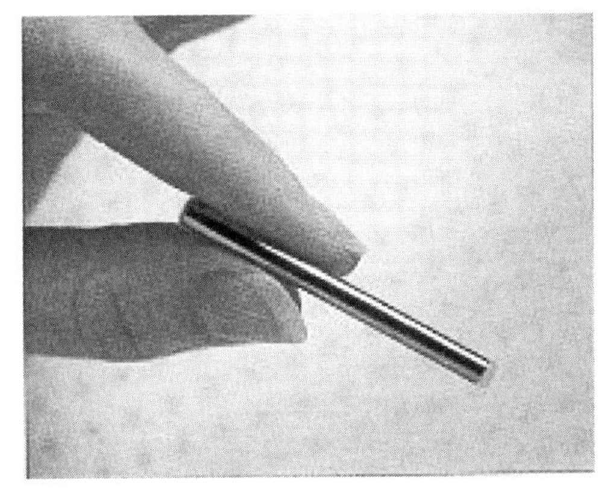

FIGURE 3.6 DUROS osmotic implant. (Reproduced with permission from Figure 1 in Ref. [65]. Copyright, 2001 Elsevier.)

in phase 2b trials [71]. Intarcia also has ω-interferon in phase 1b (25 μg/day and 50 μg/day) for Hepatitis C [72,73]. ω-interferon was formulated as a suspension and showed good stability for 2 years at 30 °C. The protein was delivered from the device at 9 μg/day and 22 μg/day for 6 months at 37 °C.

Finally, a bioerodible, micro-osmotic implant has also been constructed for the local delivery of basic fibroblast growth factor (bFGF) for tissue regeneration. The implants were micromolded and thermally assembled from PLGA sheets. The bottom layer contained the drug reservoir and microchannel array for drug release. An osmotic potential drives water into the reservoir through a semipermeable membrane. The inflated reservoir exerts pressure, delivery drug from the microarrays at 40 ng/day for 4 weeks [74].

3.4.2 Active Delivery Systems

3.4.2.1 Single Reservoir Systems Pumps generally include a drug reservoir, an actuator, and one or more valves in order to accurately control the delivery of small volumes of a drug in solution. They may operate by manual actuation, electrolysis, piezoelectric actuation, resistive heating, magnetic actuation, or by incorporation of reversible polymeric valves.

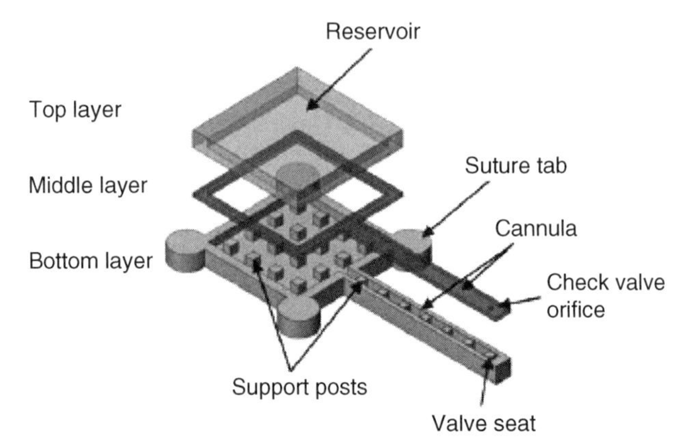

FIGURE 3.7 Manually actuated drug delivery pump for ophthalmic use. (Reproduced with permission from Figure 1 in Ref. [75]. Copyright 2009, Springer.)

A simple drug delivery pump was designed for the treatment of glaucoma, age-related macular degeneration, and diabetic retinopathy [53,75]. The pump was fabricated from three layers of polydimethylsiloxane (PDMS) using soft lithography. The device was sutured to the outside of the eye and contained a drug reservoir of ~200 μl, a check valve, and a cannula (10 mm × 1 mm × 1 mm) placed through the wall of the eye (Figure 3.7). When the pump was manually actuated (with instrument or finger) by pressing on the drug reservoir, the increase in pressure in the reservoir caused the check valve in the cannula to open, dispensing the drug.

Replenish developed an electrochemically driven drug delivery pump that allows for variable drug delivery rates. The system contains a refillable reservoir for a drug solution and a one-way check valve made of parylene. The device is surgically implanted beneath the conjunctiva with a flexible parylene cannula inserted through the eye wall. An electric current is passed between two electrodes located on the silicon in contact with the drug solution when a dose is required. The gas generated by electrolysis of the water increases the pressure on the flexible membrane of the drug reservoir, pushing drug solution out through the cannula and into the eye (Figure 3.8) [76].

ForSight Vision 4 and Genentech are in clinical trials on a refillable drug port delivery system to release Lucentis® (ranibizumab) for wet AMD. The capsule-like implant is intended for delivery to the posterior segment of the eye. One end of the capsule has an external port that can be periodically refilled by needle injection [51].

A magnetically actuated ocular implant was developed at the University of British Columbia for the treatment of diabetic retinopathy [77]. The implant consisted of a reservoir (6 mm × 550 μm) containing docetaxel, sealed with an elastic magnetic polydimethylsiloxane (PDMS) membrane (6 mm × 40 μm), and a laser-drilled orifice ($100 \times 100\ \mu m^2$). The device is designed to be surgically implanted behind the eye. Upon application of a magnetic field (255 mT), the membrane deforms, causing expulsion of drug solution (~171 ng/actuation) from the implant.

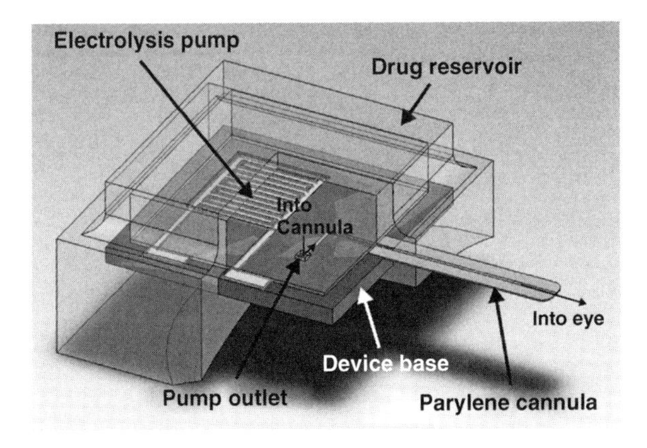

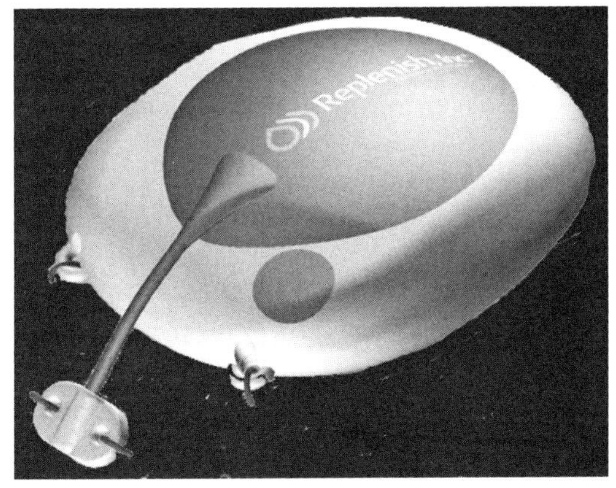

FIGURE 3.8 Replenish delivery pump for ophthalmic use. (Reproduced with permission from Figure 1 in Ref. [76]. Copyright 2008, Elsevier.)

The MIP implantable pump from Debiotech is a piezo-actuated silicon micropump for drug delivery. The pump consists of a pair of check valves and a reciprocating pumping membrane to guide liquid flow in the proper direction from a drug reservoir to the target location. The pump is fabricated from multiple bonded layers of silicon and glass, with a piezoelectric ceramic disk and titanium fluid connectors, with a flow rate of 1 ml/min [27].

Burst release devices have also been proposed for emergency care [33,78,79]. As an example, the IRD^3 (implantable rapid drug delivery device) developed at the MIT is comprised of three layers: a drug reservoir layer, a membrane layer that seals the drug reservoir, and an actuation layer where bubbles are formed. Microresistors in the actuation layer heat the drug solution, generating bubbles that rupture the membrane over the reservoir and deliver the drug from the reservoir (~20 μl vasopressin solution in 45 s). Another group developed a similar microfluidic implant (4.4 mm × 2.3 mm × 22 mm) to deliver vasopressin from a 15 μl reservoir [79].

Finally, valves made from polymer actuators have been proposed to modulate release of a drug solution from a single-reservoir "smart pill" implant [80]. The implant contains micrometer-sized polymer rings that expand and contract in response to an electrical signal transmitted through a conducting polymer, thereby allowing the modulation of flow of drug solution out of the reservoir.

3.4.2.2 Multireservoir Systems The selection of one delivery mechanism versus another is highly dependent on the treatment modality of the specific disease state and the preferred delivery profile (sustained release or pulsatile). Some treatment modalities would benefit from more than one drug reservoir. Active delivery from multireservoir implants can be actuated by electrochemical, electrothermal, or laser means [35,81–86]. The initiation of drug release is actively controlled by the application of the electrical or laser stimulus to create an opening in the sealing material. The rate of release is then passively controlled by the dissolution and diffusion characteristics of the drug formulation in the reservoir.

MIT developed arrays of gold membrane-capped drug reservoirs in silicon. When an electric potential was applied to the gold cap, the gold membrane was converted to a soluble gold salt and dissolved away, thereby exposing the drug to the surrounding environment and releasing the drug [81–84]. Although *in vivo* reproducibility of an electrochemical dissolution process can be challenging, these devices demonstrated macromolecule drug pulsatile delivery [32,85].

MicroCHIPS developed reservoir arrays where the metal membranes coating the reservoirs were opened electrothermally instead of electrochemically [86,87]. Metal membranes composed of either gold or platinum/titanium laminate were removed by resistive heating from an applied current [86]. Microchip arrays of $15 \times 15 \times 1\ mm^3$ containing 100 reservoirs capable of holding 300 nl drug formulation per reservoir have been fabricated [86]. The fully assembled devices included a drug-filled array, microprocessor, implantable battery, and wireless antenna in a titanium housing that measured $4.5 \times 5.5 \times 1\ cm^3$ with a volume of ~30 ml. Initial studies with leuprolide formulated at 100 mg/ml and filled into 300 nl reservoirs showed good stability and an *in vitro–in vivo* release rate correlation [88]. Similarly, an osteoporosis implant delivering pulsatile parathyroid hormone (PTH) was formulated at 500 mg/ml (~20 μg PTH/reservoir) and demonstrated good stability after 6 months at 37 °C [89].

On Demand Therapeutics utilizes laser activation to achieve ophthalmic drug delivery for treatment of retinal diseases, such as wet AMD or diabetic retinopathy. The preclinical product is a small, injectable rod containing multiple reservoirs that are hermetically sealed to provide drug stability from the interstitial fluid or different therapeutic agent in the adjacent reservoir. The drug-filled device is implanted into the periphery of the vitreous in the region of the pars plana using a standard intravitreal injection technique. When the ophthalmologist decides to release drug from the implant, a slit lamp and a Goldmann-mirrored lens are used to locate the implant and focus the laser beam on the selected reservoir. Upon laser activation, an opening is created that allows the drug formulation to elute into the vitreous and diffuse to the retina. The drug doses in the unopened reservoirs remain intact, such that additional drug doses can be released during subsequent visits. This

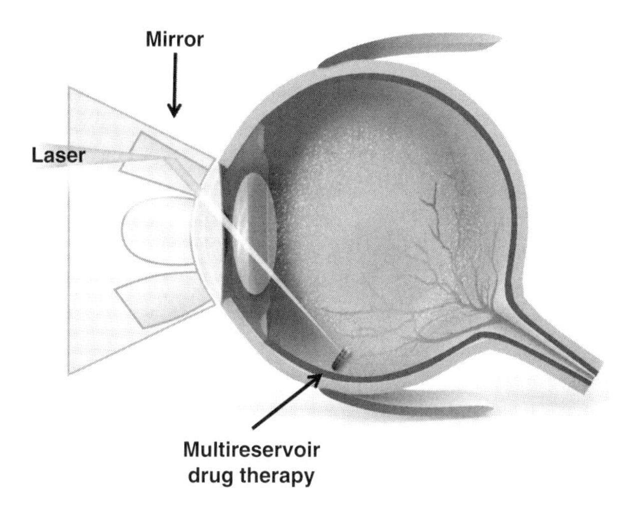

FIGURE 3.9 A laser activated system for ophthalmic a multireservoir, intravitreal implant. (Reproduced with permission from On Demand Therapeutics.) (See colour plate section.)

approach provides a less invasive dosing regimen than monthly injections (Figure 3.9) (www.ondemandtx.com).

3.5 CONCLUSIONS

The knowledge and tools enabling the development of combination product drug delivery systems has come from number of diverse fields of study including chemistry, pharmaceutics, materials science, mechanics, information technology, and microelectronics. As combination products gain maturity, several factors will impact their development success, regulatory approval, and profitability.

First, the choice of pharmaceutically active drug in need of an improved delivery paradigm should be carefully chosen. The ultimate success of a commercial product is not wholly dictated by a good technical match between the drug and the delivery method. Considerations such as time and money required to complete pivotal clinical trials, anticipated competitive products in the market at the time of projected regulatory approval, competitive pricing, and product positioning are required for a successful launch.

Second, thought should be given to supply chain issues early in development. Drug excipients should be generally regarded as safe (GRAS). Device materials that are drug- or patient-contacting should be Medical Grade with acceptable and meet USP requirements for physicochemical characteristics, biocompatibility, and leachables and extractables. This is especially important for long-term or permanent implants, where tissue adhesion or polymeric degradation products can cause inflammation. Furthermore, supplier selection should be done with care to ensure that the vendor has or will be able to provide a drug master file (DMF). Validation and verification testing of the integrated combination product in its commercial packaging should be anticipated.

Third, the combination product must demonstrate adequate shelf life stability and *in vivo* use-life stability to ensure that the final dosing period was as efficacious as the initial dosing period. For example, if the drug is released from the device for 6 months, then studies must be completed demonstrating a 2-year shelf life plus 6 months of use-life. Drug release tests used to demonstrate the lack of dose dumping, percentage of total formulation released from the delivery system, *in vivo* stability, and an *in vivo–in vitro* correlation are all critical.

Fourth, the combination product must provide a distinct advantage over current therapies, with only a small premium in price. The impact of pricing can now be felt early in the R&D process, as these budgets are increasingly curtailed. The outcome is that engineers and scientists must provide innovative products and cost-effective processes at every stage of development. This not only applies to the novel delivery system cost of goods (COGS) but also to the novel microfabrication and manufacturing lines that must be scaled up to meet commercial demands in a cost-effective manner. The risk to the investor becomes even more complex if the pharmaceutical portion of the product is a new chemical entity (NCE), instead of a previously approved compound.

Finally, development of successful drug delivery products requires forethought into the regulatory hurdles in order to receive timely market approval. A regulatory strategic plan must be developed to determine if (a) the device will be approved independent of the drug and seek approval for a 510(k) through CDRH, and/or (b) the drug is previously approved and a new delivery route is provided for a 505(b) filing through CDER. Furthermore, human factors testing should be executed to ensure that the patient and/or medical practitioner can understand the instructions for use and demonstrate that they can use the product as intended without harm to the patient, which is critical to its ultimate acceptability.

ACKNOWLEDGMENTS

Robert Langer would like to acknowledge grant support for reservoir-based, microchip drug delivery systems from the U.S. National Institutes of Health, grant #EB006365, MicroCHIPS, and On Demand Therapeutics.

REFERENCES

1. R.K. Verma and S. Garg, Drug delivery technologies and future directions. *Pharm. Technol.* 25 (2001) 1–14.
2. H. Rosen and T. Abribat, The rise and rise of drug delivery. *Nat. Rev. Drug Discov.* 4 (2005) 381–385.
3. Medical Economics Staff *Physicians' Desk Reference*, 63rd ed., Thomson Publishing, Montvale, NJ, 2009.

4. A. Ahmed, C. Bonner, and T. Desai, Bioadhesive microdevices with multiple reservoirs: a new platform for oral drug delivery. *J. Control. Release* 81 (2002) 291–306.
5. S.L. Tao and T.A. Desai, Micromachined devices: the impact of controlled geometry for cell-targeting to bioavailability. *J Control. Release* 109 (2005) 127–138.
6. K.M. Ainslie and T.A. Desai, Microfabricated implant for applications in therapeutic delivery, tissue engineering and biosensing. *Lab Chip* 8 (2008) 1864–1878.
7. K.M. Ainslie, R.D. Lowe, T.T. Beaudette, L. Petty, E.M. Bachelder, and T.A. Desai, Microfabricated devices for enhanced bioadhesive drug delivery: attachment to and small-molecule release through a cell monolayer under flow. *Small* 5 (2009) 2857–2863.
8. S.L. Tao and T.A. Desai, Microfabrication of multilayer, asymmetric, polymeric devices for drug delivery. *Adv. Mater.* 17 (2005) 1625–1630.
9. S.L. Tao and T.A. Desai, Gastrointestinal patch systems for oral drug delivery. *Drug Discov. Today* 10 (2005) 909–915.
10. E.E. Nuxoll and R.A. Siegel, BioMEMS devices for drug delivery. *IEEE Eng. Med. Biol.* Jan/Feb (2009) 31–39.
11. S.P. Sullivan, D.G. Koutsonanos, M.D.P. Martin, J.W. Lee, V. Zarnitsyn, S.O Choi, N. Murthy, R.W. Compans, I. Skountzou, and M.R. Prausnitz, Dissolving polymer microneedle patches for influenza vaccination. *Nat. Med.* 16 (2010) 915–920.
12. S. Kommareddy, B.C. Boudner, S. Oh, S.Y. Kwon, M. Singh, and D.T. O'Hagan, Dissolvable microneedle patches for the delivery of cell culture derived influenza vaccine antigens. *J. Pharm. Sci.* 101 (2012) 1021–1027.
13. A. Aroro, M. Prausnitz, and S. Mitragotri, Micro-scale devices for transdermal drug delivery. *Int. J. Pharm.* 8 (2008) 227–236.
14. M. Staples, K. Daniel, M.J. Cima, and R. Langer, Application of micro- and nano-electromechanical devices to drug delivery. *Pharm. Res.* 23 (2006) 847–863.
15. J. Matriano, M. Cormier, J. Johnson, W. Young, M. Buttery, K. Nyam, and P. Daddona, Macroflux microprojection array patch technology: a new and efficient approach for intracutaneous immunization. *Pharm. Res.* 19 (2002) 1653–1664.
16. S. Chandrasekara, J. Brazzle, and A. Frazier, Surface machined metallic microneedles. *J. Microelectromech. Syst.* 12 (2003) 281–288.
17. J. Zahn, N. Talbot, D. Liepmann, and A. Pisano, Microfabricated polysilicon microneedles for minimally invasive biomedical devices. *Biomed. Microdevices* 2 (2000) 295–303.
18. J. Zahn, A. Deshmekh, A. Pisano, and D. Liepmann, Continuous on-chip micropumping from microneedled enhance drug delivery. *Biomed. Microdevices* 6 (2004) 183–190.
19. M.A.L. Teo, C. Shearwood, K.C. Ng, J. Lu, and S. Moochhala, *In vitro* and *in vivo* characterization of MEMS microneedles. *Biomed. Microdevices* 7 (2005) 47–52.
20. D.V. McAllister, P.M. Wang, S.P. Davis, J.H. Park, P.J. Canatella, M.G. Allen, and M.R. Prausnitz, Microfabricated needles for transdermal delivery of macromolecules and nanoparticles: fabrication methods and transport studies. *Proc. Natl. Acad. Sci. USA* 100 (2003) 13755–13760.
21. B. Stoeber and D. Liepmann, Arrays of hollow out-of-plane microneedles for drug delivery. *J. Microelectromech. Syst.* 14 (2005) 472–479.
22. R. Sivamani, B. Stoeber, G. Wu, H. Zhai, D. Liepmann, and H. Maibach, Clinical microneedle injection of methyl nicotinate: stratum corneum penetration. *Skin Res. Technol.* 11 (2005) 152–156.

23. P. Griss and G. Stemme, Side-opened out-of-plane microneedles for microfluidic transdermal liquid transfers. *J. Microelectromech. Syst.* 12 (2003) 296–301.
24. N. Roxhed, C. Gasser, P. Griss, G. Holzapfel, and G. Stemme, Penetration enhanced ultrasharp microneedles and prediction in skin interaction for efficient transdermal drug delivery. *J. Microelectromech. Syst.* 16 (2007) 1429–1440.
25. H. Gardeniers, R. Luttge, E. Bereschot, M. de Boer, S. Yeshurun, M. Hefetz, R. van't Oever, and A. van den Berg, Silicon micromachines hollow microneedles for transdermal liquid transport. *J. Microelectromech. Syst.* 12 (2003) 855–862.
26. P. Van Damme, F. Oosterhuis-Kafeja, M. van der Wielen, Y. Almagor, O. Sharon, and Y. Levin, Safety and efficacy of a novel microneedle device for dose sparing intradermal influenza vaccination in healthy adults. *Vaccine* 27 (2009) 454–459.
27. U.O. Hafeli, A. Mikhtari, and D. Liepmann, *In vivo* evaluation of a microneedle-based miniature syringe for intradermal drug delivery. *Biomed. Microdevices* 11 (2009) 943–950.
28. N. Roxhed, P. Griss, and G. Stemme, Membrane-sealed hollow microneedles and related administration schemes for transdermal drug delivery. *Biomed. Microdevices* 10 (2008) 271–279.
29. L. Nordquist, N. Roxhed, P. Griss, and G. Stemme, Novel microneedle patches for active insulin delivery are efficient in maintaining glycaemic control: an initial comparison with subcutaneous administration. *Pharm. Res.* 24 (2007) 1381–1387.
30. S.P. Davis, W. Martanto, M.G. Allen, and M.R. Prausnitz, Hollow metal microneedles for insulin delivery to diabetic rats. *IEEE Trans. Biomed. Eng.* 52 (2005) 909–915.
31. S.R. Patel, A.S.P. Lin, H.F. Edelhauser, and M.R. Prausnitz, Suprachoroidal drug delivery to the back of the eye using hollow microneedles. *Pharm. Res.* 28 (2010) 166–176.
32. A.C.R. Grayson, R.S. Shawgo, A.M. Johnson, N.T. Flynn, Y. Li, M.J. Cima, and R. Langer, A BioMEMS review: MEMS technology for physiologically integrated devices. *Proc. IEEE* 92 (2009) 6–21.
33. J.H. Sakamoto, A.L. van de Ven, B. Godin, E. Blanco, R.E. Serda, A. Grattoni, A. Ziemys, A. Bouramrani, T. Hu, S.I. Ranganathan, E. De Rosa, J.O. Martinez, C.A. Smid, R.M. Chchana, S.Y. Lee, S. Srinavasan, M. Landry, A. Meyn, E. Tasciotti, X. Liu, P. Decuzzi, and M. Ferrari, Enabling individualized therapy through nanotechnology. *Pharmacol. Res.* 62 (2010) 57–89.
34. A. Nisar, N. Afzulpurkar, B. Mahaisavariya, and A. Tuantranont, MEMS-based micropumps in drug delivery and biomedical applications. *Sens. Actuat. B.* 130 (2008) 917–942.
35. R.A.M. Receveur, F.W. Linemans, and N.F. de Rooij, Microsystem technologies for implantable applications. *J. Micromech. Microeng.* 17 (2007) R50–R80.
36. Y.E. Choonara, V. Pillay, M.P. Danckwerts, T.R. Carmichael, and L.C. Du Toit, A review of implantable intravitreal drug delivery technologies for the treatment of posterior segment eye diseases. *J. Pharm. Sci.* 99 (2010) 2219–2239.
37. G.E. Sanborn, R. Anand, R.E. Torti, S.D. Nightingale, S.X. Cal, B. Yates, P. Aston, and T.J. Smith, Sustained release ganciclovir therapy for treatment of cytomegalovirus retinitis: uses of an intravitreal device. *Arch. Ophthalmol.* 110 (1992) 188–195.
38. T.J. Smith, P.A. Pearson, D.L. Blandford, J.D. Brown, K.A. Goins, J.L. Hollins, E.T. Schmiesser, P. Glavinos, L.B. Baldwin, and P. Ashton, Intravitreal sustained release ganciclovir. *Arch. Ophthalmol.* 110 (1992) 255–258.

39. P. Mruthyunjaya, D. Khalatbari, P. Yang, S. Stinnett, R. Tano, P. Ashton, H. Guo, M. Nazzar, and G.J. Jaffe, Efficacy of low release rate fluocinolone acetonide intravitreal implants to treat experimental uveitis. *Arch Ophthalmol.* 124 (2006) 1012–1018.
40. D.G. Callanan, G.J. Jaffe, D.F. Martin, P.A. Pearson, and T.L. Comstock, Treatment of posterior uveitis with a flucinolone acetonide implant. *Arch. Ophthalmol.* 126 (2008) 1191–1201.
41. E. Eljarrat-Binstock, J. Pe'er, and A.J. Domb, New techniques for drug delivery to the posterior eye segment. *Pharm. Res.* 27 (2010) 530–543.
42. F.E. Kane, J. Burdan, A. Cutino, and K.E. Green, Iluvien: a new sustained delivery technology for posterior eye disease. *Expert Opin. Drug Deliv.* 5 (2008) 1039–1046.
43. P.A. Campochiaro, D.M. Brown, A. Pearson, T. Ciulla, D. Boyer, F.G. Holz, M. Tolentino, A. Gupta, L. Duarte, S. Madreperla, J. Gonder, B. Kapik, K. Billman, and F. Kane, Long-term benefit of sustained-delivery fluocinolone acetonide vitreous inserts for diabetic macular edema. *Ophthalmology* 118 (2011) 626–635.
44. S.A. Molokhia, H. Sant, J. Simonis, C.J. Bishop, R.M. Burr, B.K. Gale, and B.K. Ambati, The capsule drug device: novel approach for drug delivery to the eye. *Vis. Res.* 50 (2010) 680–685.
45. B.D. Kuppermann, Implant delivery of corticosteroids and other pharmacologic agents. Retina 2006: Emerging New Concepts. American Academy of Ophthalmology Annual Meeting, Las Vegas, NV, November 10–11, 2006.
46. N. Kuno and S. Fujii, Ocular drug delivery for the posterior segment: a review. *Retina Today* (2012) 54–59.
47. D. McCann, Medical and dental technologies help usher in era of minimally invasive treatments. *Micro Manuf.* 2 (1) (2009) 21–23.
48. S. Mansoor, B.D. Kuppermann, and M.C. Kenney, Intraocular sustained release delivery systems for triamcinolone acetonide. *Pharm. Res.* 26 (2009) 770–784.
49. L.D. Piveteau, H. Hofmann, F. Neftel, Anisotropic nanoporous coatings for medical implants. EP Patent 1,891,988, 2008.
50. S. Verhey, P. Adostoni, K.D. Dawkins, J. Dens, W. Rutsch, D. Carrie, J. Schofer, C. Lotan, C.L. Dubois, S.A. Cohen, P.J. Fitzgerald, and A.J. Lansky, The GENESIS (Randomized, Multicenter Study of the Pimecrolimus-Eluting and Pimecrolimus/Paclitaxel-Eluting Coronary Stent System in Patients with *De Novo* Lesions of the Native Coronary Arteries) trial. *J. Am. Coll. Cardiol. Interv.* 2 (2009) 205–214.
51. D. Capodanno, F. Dipasqua, and C. Tamburino, Novel drug-eluting stents in the treatment of *de novo* coronary lesions. *Vasc. Health Risk Manage.* 7 (2011) 103–118.
52. E. Grube, J. Schofer, K.E. Hauptmann, G. Nickenig, N. Curzen, D.J. Allocco, and K.D. Dawkins, A novel paclitaxel-eluting stent with an ultrathin abluminal biodegradable polymer. *J. Am. Coll. Cardiol. Interv.* 3 (2010) 431–438.
53. S. Garg and P.W. Serruys, Coronary stents: looking forward. *J. Am. Coll. Cardiol.* 56 (2010) S43–S78.
54. C.C. Wei and D. W. Kennedy, Mometasone implant for chronic rhinosinusitis. *Med. Devices* 5 (2012) 75–80.
55. F. Martin, R. Walczak, A. Boiarski, M. Cohen, T. West, C. Casentino, and M. Ferrari, Tailoring width of microfabricated nanochannels to solute size can be used to control diffusion kinetics. *J. Control. Release* 102 (2005) 123–133.

56. G. Lesinski, S. Sharma, K. Varker, P. Sinha, M. Ferrari, and W. Carson, Release of biologically functional interferon-alpha from a nanochannel delivery system. *Biomed. Microdevices* 7 (2005) 71–79.

57. S. Pricl, M. Ferrone, M. Fermeglia, F. Amato, C. Cosentino, M.M. Cheng, R. Walczak, and M. Ferrari, Multiscale modeling of protein transport in silicon membrane nanochannels: Part 1. Derivation of molecular parameters from computer simulations. *Biomed. Microdevices* 8 (2006) 277–290.

58. J.C.T. Eijkel and A. van den Berg, Nanofluidics: what is it and what can we expect from it? *Microfluid. Nanofluid.* 5 (2005) 249–267.

59. P.M. Sinha, G. Valco, S. Sharma, X. Liu, and M. Ferarri, Nanoengineered device for drug delivery application. *Nanotechnology* 15 (2004) S585–S589.

60. A.N. Paisley and P.J. Trainer, Recent advances in the treatment of acromegaly. *Recent Adv. Investig. Drugs* 15 (2006) 251–256.

61. P. Gardner, Use of a nanopore membrane in a novel drug delivery device. *Future Drug Deliv.* (2006) 59–60.

62. H.J. Lee and M.J. Cima, An intravesical device for the sustained delivery of lidocaine to the bladder. *J. Control. Release* 149 (2011) 133–139.

63. J.C. Nickel, P. Jain, N. Shore, J. Anderson, D Giesing, H. Lee, G. Kim, K. Daniel, S. White, C. Larrivee-Elkins, J. Lekstrom-Himes, and M. Cima, Continuous intravesical lidocaine treatment for interstitial cystitis/bladder pain syndrome: safety and efficacy of a new drug delivery device. *Sci. Trans. Med.* 9 (2012) 143.

64. I.S. Tobias, H. Lee, G.C. Engelmayr, Jr., D. Macaya, C.J. Bettinger, and M.J. Cima, Zero-order controlled release of ciprofloxacin-HCl from a reservoir-based, bioresorbable and elastomeric device. *J. Control. Release* 146 (2010) 356–362.

65. J.C. Wright, S.T. Leonard, C.L. Stevenson, J.C. Beck, G. Chen, R.M. Jao, P.A. Johnson, J. Leonard, and R.J. Skowronski, An *in vivo/in vitro* comparison with a leuprolide osmotic implant for the treatment of prostate cancer. *J. Control. Release* 75 (2001) 1–10.

66. C.L. Stevenson, Formulation of leuprolide at high concentration for delivery from a one year implant, in E.J. McNally and J. E. Hastedt (Eds.), *Protein Formulation and Delivery*, 2nd ed., Marcel Dekker, Inc., New York, 2007, pp. 153–175.

67. M.J. Cukierski, P.A. Johnson, and J.C. Beck, Chronic (60 week) toxicity study of DUROS leuprolide implants in dogs. *Int. J. Toxicol.* 20 (2001) 369–381.

68. D.M. Fisher, N. Kellett, and R. Lenhardt, Pharmacokinetics of an implanted osmotic pump delivering sufentanil for the treatment of chronic pain. *Anesthesiology* 99 (2003) 929–937.

69. Durect Corp. Available at http://phx.corporate-ir.net/phoenix.zhtml?c=121590&p=irol-newsArticle&ID=459693&highlight (press release, October 16, 2003).

70. B. Yang, C. Negulescu, R. D'vas, C. Eftimie, J. Carr, S. Lautenbach, K. Horwege, C. Mercer, D. Ford, and T. Alessi, Stability of ICTA 560 for continuous subcutaneous delivery of exenatide at body temperature for 12 months, 2009 Diabetes Technology Meeting, Long Beach, CA, April 3, 2009.

71. R.R. Henry, R. Cuddig, J. Rosenstock, T. Alessi, and K. Luskey, Comparing ITCA 650, continuous subcutaneous delivery of exenatide via DUROS device, vs. twice daily exenatide injections in metformin-treated type 2 diabetes, 46th Annual Meeting of the

European Association for the Study of Diabetes (EASD), Abstract 78, September 22, 2010.

72. B. Yang, T. Alessi, C. Rohloff, R. Mercer, C. Negulescu, S. Lautenbach, J. Gumucio, M. Guo, E. Weeks, J. Carr, and D. Ford, Continuous delivery of proteins and peptides at consistent rates for at least 3 months from the DUROS® device, AAPS 2008 Meeting, Poster T3150, Atlanta, GA, November 18, 2008.
73. C.M. Rohloff, T.R. Alessi, B. Yang, J. Dahms, J.P. Carr, and S.D. Lautenbach, DUROS® technology delivers peptides and proteins at consistent rate continuously for 3 to 12 months. *J. Diabetes Sci. Technol.* 2 (2008) 461–467.
74. W.H. Ryu, Z. Huang, F.B. Prinz, S.B. Goodman, and R. Fasching, Biodegradable microosmotic pump for long-term and controller lease of basic fibroblast growth factor. *J. Control. Release* 124 (2007) 98–105.
75. R. Lo, K. Kuwahar, P.Y. Li, S. Saati, R.N. Agrawal, M.S. Humayun, and E. Meng, A passive MEMS drug delivery pump for treatment of ocular diseases. *Biomed. Microdevices* 11 (2009) 959–970.
76. P.Y. Li, J. Shih, R. Lo, S. Saati, R. Agrawal, M.S. Humayun, Y.C. Tai, and E. Meng, An electrochemical intraocular drug delivery device. *Sensors Actuat. A* 143 (2008) 41–48.
77. F.N. Pirmoradi, J.K. Jackson, H.M. Burt, and M. Chaio, On-demand controlled release of docetaxel from a battery-less MEMS drug delivery device. *Lab Chip* 11 (2011) 2744–2752.
78. N.M. Elman, H.L. Ho Duc, and M.J. Cima, An implantable MEMS drug delivery device for rapid delivery in ambulatory emergency care. *Biomed. Microdevices* 11 (2009) 625–631.
79. A.J. Chung, Y.S. Huh, and D. Erickson, A robust, electrochemically driven microwell drug delivery system for controlled vasopressin release. *Biomed. Microdevices* 11 (2009) 861–867.
80. L.M. Low, S. Seetharaman, K.Q. He, M.J. Madou, Microactuators toward microvalves for responsive controlled drug delivery. *Sensors Actuat. B.* 1 (2000) 149–160.
81. J.T. Santini, Jr., M.J. Cima, and R. Langer, A controlled-release microchip. *Nature* 397 (1999) 335–338.
82. J.T. Santini, Jr., A.C. Richards, R.A. Scheidt, M.J. Cima, and R. Langer, Microchips as implantable drug delivery devices. *Agnew. Chem., Int. Ed.* 39 (2000) 2396–2407.
83. Y. Li, H.L. Ho Duc, B. Tyler, T. Williams, M. Tupper, R. Langer, H. Brem, and M.J. Cima, *In vivo* delivery of BCNU from a MEMS device to a tumor model. *J. Control. Release* 106 (2005) 138–145.
84. Y. Li, R. Shawgo, B. Tyler, P. Henderson, J. Gobel, A. Rosenberg, P.B. Storm, R. Langer, H. Brem, and M.J. Cima, *In vivo* release from a drug delivery MEMS device. *J. Control. Release* 100 (2004) 211–291.
85. A.C.R. Grayson, R.S. Shawgo, Y. Li, and M.J. Cima, Electronic MEMS for triggered delivery. *Adv. Drug Deliv. Rev.* 56 (2004) 173–184.
86. J.H. Prescott, S. Lipka, S. Baldwin, N.F. Sheppard, Jr., J.M. Maloney, J. Coppeta, B. Yomtov, M.A. Stables, and J.T. Santini, Jr., Chronic, programmed polypeptide delivery from an implanted, multireservoir microchip device. *Nat. Biotechnol.* 24 (2006) 437–438.

87. J.M. Maloney, S.A. Uhland, B.F. Polito, N.F. Sheppard, Jr., C.M. Pelta, and J.T. Santini, Jr., Electrothermally activated microchips for implantable drug delivery and biosensing. *J. Control. Release* 109 (2005) 244–255.
88. J.H. Prescott, T.J. Krieger, S. Lipka, and M.A. Staples, Dosage form development, *in vitro* release kinetics, and *in vitro–in vivo* correlation for leuprolide released from an implantable multi-reservoir array. *Pharm. Res.* 24 (2007) 1252–1261.
89. E.R. Proos, J.H. Prescott, and M.A. Staples, Long-term stability and *in vitro* release of hPTH(1–34) from a multi-reservoir array. *Pharm. Res.* 25 (2008) 1387–1395.

4

DRUG–MATERIAL INTERACTIONS, MATERIALS SELECTION, AND MANUFACTURING METHODS

SuPing Lyu[1] and Ronald A. Siegel[2]
[1]*Cardiac Rhythm and Heart Failure, Medtronic plc, 8200 Coral Sea St NE, Mounds View, MN 55112, USA*
[2]*Departments of Pharmaceutics and Biomedical Engineering, University of Minnesota, Minneapolis, MN 55455, USA*

4.1 INTRODUCTION

Drugs and devices have been combined into integrated medical systems for two main purposes. The first purpose is to enable or improve devices that are not effective in their primary mode of action or lack critical functions. One example is steroid-eluting cardiac pacing leads [1–3]. To effectively stimulate the heart, a cardiac pacing lead must provide a minimum voltage or current to the cardiac tissue (pacing threshold). However, this minimum voltage increases shortly after implanting the lead, due to local inflammation reactions. The minimum voltage may reach a steady state value after a few months, the value varying from patient to patient. Physicians take this variation into account when they program a minimum pacing voltage for their patients. A simple option is to program a high voltage. However, since electrical energy is a product of current and voltage, battery life may be shortened. This problem is solved by attaching a small component that delivers a steroidal anti-inflammatory drug to the tissue in the vicinity of the lead electrode. The required minimum pacing voltage then becomes stable at a low value, making pacemaker programming simple and reliable. Oral glucocorticosteroids have been reported as effective in lowering thresholds, but are not an option for long-term therapy due to undesirable side effects

Drug–Device Combinations for Chronic Diseases, First Edition. Edited by SuPing Lyu and Ronald A. Siegel.

and patient compliance issues [4]. Direct release of steroid into the neighboring tissue permits very low doses to be administered over long periods of time without side effects.

There are several other examples where device patency is improved by local administration of drugs that are combined with the device. The performance of coronary stents, used to open clogged coronary arteries, is improved by coating the stents with drugs that slowly elute drugs (see Chapter 8) [5–19]. With bare metal stents, about 20—40% of the patients experienced renarrowing, or restenosis of vessels following stent insertion [5,15,18,19]. With the addition of drug coatings, the restenosis rate has been reduced to a single digit percentage.

Vascular grafts are an excellent product to replace diseased vessels, or to access blood in hemodialysis patients. However, blood clotting often occurs at the graft surface. Heparin coated vascular grafts offer an effective nonthrombogenic surface (see Chapter 6) [20]. Here heparin, which is normally injected into the blood circulation for systemic antithrombogenic effect, is covalently attached to the inner surface of the vascular grafts, and provides antithrombogenic activity locally. This is an example where the drug is effective but need not be eluted.

Bacterial infections related to medical devices, while relatively rare, are inconvenient and expensive to treat. Recent evidence of increased mortality rates of patients with cardiac implantable electronic devices with infection has emerged. Regardless of whether bacteria enter the patient through a nonsterile device or through another route (e.g., oral surgery), the device provides a platform for bacterial colonization. Once an organized colony, or biofilm, has formed, the infection is difficult to treat, often requiring a second surgery to remove the device [21]. This is because bacteria in biofilms become resistant to the drugs that are effective in treating circulating or planktonic bacteria of the same species. The bacterial colonies may lie dormant for years before activating. While traditional biomaterials cannot prevent the formation of a biofilm, addition of an antimicrobial drug to an implanted device may kill the bacteria before a biofilm is formed on the surface of the device [22–24].

A final example where a locally delivered drug improves device patency is the Infuse® Bone Graft/LT-Cage® Lumbar Tapered Fusion Device. This product is composed of two parts—a tapered spinal fusion case (the LT-Cage Lumbar Tapered Fusion Device) and a bone graft alternative (Infuse Bone Graft). The tapered fusion cage is intended to restore the degenerated disc space to its original height. The graft component is used to fill the cage. It consists of two parts—a solution containing recombinant human bone morphogenetic protein 2 (rhBMP2) and an absorbable collagen sponge (ACS), which the physician uses to carry the drug. During surgery, the onsite reconstituted protein solution is soaked into the ACS, forming the graft. When this product is used with an fusion cage, the infusion rate between neighboring lumbar vertebrae is comparable to that seen when an autograft is used [25–27]. The latter procedure involves a second or larger incision that may be painful and/or take long time to heal.

The second reason to combine a drug with a device is to use the device to store and deliver the drug to the right tissue site with the right dosing pattern, which is practically difficult to achieve with conventional drug administration routes. This approach can improve disease management, increase the patient's quality of life, and reduce systemic toxicity of the drug. One example is using insulin to manage diabetes. The conventional approach is to administer an insulin bolus by injection before meals based on the amount of food consumed. Blood glucose is measured with finger-stick blood samples, test strips, and portable meters. An advanced insulin delivery system, for example, the MiniMed Paradigm® Reveal™ system and Dexcom G4 Platinum system, combines a built-in continuous glucose monitoring technology with an insulin delivery pump that delivers both bolus and basal insulin to manage blood glucose levels after eating and during the fasting state [28–30]. The system displays glucose trends and provides a predictive warning, which allows patients to avoid accidents involving high and low glucose levels. The MiniMed system has a drug reservoir with a capacity of 2–3 days of insulin and a transcutaneous electrode that monitors glucose levels for 3–6 days.

A second example is intrathecal drug delivery for pain or muscle spasticity management. Implantable pump technologies that are composed of a pumping mechanism, a drug reservoir, a catheter, and a controller allow for local delivery of concentrated pharmaceuticals to the subarachnoid space of the central nervous system, which is filled with cerebrospinal fluid [31–33]. The amount of drug can be precisely dosed and released to the target at the right time. The drug reservoir can be refilled before it depletes.

Birth control drug delivery systems such as Nexplanon® and Mirena® (see below) are a third example [34–37]. Nexplanon is a flexible elastomer rod that is implanted underneath the skin and releases etonogestrel for 3 years. During this period, the patient can depend on the implant to provide contraception. The depleted device can be removed and a new one is implanted if needed.

A fourth example of local drug delivery is a bioresorbable polymer wafer loaded with chemotherapeutic drug for treating brain cancer [38–41]. Local release improves treatment efficacy by increasing the target tissue dose while minimizing the systemic toxicity that would occur with oral of IV delivery of the same agent. Different from Mirena that needs to be removed, this device is absorbed into the tissues.

Table 4.1 provides a partial list of commercially available drug–device combination products. The increased success of such combination products has been recognized as a trend in the medical device industry [42], and FDA has recognized the need to better understand and regulate such combinations.

4.2 DRUG–DEVICE INTERACTIONS AND TESTS

4.2.1 Testing Considerations

While medical devices are tested for their functionalities, biocompatibility and biostability, drugs are tested for their bioavailability, pharmacologic efficacy, and safety. Drug–device combination products must be tested for all of these attributes. These qualities must persist throughout manufacturing, storage, and use.

Medical device tests are concerned with device functions, biocompatibility, and biostability for the duration described in the quality statements. Medical devices are

TABLE 4.1 Marketed Drug–Device Combination Products

Product	Drug-Contacting Material	Drug	References
Steroid releasing cardiac pacing lead	Silicone, polyurethane	Dexamethasone and derivatives	[1–4]
Drug-eluting stent	EVA/PBMA, SIBS, Biolinx, PVDF-HFP & PBMA	Rapamycin, paclitaxel, zotarolimus, everolimus	(see Chapter 8) [5–19]
Antithrombosis vascular graft	CBAS	Heparin	(see Chapter 6) [20,43]
Infection reduction coating	Silicone	Rifampin, minocycline, clindamycin	[21–24]
Spinal fusion	ACS	rhBMP2	[25–27]
Insulin delivery pump		Insulin	[28–30]
Intrathecal drug delivery pump	Radiopaque silicone	Morphine, baclofen	(see Chapter 7) [31–33]
Contraceptive implants	EVA, $BaSO_4$-loaded PE, silicone	Etonogestrel, levonorgestrel	[34–36]
Brain cancer drug delivery	P(CPP-SA)	Carmustine	[38–41]

SIBS: poly(styrene–isobutylene–styrene); CBAS: CARMEDA BioActive Surface; P(CPP-SA), copolymer of poly[bis(*p*-carboxyphenoxy) propane] anhydride and sebacic acid (PCPP-SA); ACS: absorbable collagen sponge (type I); EVA: ethylene-vinyl acetate copolymer; PBMA: poly *n*-butyl methacrylate; PVDF–HFP: poly(vinylidenefluoride-hexafluoropropylene); rhBMP2: recombinant human bone morphogenetic protein 2.

designed to perform diagnosis or produce therapeutic effects by mechanical, electrical, or transport actions. These functions are built into a device through design and manufacturing, confirmed with design verification tests, and validated with preclinical and/or clinical studies. Biocompatibility depends on actions of the device and its degradation products on the patients during the use. Biocompatibility, while not specifically related to the device's primary mode of action, is an essential quality that is influenced by the materials, processing, design, and sterilization method. Biocompatibility can also be influenced by materials and agents that are not part of the product but are used with the product during the product life cycle. Strictly speaking, biocompatibility should be tested with the finished product but not with the materials. The tests can be very time consuming and costly. Practically, as long as materials, processing, sterilization methods, and use conditions are the same, components can be used for biocompatibility assessment in lieu of the finished product. Detailed testing procedures for biocompatibility can be found in ISO10993. The main questions to answer include regarding toxicity, hemocompatibility, allergic reactions and sensitization, and local effects of implants on the tissues, including inflammation, tissue remodeling, etc.

The biostability of a device refers to its ability to resist possible adverse effects engendered by biological, chemical, or mechanical processes that impinge on the devices when inside the patient. These processes can adversely affect the component materials, shapes, or relationships (e.g., connections), potentially degrading the functions and biocompatibility of the device and shortening the lifetime of use. Biostability depends on device design, manufacturing, intended functions, and use conditions. Different from biocompatibility tests, biostability tests are designed and performed case by case. Designing and validating the testing methods for device biostability are part of risk management activities and often constitute a large part of the product development effort. Although many tests are performed before a product is approved for market release, none of these can completely predict the performance of the product in patients. Real-time performance of a product is learned by conducting postmarket surveillance. Design and manufacturing of devices are regulated according to Federal Regulation 21CFR820.

Drugs are studied by testing their pharmacokinetic, pharmacodynamic, and toxicity profiles. Pharmacokinetics deals with the effects of the body on the drug, including drug absorption into the tissue, distribution among the organs, metabolism, and excretion. In a sense, this is analogous to the biostability of medical devices. The physical state of the drug (e.g., amorphous versus crystalline), the dosage form, excipients, dose, dosing interval, and delivery route all affect pharmacokinetics.

Pharmacodynamics deals with the effect of the drug on the body, and is analogous to device function. Most drugs work by binding to their targeted receptors or enzymes. The percentage of receptors bound, and hence drug effect is a function of the free drug concentration in the biological fluids (blood, interstitial fluid, or cytoplasm) in contact with the receptors. Therapeutic effectiveness (ability to bind the maximum percentage of targets), potency (ability to achieve effectiveness at low drug concentration), therapeutic index (ratio of toxic concentration to minimally effective concentration), and safety margin are the subjects of study.

Toxicology studies of a drug are analogous to biocompatibility studies performed on a device. They are carried out to demonstrate safety before introducing the drug into humans in clinical trials. Preclinical *in vitro* tests and animal studies must be carried out before human trials. Detailed guidance is provided in International Conference on Harmonization (ICH) Guidelines for Toxicological and Pharmacological Studies. Postmarket surveillance is also carried out for drugs to further understand their effectiveness and toxicity. Small molecular weight drug product manufacturing is regulated according to 21CFR210 and 21CFR211 while biological-based drugs are regulated according to 21CFR210, 21CFR211, 21CFR600, and 21CFR610. Reference [44] provides a nice review.

For a drug–device combination product, all of these aspects are considered. Most of the drug–device combination products use drugs that have already been approved as pharmaceutical products by regulatory agencies. This provides great advantages for developers to achieve efficacy targets, reduce risks of having side effects, and simplify manufacturing process development efforts. However, combining a drug with a device means that the drug may be formulated with excipients that are different from those in its original formulations, manufactured via extra and different processes and applied to the patients via different routes. Moreover, the drug's intended effect (mode of action) may differ when used with a device than when administered in its conventional forms, and the relevant pharmacodynamics markers may have to be revised. Finally, release of drug from the device constitutes a new route of delivery, which may alter its stability and ADMET (administration, distributions, metabolism, elimination, and toxicity) properties. Therefore, all tests are very likely needed. It is even more challenging that the tests are not a simple sum of those for the drug and the medical device alone. New dimensions need to be considered.

Clearly, the synergy between drug and device makes the drug–device combination product successful and attractive. The drug can enhance the device's functions and the device can better deliver the drug. These synergistic interactions are well recognized during the research and development of combination products. However, there are also interactions between the drug and the materials and components of the device that are not seen when the drug and device are used alone. These interactions may not always be helpful, and may cause product quality issues. Some examples will be given in the next section.

In drug–device combination products, the device "participates" in drug storage and release. The drug can interact with the components of the device by direct or indirect contact. Interactions can occur during the entire product life cycle from manufacturing, sterilization, shipping, and storage to use in the operating room, use in the patient, and during explantation. Interactions can be facilitated by other factors during the process. For example, when a product is implanted, the body temperature and tissue fluids may accelerate interactions. The patient's motion can agitate the drug/device system, altering pharmacokinetics and/or pharmacodynamics over time. Table 4.2 lists factors that can impact the drug–device interactions in each phase.

4.2.2 Functional Interactions between a Drug and Device

Interactions between a drug and device can impact the therapies negatively, and some interactions may not be revealed in early testing. The late thrombosis associated with certain drug-eluting stents is one example. Since Johnson & Johnson released its first drug-eluting stent (Cypher™), other companies have released their own products. All of them show effective reduction of restenosis, that is, renarrowing of the vessel lumen after implantation of the stents. However, since 2004, it has been reported that some patients with implanted drug-eluting stents experience blood clotting after stopping supplementary anticlotting drugs 2 years later [45–48]. This side effect occurs more often with drug-eluting stents than with bare metal stents although the rates of both are very low. These results prompted FDA to organize a panel to review the results [49]. It was found that increased rates of stent thrombosis occurred 1 year after implantation, even with on-label use. However, it was concluded that the risk of thrombosis does not outweigh their advantages of drug-eluting stents over bare metal stents in reducing the rate of repeated revascularization when the former are used for their approved indications [49]. The root cause of the late thrombosis has been explored for years since then.

The research group headed by Vironmi performed pathology studies of some late thrombosis cases where Cypher and Taxus™ were used [50]. It was found that

TABLE 4.2 Factors in Each Phase of Product Life Cycle That Could Impact the Drug, Device, and Their Interactions

Phase	Factors to be Considered
Manufacturing	• Raw materials and components • Process temperatures and time • Environment (e.g., clean room, light, solvent, and moisture) • Containers, working surface, and tools • Process and processing aids • Mechanical operation, handling, and so on • Cleaning
Sterilization	• Chemical actions • Temperature • Time
Shipping and storage	• Temperature and thermal shock • Time • Moisture, light, mechanical action, and so on • Packaging material and contaminants • Mechanical action
Implantation	• On-site formulation procedure • Mechanical handling
In patients	• Drug liberation, absorption, and distribution • Blood circulation • Blood contact: water, lipid, O_2, CO_2, urea, glucose, fatty acid, and so on • Inflammation reactions to device • Thrombosis due to device and drug • Bacterial colonization on devices • Tissue mechanical action • Tissue proliferation and remodeling • Interactions with other therapies of the patient
Explantation	• Mechanical handling • Tissue removal

endothelialization was delayed. The authors argued that while sirolimus or paclitaxel reduce neointimal formation, those drugs also impair the normal healing processes of the injured arterial wall. After the added antithrombotic drug was stopped, clotting occurred due to the unhealed vessel wall. Rates of endothelialization after stent implantation seem to vary among currently used animal models. It is well known that the endothelialization process in pig coronary arteries is faster than in humans. Therefore, animal study data may not provide a correct reference for human use.

Kolachalama et al. of MIT studied this problem by computational fluid dynamic modeling [51]. The results elucidated the influence of luminal blood flow on arterial drug deposition and distribution. Blood flow over struts imposes recirculation zones distal and proximal to the implanted stent, which asymmetrically alters the arterial drug distribution and availability for tissue absorption. The disparities in sizes and drug concentrations in the two recirculation zones are functions of local transient flow and strut geometry that are intrinsic to the arterial anatomy, stent design, and stent embedment in the tissue determined by the deployment procedure, and blood flow. While the drug and drug coating of the stent probably impose a primary impact on tissue absorption of the eluted drug, stent design and stent deployment may influence drug absorption as well. In a pharmacokinetic study of the Promus Element™ stent (Boston Scientific), the measured drug distribution in tissues distal, proximal, and adjacent to the stent supported this prediction regarding disparities (see Chapter 7).

Regardless of what mechanisms underlie late-term drug–device interactions, it is clear that the drug's therapeutic effects, drug release from the device, and device's biocompatibility could have complex interplays that affect the response of the body. While more extensive clinical studies may arguably help to discover those interactions, the most effective approach is to conduct postmarket surveillance.

4.2.3 Physical Interactions

A drug can change the T_g (glass transition temperature) and T_m (melting point) of a polymer and hence its mechanical properties. The polymer may become tacky if its T_g is reduced below room temperature or if its crystallinity is significantly reduced. The polymer may become rigid if its T_g is increased. If a plasticizer, for example, polyethylene glycol (PEG), is used, it can change the T_g of the polymer. Moreover, PEG can elute over time and the polymer matrix may become rigid in the patient over time.

A drug's release properties can change due to its interactions with the polymer [52]. During storage or following implantation, the drug may change from the amorphous form to a crystalline form, or change from one crystalline form to another, that is, a different polymorphic form. In one example, a drug and a polymer are dissolved together in a solvent. After the solvent is quickly evaporated, an amorphous dispersion of drug in the polymer forms. Such a dispersion can also be formed by melt processing of drug and polymer, followed by rapid quenching. Such dispersions are not thermodynamically stable if the resulting drug concentration is higher than its solubility in the polymer. The amorphous drug can then undergo a slow crystallization or phase separation. The former process can lead to varying amounts of drug crystallinity during storage, and the latter can lead to an increased drug domain size or surface accumulation. Both of these processes can cause changes in drug release profiles. If the dispersion is completely dry, the drug may be stable on the shelf. However, if the product contacts moisture or other chemicals that can enhance the mobility of the drug molecules, recrystallization can occur.

The release rate of the drug is determined by the form it takes as a solid and the dispersion morphology in the polymer matrix. Lack of control of the solid form can lead to unpredictable or undesirable drug release rates. An example of this phenomenon is provided by the drug podophyllotoxin dispersed in polycaprolactone (PCL). In a study of this system, the two components were dissolved in tetrahydrofuran (THF) and cast as a film. Differential scanning calorimetry (DSC) indicated that pure podophyllotoxin's melting point, T_m is 186 °C, while neat PCL has a T_g at −40 °C and

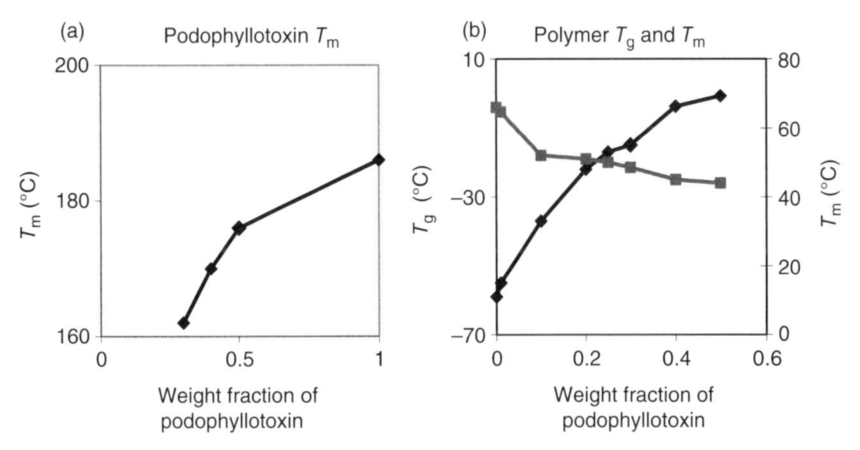

FIGURE 4.1 Thermal properties of podophyllotoxin–polycaprolactone mixture indicating the miscibility of the two components. (a) T_m of podophyllotoxin. (b) T_g (diamond) and T_m (squares) of PCL. (Reproduced with permission from Medtronic plc.)

T_m at 65 °C. After the two components were mixed, the mixture did not display a podophyllotoxin melting peak until the podophyllotoxin composition exceeded 20 wt%. Above that concentration, podophyllotoxin melting appeared but at a reduced T_m compared to that of 100% podophyllotoxin, as shown in Figure 4.1a. The T_g of the PCL, increased from −40 to 0 °C when podophyllotoxin content increased from 0 to 50 wt% (Figure 4.1b). On the other hand, the T_m of PCL decreased with increasing podophyllotoxin wt%. These results indicate that podophyllotoxin and PCL were miscible and formed one phase when podophyllotoxin in PCL is <20 wt%. Above that composition, there were two phases, one being podophyllotoxin and the other being podophyllotoxin dissolved in PCL. The continuously increasing T_g and decreasing T_m of PCL indicated that the podophyllotoxin/PCL phase may have exhibited continuous composition changes.

At room temperature and in the dry state, this system was stable, as shown in Figure 4.2a. When the system was soaked in water at room temperature for 1 h the podophyllotoxin crystallized from the PCL matrix and formed needle crystals (Figure 4.2b). When soaked in lipid oil for 1 h, needle crystals also formed, but were shorter (Figure 4.2c). While the patterns may be visually appealing, their appearance suggests that this system would be subject to transformations in the presence of body water and fat, and is likely not suitable for drug delivery, as the drug in the needles would have different rates of liberation into the tissue fluids compared to drug dispersed in the polymer matrix.

Many drugs are crystalline. Crystallization behavior changes in polymer matrices under different conditions especially when the drug and polymer are miscible. If drug and polymer are immiscible, the change of thermal properties due to the interactions may not be significant. In such case, thermal properties of the mixture may not be distinguishable from the un-mixed components. The PROMUS Element® stent (Boston Scientific) coating system is an example, where everolimus is dispersed in

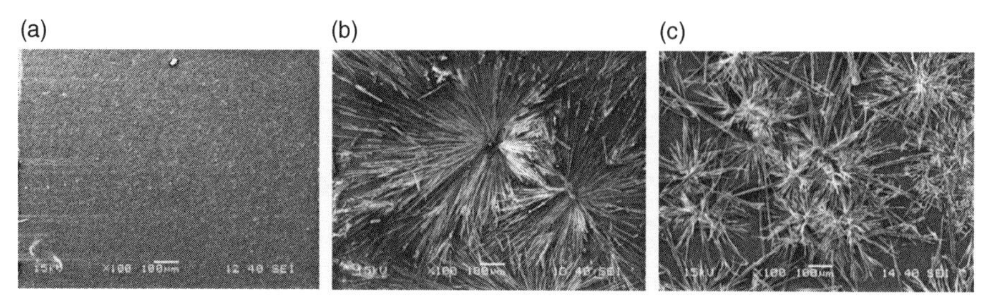

FIGURE 4.2 (a) Podophyllotoxin–polycaprolactone (50/50 wt) dry, (b) soaked in water for 1 h, and (c) soaked in lipid oil for 1 h. (Reproduced with permission from Medtronic plc.)

poly(vinylidenefluoride-hexafluoropropylene) (PVDF-HFP) and has a distinguishable thermal transition (see Chapter 7). In this case, any solid-state transformations are due to the drug alone.

4.2.4 Chemical Interactions

Chemical changes of drugs or excipients usually are not favorable, but are not uncommon either. If a product contains certain amounts of degradation products, the product developers must generate or find enough evidence that the product is safe and efficacious and the amount of degradation is under control through the manufacturing process in order for the product to be used in patients. Otherwise, the formulation needs to be adjusted. For example, rifampin and minocycline could degrade under extreme conditions such as low and high pH, high temperature, oxidative environment, etc. Reference 53 reports forced degradation of minocycline. However, those drugs have been widely used in medical devices such as antimicrobial meshes [24,54], hydrocephalus shunts [55,56], and central venous catheters [22,23,57] where the products have been demonstrated being safe and effective in reducing device-related infection.

Chemical changes of drugs can happen at any phase of a product's life cycle from manufacturing, sterilization, shipping under extreme temperature and humidity conditions, or storage. Each of these can impact the success of product development. In Ref. [58], a rifampin-loaded silicone sheet with thickness ranging from 0.1 to 0.4 mm was sterilized by steam, ethylene oxide (ETO), gamma irradiation, and electron beam. It was found that drug recovery after sterilization by these methods was 20–30%, 60–80%, 70–80%, and >90%, respectively. Minocycline-loaded silicone sheets were sterilized as well. The recovery was 0, 85–100%, 70–90%, and >90%, respectively. Thus, both drugs exhibited a low recovery from steam and ETO methods and a fairly high recovery with the electron beam and gamma methods. Similar results with rifampin were observed in another patent application [59]. If these drugs are incorporated into a device that can be sterilized with gamma irradiation or electron beam, then the whole product can be sterilized in one step. However, if the device part can be sterilized only with ETO or steam (e.g., bulky metallic parts), terminal sterilization of the product would be challenging. Sometimes, it may not be the best option to sterilize drugs and devices together. Instead, it may be easier to sterilize them separately. Similar considerations apply to packaging, storage, and shipping.

The presence of drug can change the rate of polymer degradation. The hydrolytic degradation rate of a polymer depends on pH inside the sample. This fact has been demonstrated with polylactide, in which the acidic degradation products accelerate the degradation rate (the so-called *autoacceleration*). The degradation rate of large samples is faster than that of small samples because the degradation products cannot easily diffuse out of the sample, and changes of the inside pH are more substantial in the larger samples [60]. The slowest degradation of PLA occurs at pH = 4 [61]. Some drugs buffer the internal pH of a sample near that value, in which case sample degradation can be slowed.

Polymer hydrolysis also depends on the water content in the polymer. A highly water-soluble drug increases the water concentration in the polymer, sometimes leading to accelerated degradation. Combining these factors, the impact of the drugs on polymer degradation can be complicated, as reported in Ref. [62]. When the degradation properties of the polymer are used to control drug delivery, the interplay between drug and polymer needs to be well evaluated because it may impact the pharmacology as well as the biocompatibility of the system.

4.2.5 Corrosion of Device Due to Drug

In a pump-enabled drug delivery device, the drug may interact with the device components that it directly contacts, for example, the drug reservoir, catheters, valves, and so on. Metal corrosion, polymer swelling, polymer degradation, etc., on are typical consequences that are evaluated by product developers. If there is mechanical action, the interactions can be complicated. For example, the polymer may creep under mechanical loading. In the presence of a drug or a solvent, the creep process can be accelerated. Chemicals in the blood such as oxygen, carbon dioxide, urea, etc. can penetrate through a catheter wall and participate in interactions.

Device–drug interactions may happen between drug and the components of the device that are not in direct contact. One example is the recent FDA-documented case concerning the Synchromed II Intrathecal drug pump [63]. This device stores the drug inside the pump and delivers it through a silicone catheter. It was demonstrated that the silicone catheter was stable. There is no report documenting changes of the drug due to contact with silicone. However, based on the document from the manufacturer's Implantable Systems Performance Registry (ISPR), the overall failure rate of the SynchroMed II pump at 78 months postimplant is 2.4% when used to dispense approved drugs, and 7.0% when used to dispense unapproved drugs [63,64]. After examination, it was found that "corrosive agents (e.g., chloride ion and sulfate ion) originating from drug formulations can permeate through the internal pump tubing and initiate corrosion of internal components." Long-time accumulation could result in motor stall although the pump motor was not in direct contact with the drug. This example suggests that drug–material interactions could happen among components without direct contact, which can impact the biostability of the products.

4.3 MATERIAL SELECTION AND DEVELOPMENT

The required properties of a candidate material for a medical device or drug–device combination product are not limited to those that allow the products to have critical functionalities. Criteria related to product design, supply, manufacturing, and use conditions in the patient must also be considered. For example, although supply is not an intrinsic property of a material, it is the supplier who determines the quantity and quality of the materials used in the products made for the end users. As already mentioned in Section 4.1, biocompatibility, biostability, pharmacokinetics, pharmacodynamics, toxicity, and functions all need to be considered for combination products. Then, the selection of materials needs to be checked against those points by considering material properties, use conditions of the device in the patients, and manufacturing processes. Here, we focus on materials and manufacturing.

4.3.1 Solubility Estimation

Knowing the solubility of drug in the material(s) controlling drug release is crucial to success in developing delivery systems. Different situations require different solubilities. For example, in steroid-eluting leads, the solubility of dexamethasone sodium phosphate in silicone is low. Such low solubility allows a very slow elution rate such that the drug lasts over 10 years (see Chapter 5). For drug delivery catheters, in which drug is administered through the catheter lumen, drug solubility in the catheter material should be extremely low so that the drug is not trapped in the catheter wall or lost via permeation through the catheter. Drug solubility in the polymer matrix in the stent coating varies. High solubility of drug in the polymer coating is a target for some cases but not a primary target for others.

Solubility of an amorphous drug in an amorphous polymer is determined by the change in the normalized free energy due to mixing of the two components,

$$\Delta G_{\mathrm{mix}} = \Delta H_{\mathrm{mix}} - T\Delta S_{\mathrm{mix}} \tag{4.1}$$

If free energy decreases due to mixing ($\Delta G_{\mathrm{mix}} < 0$), then high solubility of the drug in the polymer can be achieved. The entropy change, ΔS_{mix}, is always positive and drives mixing. The enthalpy change, ΔH_{mix}, is usually positive, which does not favor mixing. In this case, small or near zero enthalpy change is required for achieving good mixing. However, ΔH_{mix} can be negative when there exists a special attractive interaction between the two components. An example is an anionic component (e.g., dextran sulfate) mixed with cationic components (e.g., polyethylenimine) [20,43] (see Chapter 6). In such a case, the enthalpy change can be negative and it drives mixing. When van der Waals interactions are dominant, the normalized enthalpy of mixing can be estimated according to Eq. (4.2):

$$\Delta H_{\mathrm{mix}} = \upsilon_d(\delta_d - \delta_p)^2\varphi_d(1 - \varphi_d) \tag{4.2}$$

where ϕ_{d} is the volume fraction of drug dissolved in the polymer, υ_{d} is the molar volume, and δ_{d} and δ_{p} are the solubility parameters (square roots of cohesive energy densities) of the drug and polymer, respectively. In order to have a small mixing enthalpy, the solubility parameters of the two components should be similar. Solubility parameters can be estimated with molecular dynamics simulation or by simple estimation based on additive group contributions. Many estimation methods are presented in the book by Van Krevelen [65].

The normalized entropy term is estimated using Flory–Huggins theory, according to which

$$\Delta S_{\mathrm{mix}} = -\mathrm{R}\left[\phi_{\mathrm{d}} \ln \phi_{\mathrm{d}} + (1-\phi_{\mathrm{d}}) \ln (1-\phi_{\mathrm{d}})/P\right] \tag{4.3}$$

where P is the ratio of (average) polymer chain volume to the volume of the drug molecule. It can be shown that equilibrium will occur where ΔG_{mix} is minimized, and the result is

$$\ln \phi_{\mathrm{d}} + (1-1/P)(1-\phi_{\mathrm{d}}) + (\upsilon_{\mathrm{d}}/\mathrm{RT})(\delta_{\mathrm{d}}-\delta_{\mathrm{p}})^2(1-\phi_{\mathrm{d}})^2 = 0 \tag{4.4}$$

Once ϕ_{d} is solved for, solubility of drug in the polymer is calculated as $S_{\mathrm{d}} = \rho_{\mathrm{d}}\phi_{\mathrm{d}}$, where, $\rho_{\mathrm{d}} = \mathrm{MW}/\upsilon_{\mathrm{d}}$ is the drug's density.

For a crystalline drug in an amorphous polymer, the solubility is determined by the free energy change due to dissolution of crystalline solid and the free energy of mixing [66]:

$$\Delta G_{\mathrm{diss}} = \Delta G_{\mathrm{melt}} + \Delta G_{\mathrm{mix}} = \Delta H_{\mathrm{melt}}(1 - T/T_{\mathrm{melt}}) + \Delta H_{\mathrm{mix}} - T\Delta S_{\mathrm{mix}} \tag{4.5}$$

With this added term,

$$\frac{\Delta H_{\mathrm{melt}}}{R}\left(\frac{1}{T}-\frac{1}{T_{\mathrm{melt}}}\right) + \ln \phi_{\mathrm{d}} + (1-1/P)(1-\phi_{\mathrm{d}}) + (\upsilon_{\mathrm{d}}/\mathrm{RT})(\delta_{\mathrm{d}}-\delta_{\mathrm{p}})^2(1-\phi_{\mathrm{d}})^2 = 0 \tag{4.6}$$

4.3.2 Polymer Matrix for Drug Delivery Depots

Drug–eluting stents (DES) (Chapter 8) are among of the successful drug/device combination products [5–19,67–69]. In the early days of the product development of DES, requirements including drug selection and drug release rates were not clearly defined. There were only a few generic criteria for the polymer carrier, or coating. The most obvious requirements were that the polymer should be able to load a drug over a reasonable range, control the release rate within a broad range, and that the polymer and the drug to be used would not have adverse interactions. On top of these requirements, the polymer coating should present a biocompatible surface. It should be easily processed and stored, and be readily produced on a commercial scale. The drug can be hydrophilic or hydrophobic, and miscible or immiscible with the polymer. Finally, the polymer should be able to control the rate of drug release over specified time periods. The task of selecting and developing a polymer system meeting all these requirements was challenging.

One approach was to make copolymers composed of hydrophilic and hydrophobic monomers. By adjusting the ratio of the two monomers, one could produce a range of copolymers meeting specified requirements. This approach was used for Taxus™ drug-eluting stents (styrene–isobutylene–styrene block copolymer, or SIBS). However, a practical challenge is that synthesizing a block copolymer with well-controlled composition and chain configuration (monomer sequence) is time consuming. Synthesizing a new copolymer, even with the same monomers as before but with different monomer ratio, requires new processes and characterizations. If a random copolymer is targeted, the sequence of the monomers is determined by the nature of the copolymerization kinetics (reactivity ratios) of the two monomers. The sequence can vary over the reaction process when the reaction is performed in batch. Better control can be achieved by careful addition of new monomers as the polymerization proceeds.

Another approach is to blend two polymers that, by themselves, are at the two ends of the desired property range. For example, one polymer is hydrophilic and other hydrophobic. Alternatively, one polymer has a high glass T_g so that at body temperature it is glassy and releases drug slowly, while the other has a low T_g so that it is rubbery and releases drug quickly. The two polymers can both be hydrophobic or hydrophilic. By mixing them together, a series of blends can be achieved, which provides a broad range of drug release rates and drug loadings. Cypher® drug-eluting stents used a blend of ethylene vinyl acetate copolymer (EVA) and polybutylmethacrylate (PBMA) (This was the first DES approved by FDA for market release). However, not every pair of polymers can be blended to achieve this purpose. Only polymers that are thermodynamically miscible may provide the possibility to achieve adjustable drug loading and drug release rates. Actually, the miscibility between the two Cypher stent coating polymers has not been clearly documented. In Ref. [70] and a series of patent applications [71–74], a number of miscible polymers and immiscible polymers were tested by one of the present authors (Lyu) and his colleagues. The results demonstrated that it was possible to achieve adjustable drug release rates by making miscible polymer blends. If the polymers that are miscible also have a history of applications in implantable medical devices, the predicate applications are valuable references for discussing the biocompatibility and biostability questions.

Most polymers are thermodynamically immiscible due to their high molecular weight. This limitation has driven researchers to develop a third generation concept of blending miscible copolymers. Take two polymers, call them S and F, that can provide drug release rates at the two ends of a desired range (S for slow and F for fast). If S and F are immiscible, then introduce a third implantable component, call it B, and copolymerize it with S and F individually, yielding two copolymers, that is, SB and FB. Because these two copolymers have a common component B, they should have an increased probability of being miscible. Blending SB and FB can result in a series of blends SB/FB that may provide adjustable drug release rates within a broad range bounded by the release rates of S and F (Figure 4.3).

This idea turned out to be very successful and is documented in U.S. Patent 8,088,404 [73]. Polybutylmethacrylate, a hydrophobic polymer, was chosen as

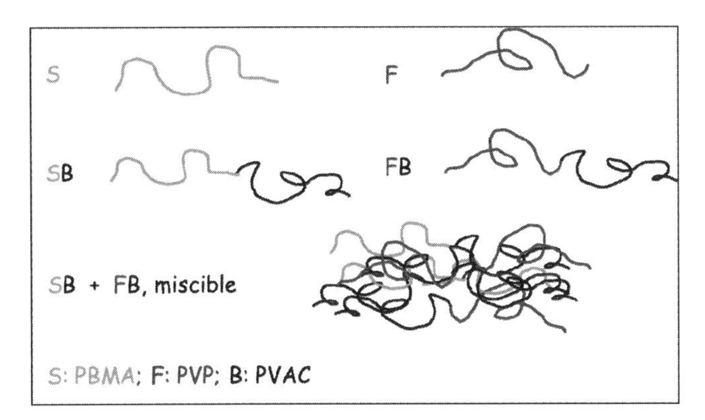

FIGURE 4.3 Miscible polymer blends made of SB and FB copolymers. This schematic is for demonstrating the concept, and not necessarily represents chain sequence or configuration of copolymers. (Reproduced with permission from Medtronic plc.) (See colour plate section.)

S. Polyvinylpyrrolidone (PVP), a hydrophilic polymer, was chosen as F. Polyvinyl acetate (PVAC) was chosen as B. Two copolymers, PVP–PVAC and PBMA–PVAC, were synthesized. To fine-tune the hydrophobicity of PVP–PVAC, polyhexylmethacrylate (PHMA), which has lower T_g and higher hydrophobicity was added, resulting in a terpolymer, PVAC–PVP–PHMA. Extensive examples and studies were reported in the patent with these copolymers. The results demonstrated that these polymers could form stable blends and provide a broad range of drug release rates [74].

One composition of this broad range of formulation invention is worth mentioning. This formulation is referred to as Biolinx [75], which is a blend of three components: PVAC–PHMA–PVP (A), PVAC–PBMA (B), and PVP (C). The estimated solubility parameters of the components A and B are 18.0 and 17.9 $(J/cm^3)^{1/2}$, respectively. Glass transition measurements indicated that these two polymers are miscible [74]. This polymer system was used to load and control the release of zotarolimus, a tetrazole-containing macrocyclic immunosuppressant. The solubility parameter of the drug is 17.8 $(J/cm^3)^{1/2}$, which is close to that of the polymer system. This good match of the solubility of the polymers and drug allowed us to achieve sustained release of zotarolimus [74,75]. This whole idea helped to advance the development of a product, the Endeavor Resolute drug-eluting stent [10]. Stents coated with this polymer–drug system are reported to have a very low target lesion revascularization (2.9%) and low definite and probable stent thrombosis (<0.1%) at the 1 year time point in a U.S. clinical trial [16]. This is an example of how to develop a polymer coating material by combining first principles of physical chemistry with application requirements.

4.3.3 Materials for Drug Delivery Catheters

A well-known example of drug delivery pump therapy is the use of intrathecal drug delivery pumps for pain and muscle spasticity management. These pumps deliver, when needed, morphine or baclofen directly into the subarachnoid space of the central nervous system, which contains cerebrospinal fluid [21]. In these systems, a drug delivery catheter connects the drug reservoir and the tissue that is targeted to deliver the drug. Ideally, a drug delivery catheter should have the following characteristics:

- The catheter materials should not have adverse interactions with the drug that could either degrade the drug or degrade the materials in terms of stability and biocompatibility. Some drug solutions have acidic or basic pH, which may adversely affect the materials. Interactions between drug and materials need to be evaluated.
- Drug and excipients should not permeate through the catheter wall significantly such that the amount of drug delivered is affected or the drug interacts with device components that are not in direct contact.
- Components of the tissues such as O_2, CO_2, lipids, free radicals, etc. may diffuse into the catheters. Undesirable interactions of these components with the drug or polymer should be avoided.
- The catheter is mechanically stiff enough such that pumping pressure does not cause volume increase, and at the same time flexible enough that it does not irritate contacting tissues.
- The catheter should resist kinking, wrinkling, twisting, fatigue, and other mechanical conditions.
- There must be no back flow of tissue fluids into the catheter.

Silicone rubber has good chemical stability in biological environments. It has a good track record of use in human implants. Caution is required that polydimethylsiloxane (PDMS), the building chains of silicone rubber, can react with strong acids, although this is not a significant issue for silicone rubber due to its cross-linking structures.

4.4 MANUFACTURING METHODS FOR COMBINING DRUGS INTO DEVICES

Special manufacturing processes of combination products are needed to incorporate drugs into devices. The drugs and devices may be already market-released products, or equivalents to existing products. In this sense, manufacturing a combination product is reformulating the drug with the "device" instead of making a new drug. The new product has the ability to store the drug and deliver it to the right location with the right dosing pattern. Clearly, the manufacturing process must comply with related regulatory rules. However, the process itself often involves technological innovation. In this section, processes that are commonly used in manufacturing combination products are reviewed from the technological point of view. Compliance with regulatory rules is not discussed here. In general, manufacturers must demonstrate good process control by showing that the drug contents, content uniformity, and *in vitro* drug release meet certain requirements.

4.4.1 Dip Coating

Dip coating is a simple and flexible method. It can be used to build a thick coating over a device with complicated geometry. Often, it is used to build prototype samples for proof of concept or feasibility tests. Chemical reaction processes such as grafting can be used throughout the dipping process due to its flexibility. If more than two materials are complementary to each other (e.g., anionic and cationic) and need to be coated on the same surface, a layer-by-layer method may be used. For example, the Carmeda® Bioactive Surface (CBAS) for heparin coating is built via layer-by-layer technology (see Chapter 6) [43]. This coating is applied to Gore's ePTFE surface and has been proved to be nonthrombogenic, biocompatible, and hemocompatible. The dip coating process is an art. The coating thickness and quality depend on the concentration and viscosity of the solutions, wetting of the solution on the surfaces, angles and speed of dipping in and taking out, time of dipping and drying, number of cycles, and so on. Webs may form in small openings or corners of the mandrel from dip coating.

Dip coating can be performed in different ways, for example, applying solution droplets to the surface instead of dipping the surface in the solution. If the device to be coated has a circular form, it can be rotated over a mandrel to achieve uniform coating thickness and remove extra solution. Cordis, the maker of Cypher stents, has a patent describing how to coat stents with drugs using this technology [76].

A disadvantage of dip coating involves the handling of solvent. First, dissolving the drugs and polymers and maintaining concentrations at the targeted values requires effort. Second, if degradation or adverse interactions are a concern, caution should be taken when using solvents, especially during storing and operating. This is because chemical and physical interactions are faster in solutions than in solids. Third, contaminants can be introduced by solvents. Fourth, drying or removing solvents can be a challenging step. Finally, if the components of the system are not fully miscible (e.g., drug–polymer or polymer–polymer), phase ripening (or Ostwald ripening) and migration of the minor components to the surface could happen, which can lead to variation in drug release rates.

4.4.2 Spray Coating

A nice literature review of spraying technology in manufacturing drug-eluting stents can be found in Ref. [77]. In spray coating, drug and excipients are dissolved in a solvent. The solution is sprayed out from a nozzle and breaks up into droplets, which deposit on the surface. After the solvent evaporates, a thin layer of polymer coating forms on the surface. The sprayability of a solution is related to its surface energy. The evaporation rate of a solvent is related to its boiling point. The coating thickness and its variation depend on how accurately each layer is sprayed and how many layers are sprayed.

In spray coating, the solution droplet size and distribution are critical to the quality of coating. Fundamentally, the droplet size is determined by a competition between the kinetic energy of the jet and the surface tension of the liquid. Lord Rayleigh's initial work on this problem has been extended in many directions [78,79]. For ultrasonic atomizing spray coating, Lang et al. proposed the following equation to predict the median size of droplet (D_{50}) [80]:

$$D_{50} = 0.34\left(\frac{8\pi\Gamma}{\rho f^2}\right)^{1/3} \tag{4.7}$$

where Γ, ρ, and f are surface tension, liquid density, and vibration frequency, respectively. Viscosity of the liquid jet is not considered in this equation. In production, in order to build coating thickness faster, it is beneficial to have a high polymer concentration. However, a higher polymer concentration will lead to a higher viscosity of the spraying solution, which can make it difficult to achieve a high flow rate for spraying. Therefore, a balance among the parameters is needed.

A simple spray coating process is to load a drug–polymer solution and pump it out through a nozzle. As the fine liquid stream is ejected out of the nozzle, it breaks up into droplets. A device is placed in the droplet cloud and the surface is coated. In order to produce an even and consistent coating thickness, the device rotates and moves back and forth in the cloud. There are many advanced spray processes. One of them is gas-assisted atomization spraying. Here the liquid is forced out of a nozzle with a gas flow in order for the liquid droplets to be further homogenized. In another advanced spray process, the nozzle vibrates at an ultrasonic frequency and the droplets are further refined.

In spray coating, a lot of the solution is wasted because the target surface is smaller than the droplet cloud. Electrical spraying can be used to improve the efficiency of catching droplets by the target. Spray coated surfaces can also have defects such as webbing, surface roughness, and narrow regions that are shaded from the spray. Because solvents are used, spray coating technologies have similar advantages and drawbacks as those associated with dip coating.

Boston Scientific, Abbott, and Medtronic have published various spray coating techniques that have been used in manufacturing drug-eluting stents [67,69,74].

4.4.3 Impregnation

Impregnation is a process in which drug is loaded into the polymer by swelling the latter in a solution in which the drug is dissolved. The solvent is then removed. A typical process is to dissolve the drugs in a solvent, place the device in the solution for a controlled time to load the drug, and dry the device by vaporizing the solvents. The concentration of drug in the solution and duration of soaking are controlled such that the total amount of drug entering the tubing meets the specification. Choice of solvent is a key factor in making satisfactory devices. The solvent chosen should be able to dissolve enough drug to be loaded into the polymer. It should be able to swell the polymer but not dissolve it. Often, cross-linked polymers such as silicone rubber are loaded by impregnation. Some anti-infection catheters containing, for example, rifampin and minocycline, are produced by impregnation [81,82].

4.4.4 Extrusion

Nexplanon®, a subdermal contraceptive rod implant made by Merck & Co., is produced by extrusion [83]. The rod, about 2 mm in diameter and 4 cm in length, is made of ethylene-vinyl acetate copolymer (EVA) and contains etonogestrel. The therapy is claimed to be effective for 3 years [20]. There are many advantages of using the extrusion method. It is a continuous process that allows for high yield and high production rate with consistent product quality. There is no solvent involved, which significantly simplifies the process by eliminating the dissolution, drying, and purification steps. Degradation of drug and excipients can be reduced because degradation usually is slower in the solid form than in solution. Dosage forms such as pellets, rods, fibers, sheets, films, and so on can be produced by extrusion.

The extrusion process includes premixing the ingredients (also called dry mixing), melt mixing, and shaping [83]. In principle, these processes are similar to those used in commercial polymer melt blending. For example, premixing can be accomplished using multiple feeders, each of which feeds a different ingredient into the extruder. Twin screw extruders may be used for melt mixing. A single extruder or meter pump (gear pump) in combination with various dies (strand die, sheeting die, tubing dies etc.) can be used to shape the mixture into predose forms such as strands and sheets. Various cutting technologies, for example, mechanical cutting and laser cutting, are used to make the final dose forms.

However, there are a few differences between the drug/polymer mixtures and nonmedical commercial polymers, requiring specific modifications of the extrusion processes. First, processes for medical devices must be carried out under cGMP, which influences the selection of equipment, use of processing additives, environment control, qualification, cleaning, change control, etc. Second, extrusion is a continuous process, and the drug concentration in the extrudate (strand or sheet) is a function of the concentration of the feed. The volume of the extruder cavity can serve as a buffer that smooths variation, but the drug concentration in the extrudate still depends on the feed history. Constant feed rates of drug, polymer, and other excipients over time are, therefore, critical to achieve consistent drug concentration in the extrudate.

Constant feed rates for drugs can be challenging to achieve for several reasons. First, drugs usually are in powder form. Powder flow and feeding are more difficult to control than liquids and pellets, especially at low rates. This is because the resistance of powder flow depends on the friction between solid drug particles, particle aggregation, and particle jamming (a large-scale particle stall due to friction and bridging). Therefore, flow resistance varies dramatically. Second, extrusion of drug products usually is slow compared to other commercial products. For example, each Nexplanon dose is about 2 mm in diameter and 4 cm in length, which is about 0.126 cm^3 in volume. If extrusion runs at 100 feet/min (equivalent to 760 dose/min), only 100 cm^3 of material is needed to feed each minute. If drug, excipients, and polymer are fed separately, individual feed rates are even lower.

Typical pellet or powder feeders do not have the best precision at low feed rates. Some feeders are designed for low-rate feeding of powder materials, but they are product specific. It is challenging to control feed rate consistency. To achieve constant drug concentration in the extrudate, drug, polymer, and all the excipients can be premixed together and fed through one feeder. In this case, the drug concentration in extrudate depends on uniform premixing and minimized particle settling during feeding. Another approach is to repelletize the extrudate, homogenize the whole batch, and redo extrusion. Sometimes, a high drug concentration master batch can be made, which is then blended with polymer to produce the targeted extrudate. If the polymer is a gum, it will need a gum feeder. The drug powder can be compounded in the gum ahead of extrusion.

In the final product, total amount of drug, that is, the dose, must be tightly controlled. In addition to keeping the drug concentration consistent, the device dimensions must be controlled. For example, if the diameter of a rod varies by 5%, the total amount of drug can vary by a factor of 10%. To achieve a consistent diameter of the rod, a metering pump and a high precision puller are needed. To ensure uniform length, precision cutting technologies are needed. Mechanical cutting has been widely used. Laser cutting, especially with femtosecond lasers, has been increasingly used due to its high precision and low thermal damage.

Extrusion runs at a temperature at which the polymer is a melt and has low-enough viscosity. It is obvious that the drug should not chemically degrade or have polymorphism changes at the extrusion temperature during processing. Careful testing for chemical degradation and polymorphism changes are necessary. Even with the same drug, if the polymer or excipients change then testing may need to be repeated. If the extrusion temperature is too high, plasticizers may be used to reduce the processing temperature. There are a few commonly used plasticizers, as listed in Table 4.3.

Drug elution rate is another parameter that must be controlled in a combination device. If drug is presented as particles dispersed in a polymer matrix, elution rate can also be related to the drug particle size and dispersion uniformity. It is critical to note that if a twin screw extruder is used to compound drug and polymer, the screws may

TABLE 4.3 Excipients Used in Medical Device Polymer–Drug Systems (R120)

Type	Material
Antioxidation	• Irganox®, butylated hydroxyanisole (BHA), butylated hydroxytoluene (BHT) • Sodium thiosulfate • Vitamin E
Solubility enhancement	• EUDRAGIT® poly(meth)acrylates • Polyethylene glycol (PEG) • Dextrose
Plasticizer	• Poly(ethylene glycol) dimethyl ether (PEG-DME) • Sugar alcohol • Triethylcitrate • Dibutyl sebacate • Diethyle phthalate (use of this type additives has been reduced) • Glycerol • Menitol
Preservative	• *meta*-Cresol

grind the drug particles, resulting in reduction of particle sizes or particle size distribution changes. The elution rate may then be increased. If a single screw extruder is used, the grinding effect may not be significant but mixing may not be as good as with a twin screw extruder.

4.4.5 Molding

Molding can be used for the drug and polymer when the drug can tolerate the molding temperature in the presence of the polymer and its additives. Drug content and variation are controlled by the mold cavity volume and drug concentration. Molding is a batch process, but there is no need to do postoperations such as cutting for extrusion. As in extrusion, the premixture is a key factor in controlling drug content variation. In addition, the mold size has significant effect on the dose sizes as well. In a compounding process where particles are dispersed in continuous matrices, the particles tend to stay away from the surface, leaving a thin layer of matrix covering the surface of the molded parts. During drug release, this layer acts as a barrier. The layer thickness is a function of molding parameters such as flow rate of the melt during injection. If the layer is trimmed from the primary molded part, the revealed fresh surface may release drug with an initial burst. Processing temperature and feed control have similar impact on the products as those with extrusion processes.

Melt molding is usually operated at high temperatures that may not be suitable for certain drugs. Powder molding may be a method to consider. Here a polymer and drug are ground into fine powder and dry mixed, filled in a mold cavity, heated to a temperature above the T_g of polymer, and compressed into a solid form. The heating temperature can be substantially lower than needed for melt injection molding. In U.S. Patent 8,637,064 [84], a method is described to use powder molding to make a composite composed of polylactide (PLA) and animal bone powder. High-strength polylactide usually has a high molecular weight. Accordingly, its melt processing temperature can be about 180–210 °C. This temperature is not suitable for animal bone powder. By grinding the PLA into fine powder, the process temperature can be reduced to 70–80 °C, at which dry bone powder may survive for a short period of time. If solvent is used as binder for PLA particles, the molding temperature can be further reduced, but the solvent must also be compatible with the bone powder and can be easily removed.

4.4.6 Reservoir Filling

In pump-enabled drug delivery technologies, such as insulin pump therapy and intrathecal drug pump therapy, a solution of drug and excipients is filled in a cartridge or reservoir that is connected to the delivery catheter. Controlled delivery is performed with a syringe pump acting as the cartridge or a roller pump acting on the catheter. The delivery accuracy is guaranteed by the precision of the catheter dimension and material properties (see above), the pumping mechanism, and time control. Pump and catheter designs, including materials and mechanical structures, are critical to the therapies. Cartridge replacement or reservoir refilling are usually possible so the pump can be reused by the same patient.

Formulation of the drug solution is mostly based on pharmaceutical considerations. However, the compatibility of the drug solution with the pump and catheter must be well considered to avoid adverse effects on the drug due to the delivery mechanism, as well as adverse effects of the drug solution on the pump and catheters. Mechanical movement of the patient and tissue fluids, biological components in the tissues (e.g., O_2 and CO_2), and body temperature all need to be considered when evaluating the interactions. In certain cases, temperature adjustment and degassing of the drug solutions need to be done prior to refilling.

It is interesting to note that the interaction of macromolecular drugs such as insulin has led to modification of the molecule. The insulin derivative Lispro, which is produced by recombinant methods, has a lysine base switched with a proline in its primary structure, which leads to a form of the folded protein that resists aggregation and binding to surfaces. This derivative, whose action is the same as that of native insulin, does not clog catheter lines, which was a historical problem with insulin pumps before the modified molecule was introduced.

4.5 SUMMARY

Drug–device combination products represent a trend of biomedical devices because they can offer synergistic functions. Success of development of a drug–device combination depends on proper design, manufacturing, and use of the product, which are based on thorough understanding of the drug, device, and drug–device interactions underlying the synergy. However, the drug and the device may interact with each other in ways that are not favorable. To avoid adverse effects, those interactions have to be carefully considered and sometimes mitigated in product design and testing, drug and material selection, manufacturing, and storage.

The manufacturing considerations treated in this chapter are all based on the availability of suitable materials and drugs for the task at hand. It is of course easiest to work with already existing and tested components, given the points made regarding interactions. However, we have presented two examples where innovation in materials and drugs may lead to better and easier-to-manufacture devices. The first example was the development of copolymers that enhance blending of components with different properties, while the second was the modification of a molecule (insulin) to make it more compatible with the rigors of delivery from a device. Since each innovation requires its own development, it is likely that the most common pathway will use "tried and true" components, while component innovation will be included only when no easy solutions can be found by working with existing tool boxes.

REFERENCES

1. K.B. Stokes, G.A. Bornzin, and W.A. Wiebusch, A steroid-eluting, low-threshold, low-polarizing electrode, in K. Steinbach (Ed.), *Cardiac Pacing*, Steinkopff Verlag, Darmstadt, 1983, pp. 369–376.

2. H. Mond and K.B. Stokes, The electrode-tissue interface: the revolutionary role of steroid-elution. *Pacing Clin. Electrophysiol.* 15 (1992) 95–107.
3. Gammage, R. Lieberman, R. Yee, et al., Multi-center clinical experience with a lumenless catheter-delivered, bipolar, permanent pacemaker lead: implant safety and electrical performance. *Pacing Clin. Electrophysiol.* 29 (2006) 858–865.
4. D.S. Beanlands, Y. Akyurekli, and W.J. Keon, Prednisone in the management of exit block, in Proceedings of the Sixth World Symposium on Cardiac Pacing, Montreal, 1979.
5. J.E. Sousa, M.A. Costa, A.C. Abizaid, et al., Sustained suppression of neointimal proliferation by sirolimus-eluting stents: one-year angiographic and intravascular ultrasound follow-up. *Circulation* 104 (2001) 2007–2011.
6. CYPHER™ Sirolimus-Eluting Coronary Stent on RAPTOR® Over-the-Wire Delivery System or RAPTORRAIL® Rapid Exchange Delivery System: P020026. Available at http://www.accessdata.fda.gov/scripts/cdrh/cfdocs/cftopic/pma/pma.cfm?num=p020026 (accessed April 6, 2014).
7. TAXUS™ Express 2™ Paclitaxel-Eluting Coronary Stent System (Monorail and Over-the-Wire): P030025. Available at http://www.accessdata.fda.gov/scripts/cdrh/cfdocs/cftopic/pma/pma.cfm?num=p030025 (accessed April 6, 2014).
8. Endeavor ® Zotarolimus: Eluting Coronary Stent on the Over-the-Wire (OTW), Rapid Exchange (RX), or Multi Exchange II (MX2) Stent Delivery System: P060033. Available at http://www.accessdata.fda.gov/scripts/cdrh/cfdocs/cfTopic/pma/pma.cfm?num=P060033.
9. XIENCE nano™ Everolimus Eluting Coronary Stent System: P070015/S054. Available at http://www.accessdata.fda.gov/scripts/cdrh/cfdocs/cftopic/pma/pma.cfm?num=p070015s054 (accessed April 6, 2014).
10. Resolute MicroTrac Zotarolimus-Eluting Coronary Stent System (Resolute MicroTrac) and Resolute Integrity Zotarolimus-Eluting Coronary Stent System (Resolute Integrity): P110013. Available at http://www.accessdata.fda.gov/scripts/cdrh/cfdocs/cfTopic/pma/pma.cfm?num=P110013 (accessed April 6, 2014).
11. M.C. Morice, P.W. Serruys, J.E. Sousa, J. Fajadet, E.B. Hayashi, M. Perin, A. Colombo, G. Schuler, P. Barragan, G. Guagliumi, F. Molnàr, R. Falotico, M.C. Morice, P.W. Serruys, J.E. Sousa, et al., A randomized comparison of a sirolimus-eluting stent with a standard stent for coronary revascularization. *N. Engl. J. Med.* 346 (2002) 1773–1780.
12. J.W. Moses, M.B. Leon, et al., Sirolimus-eluting stents versus standard stents in patients with stenosis in a native coronary artery. *N. Engl. J. Med.* 349 (2003) 1315–1323.
13. G.W. Stone, M. Midei, W. Newman, M. Sanz, J.B. Hermiller, J. Williams, N. Farhat, K.W. Mahaffey, D.E. Cutlip, P.J. Fitzgerald, P. Sood, X. Su, and A.J. Lansky, Comparison of an everolimus-eluting stent and a paclitaxel-eluting stent in patients with coronary artery disease: a randomized trial. *JAMA* 299 (2008) 1903–1913.
14. G.W. Stone, M. Midei, W. Newman, M. Sanz, J.B. Hermiller, J. Williams, N. Farhat, R. Caputo, N. Xenopoulos, R. Applegate, P. Gordon, R.M. White, K. Sudhir, D.E. Cutlip, and J.L. Petersen, Randomized comparison of everolimus-eluting and paclitaxel-eluting stents: two-year clinical follow-up from the Clinical Evaluation of the Xience V Everolimus Eluting Coronary Stent System in the Treatment of Patients with *de novo* Native Coronary Artery Lesions (SPIRIT) III trial. *Circulation* 119 (2009) 680–686.

15. J. Fajadet, W. Wijns, G.-J. Laarman, K.-H. Kuck, J. Ormiston, T. Münzel, J.J. Popma, P.J. Fitzgerald, R. Bonan, and R.E. Kuntz, Randomized, double-blind, multicenter study of the endeavor zotarolimus-eluting phosphorylcholine-encapsulated stent for treatment of native coronary artery lesions: clinical and angiographic results of the ENDEAVOR II trial. *Circulation* 114 (2006) 798–806.
16. A.C. Yeung, M.B. Leon, A. Jain, T.R. Tolleson, D.J. Spriggs, B.T. McLaurin, J.J. Popma, P.J. Fitzgerald, D.E. Cutlip, J.M. Massaro, and L. Mauri, Clinical evaluation of the Resolute zotarolimus-eluting coronary stent system in the treatment of *de novo* lesions in native coronary arteries: the RESOLUTE US clinical trial. *J. Am. Coll. Cardiol.* 57 (2011) 1778–1783.
17. J. Kunz and M. Turco, The DES landscape in 2011. *Cardiac Interv. Today* Jan/Feb 2012 48–52.
18. T. Kimura, H. Yokoi, Y. Nakagawa, T. Tamura, S. Kaburagi, Y. Sawada, Y. Sato, H. Yokoi, N. Hamasaki, H. Nosaka, and M. Nobuyoshi, Three-year follow-up after implantation of metallic coronary–artery stents. *N. Engl. J. Med.* 334 (1996) 561–566.
19. M.C. Morice, P.W. Serruys, J.E. Sousa, et al., A randomized comparison of a sirolimuseluting stent with a standard stent for coronary revascularization. *N. Engl. J. Med.* 346 (2002) 1773–1780.
20. P.C. Begovac, R.C. Thomson, J.L. Fisher, A. Hughson, and A. Gallhagen, Improvements in GORE-TEX(R) vascular graft performance by Carmeda® BioActive Surface heparin immobilization. *Eur. J. Vasc. Endovasc. Surg.* 25 (2003) 432–437.
21. M.R. Sohail, C.A. Henrikson, M.J. Braid-Forbes, K.J. Forbes, and D.J. Lerner, Mortality and cost associated with cardiovascular implantable electronic device infections. *Arch. Intern. Med.* 171 (2011) 1821–1828.
22. R.O. Darouiche, I.I. Raad, S.O. Heard, J.I. Thornby, O.C. Wenker, A. Gabrielli, J. Berg, N. Khardori, H. Hanna, R. Hachem, R.L. Harris, and G. Mayhall, A comparison of two antimicrobial-impregnated central venous catheters. *N. Engl. J. Med.* 340 (1999) 1–8.
23. T. Elliott, Role of antimicrobial central venous catheters for the prevention of associated infections. *J. Antimicrob. Chemother.* 43 (1999) 441–446.
24. H.L. Bloom, L. Constantin, D. Dan, et al., Implantation success and infection in cardiovascular implantable electronic device procedures utilizing an antibacterial envelope. *Pacing Clin. Electrophysiol.* 34 (2011) 133–142.
25. InFUSE™ Bone Graft/LT-CAGE™ Lumbar Tapered Fusion Devices: P000058. Available at http://www.accessdata.fda.gov/scripts/cdrh/cfdocs/cftopic/pma/pma.cfm?num=P000058 (accessed April 6, 2014).
26. J.K. Burkus, et al., Anterior lumbar interbody fusion using rhBMP-2 with tapered interbody cages. *J Spinal Disord.* 15 (2002) 337–349.
27. R.W. Fu, S. Selph, M. McDonagh, K. Peterson, A. Tiwari, R Chou, and M. Helfand, Effectiveness and harms of recombinant human bone morphogenetic protein-2 in spine fusion: a systematic review and meta-analysis. *Ann. Intern. Med.* 158 (2013) 890–902.
28. D.B. Keenan, R. Cartaya, and J.J. Mastrototaro, Accuracy of a new real-time continuous glucose monitoring algorithm. *J. Diabetes Sci. Technol.* 4 (2010) 111–118.
29. MiniMed 530G System: P120010. Available at http://www.accessdata.fda.gov/scripts/cdrh/cfdocs/cftopic/pma/pma.cfm?num=p120010 (accessed April 6, 2014).

30. Dexcom G4 PLATINUM Continuous Glucose Monitoring System: P120005. Available at http://www.accessdata.fda.gov/scripts/cdrh/cfdocs/cftopic/pma/pma.cfm?num=p120005 (accessed April 6, 2014).
31. Medtronic® Synchromed™ Pump & Infusion System: P860004. Available at http://www.accessdata.fda.gov/scripts/cdrh/cfdocs/cfpma/pma.cfm?id=23090 (accessed August 12, 2014).
32. R. Gilmartin, D. Bruce, B.B. Storrs, et al., Intrathecal baclofen for management of spastic cerebral palsy: multicenter trial. *J. Child Neurol.* 15 (2000) 71–77.
33. Prometra® Programmable Infusion Pump System: P080012. Available at http://www.accessdata.fda.gov/scripts/cdrh/cfdocs/cftopic/pma/pma.cfm?num=p080012 (accessed August 12, 2014).
34. B. Winner, J.F. Peipert, Q.H. Zhao, C. Buckel, T. Madden, J.E. Allsworth, and G.M. Secura, Effectiveness of long-acting reversible contraception. *N. Engl. J. Med.* 366 (2012) 1998–2007.
35. L. Makarainen, A. van Beek, L. Tuomivaara, B. Asplund, and H.C. Bennink, Ovarian function during the use of a single contraceptive implant; Implanon compared with Norplant. *Fertil. Steril.* 69 (1998) 714–721.
36. L. Bahamondes, C. Monteiro-Dantas, X. Espejo-Arce, et al., A prospective study of the forearm bone density of users of etonogestrel and levonorgestrel-releasing contraceptive implants. *Human Reprod.* 21 (2006) 466–470.
37. Mirena (levonorgestrel-releasing intrauterine system). Application No. 021225. http://www.accessdata.fda.gov/drugsatfda_docs/nda/2000/21-225_Mirena.cfm (accessed April 6, 2014).
38. M.A. Moses, H. Brem, and R. Langer, Advancing the field of drug delivery: taking aim at cancer. *Cancer Cell* 4 (2003) 337–341.
39. R. Langer, New methods of drug delivery. *Science* 249 (1990) 1527–1532.
40. N.A. Peppas and R. Langer, New challenges in biomaterials. *Science* 263 (1994) 1715–1720.
41. S. Valtonen, U. Timonen, P. Toivanen, H. Kalimo, L. Kivipelto, O. Heiskanen, G. Unsgaard, and T. Kuurne, Interstitial chemotherapy with carmustine-loaded polymers for high-grade gliomas: a randomized double-blind study. *Neurosurgery* 41 (1997) 44–49.
42. A. Shmulewitz, R. Langer, and J. Patton, Convergence in biomedical technology. *Nat. Biotechnol.* 24 (2006) 277–280.
43. O. Larm, R. Larsson, and P. Olsson, A new non-thrombogenic surface prepared by selective covalent binding of heparin via a modified reducing terminal residue. *Biomater. Med. Devices Artif. Organs* 11 (1983) 161–173.
44. R. Ng, *Drugs from Discovery to Approval*, 2nd ed., Wiley-Blackwell, New Jersey, 2009.
45. R. Virmani, G. Guagliumi, A. Farb, et al., Localized hypersensitivity and late coronary thrombosis secondary to a sirolimus-eluting stent: should we be cautious? *Circulation* 109 (2004) 701–705.
46. E.P. McFadden, E. Stabile, E. Regar, et al., Late thrombosis in drug-eluting coronary stents after discontinuation of antiplatelet therapy. *Lancet* 364 (2004) 1519–1521.
47. E. Camenzind, P.G. Steg, and W. Wijns, A metaanalysis of first generation drug eluting stent programs. Abstract presented at the World Congress of Cardiology 2006, Barcelona, September 2–5, 2006.

48. P. Wenaweser, K. Tsuchida, S. Vaina, et al., Late stent thrombosis following drug-eluting stent implantation: data from a large, two-institutional cohort study. Abstract presented at the World Congress of Cardiology 2006, Barcelona, September 2–5, 2006.
49. A. Farb and A.B. Boam, Stent thrombosis redux: the FDA perspective. *N. Engl. J. Med.* 356 (2007) 984–987.
50. M. Joner, A.V. Finn, A. Farb, E.K. Mont, F.D. Kolodgie, E. Ladich, R. Kutys, K. Skorija, H.K. Gold, and R. Virmani, Pathology of drug-eluting stents in humans delayed healing and late thrombotic risk. *J. Am. Coll. Cardiol.* 48 (2006) 193–202.
51. V.B. Kolachalama, A.R. Tzafriri, D.Y. Arifin, and E.R. Edelman, Luminal flow patterns dictate arterial drug deposition in stent-based delivery. *J. Control. Release* 133 (2009) 24–30.
52. J. Tao, Y. Sun, G.G.Z. Zhang, and L. Yu, Solubility of small-molecule crystals in polymers: D-mannitol in PVP, indomethacin in PVP/VA, and nifedipine in PVP/VA. *Pharm. Res.* 26 (2009) 855–864.
53. N. Jain, G.K. Jain, F.J. Ahmad, and R.K. Khar, Validated stability-indicating densitometric thin-layer chromatography: application to stress degradation studies of minocycline. *Anal. Chim. Acta* 599 (2007) 302–309.
54. Medtronic TYRX AIGIS_{RX}® device. Available at http://www.accessdata.fda.gov/cdrh_docs/pdf13/K130943.pdf (accessed October 22, 2014).
55. D. Sciubba, R. Stuart, M. McGirt, G. Woodworth, A Samdani, B Carson, and G. Jallo, Effect of antibiotic-impregnated shunt catheters in decreasing the incidence of shunt infection in the treatment of hydrocephalus. *J. Neurosurg.* 103 (2005) 131–136.
56. S. Parker, W. Anderson, S Lilienfeld, J. Megerian, and M. McGirt, Cerebrospinal shunt infection in patients receiving antibiotic-impregnated versus standard shunts: a review. *J. Neurosurg. Pediatr.* 8 (2011) 259–265.
57. A.L. Casey, L.A. Mermel, P. Nightingale, and T.S.J. Elliott, Antimicrobial central venous catheters in adults: a systematic review and meta-analysis. *Lancet Infect. Dis.* 8 (2008) 763–776.
58. D. Judd and W. Ferris, Sterilized minocycline and rifampin-containing medical device, U.S. Patent Application 2008/0075628 A1, filed September 27, 2006.
59. W.J. Bertrand, P. Marek, and D. Amery, Adjusting drug loading in polymeric materials, U.S. Patent Application 20110237687 A1, filed May 25, 2010.
60. M. Vert, J. Maudult, and S. Li, Biodegradation of PLA/GA polymers: increasing complexity. *Biomaterials* 15 (1994) 1209–1213.
61. S.P. Lyu and D. Untereker, Degradability of polymers for implantable biomedical devices. *Int. J. Mol. Sci.* 10 (2009) 4033–4065.
62. S.J. Siegela, J.B. Kahna, K. Metzgera, K.I. Winey, K. Wernerd, and N. Dand, Effect of drug type on the degradation rate of PLGA matrices. *Eur. J. Pharm. Biopharm.* 64 (2006) 287–293.
63. Medtronic SynchroMed II Implantable Drug Infusion Pump and SynchroMed EL Implantable Drug Infusion Pump. Available at http://www.fda.gov/MedicalDevices/Safety/ListofRecalls/ucm333231.htm (accessed April 6, 2014).
64. Medical Device Safety Notification: November 2012. Use of Unapproved Drugs with the SynchroMed Implantable Infusion Pump. Available at http://professional.medtronic.com/pt/neuro/idd/ind/product-advisories/NOVEMBER-2012-SYNCHII (accessed April 6, 2014).

65. D.W. van Krevelen, *Properties of Polymer*, 3rd ed., Elsevier, New York, 1990.
66. A.S. Michaels, P.S.L. Wong, R. Prather, and R.M. Gale, A thermodynamic method of predicting the transport of steroids in polymer matrices. *AlChE J.* 21 (1975) 1073–1080.
67. K. Kamath, S. Nott, L. Pinchuk, and M. Schwarz, Drug delivery compositions and medical devices containing block copolymer, U.S. Patent Application 2002107330 A1, filed December 12, 2000.
68. S. Venkatraman and F. Boey, Release profiles in drug-eluting stents: issues and uncertainties. *J. Control. Release* 120 (2007) 149–160.
69. Y.M. Chen, J. van Sciver, J. McCabe, and A. Garcia, Methods and apparatus for coating stents, U.S. Patent 8367150 B2, filed June 15, 2007.
70. S.P. Lyu, R. Sparer, C. Hobot, and K. Dang, Adjusting drug diffusivity using miscible polymer blends. *J. Control. Release* 102 (2005) 679–687.
71. S.P. Lyu, R. Sparer, C. Hobot, and K. Dang, Active agent delivery system including a poly (ethylene-*co*-(meth)acrylate) medical device and method, EP Patent 1531875, filed August 13, 2003.
72. R. Sparer, C. Hobot, S.P. Lyu, K. Dang, and P.W. Cheng, Active agent delivery system including a hydrophobic cellulose derivative, EP Patent 1539265, filed August 13, 2003.
73. K. Udipi, P. Cheng, M. Chen, and S.P. Lyu, Biocompatible controlled release coatings for medical devices and related methods, U.S. Patent 8,088,404, filed December 6, 2004.
74. K. Udipi, M. Chen, P. Cheng, K. Jiang, D. Judd, A. Caceres, R.J. Melder, and J.N. Wilcox, Development of a novel biocompatible polymer system for extended drug release in a next-generation drug-eluting stent. *J. Biomed. Mater. Res.* 85A (2008) 1064–1071.
75. A. Hezi-Yamit, C. Sullivan, J. Wong, L. David, M. Chen, P. Cheng, D. Shumaker, J.N. Wilcox, and K. Udipi, Impact of polymer hydrophilicity on biocompatibility: implication for DES polymer design. *J. Biomed. Mater. Res.* 90A (2009) 133–141.
76. D.M. Wiktor, Intravascular radially expandable stent and method of implant, U.S. Patent 4,886,062, filed October 19, 1987.
77. D. Douroumis and I. Onyesom, Novel coating technology of drug eluting stents, in M. Zilberman (Ed.), *Studies in Mechanobiology Tissue Engineering and Biomaterials*, Vol. 8, Active Implants and Scaffolds for Tissue Regeneration, Springer, Berlin, 2011, pp. 87–125.
78. M. Gañán-Calvo, Generation of steady liquid microthreads and micron-sized monodisperse sprays in gas streams. *Phys. Rev. Lett.* 80 (1998) 285–288.
79. S.P. Lin and R.D. Reitz, Drop and spray formation from a liquid jet. *Annu. Rev. Fluid Mech.* 30 (1998) 85–105.
80. R.J. Lang, Ultrasonic atomization of liquids. *J. Acoust. Soc. Am.* 34 (1962) 6–8.
81. D.J. Schuerer, J.E. Mazuski, T.G. Buchamn, et al., Catheter-related bloodstream infection rates in minocycline/rifampin vs chlorhexidine/siliver sulfadiazine-impregnated central venous catheters: results of a 46 month study. *Crit. Care Med* 36 (Suppl. 12), (2008) A199–A208.
82. H. Hanna, R. Benjamin, I. Chatzinikolaou, B. Alakech, D. Richardson, P. Mansfield, T. Dvorak, M.F. Munsell, R. Darouiche, H. Kantarjian, and I. Raad, Long-term silicone central venous catheters impregnated with minocycline and rifampin decrease rates of catheter-related bloodstream infection in cancer patients: a prospective randomized clinical trial. *J. Clin. Oncol.* 22 (2004) 3163–3171.

83. A. Almeida, B. Claeys, J. Remon, C. Vervaet, Hot-melt extrusion developments in the pharmaceutical industry, in D. Douroumis (Ed.), *Hot-Melt Extrusion: Pharmaceutical Applications*, Advances in Pharmaceutical Technology, John Wiley & Sons, Inc., Hoboken, 2012.
84. S.P. Lyu, C. Hobot, S. Haddock, B. Loy, R. Sparer, K.M. Kinnane, and J.M. Gross, Compression molding method for making biomaterial composites, U.S. Patent 8,637,064 B2, filed September 20, 2006.

PART II

PRODUCTS

5

STEROID-RELEASING LEAD

Rick McVenes and Ken Stokes
CRHF, Medtronic, Inc., Mounds View, MN 55112, USA

5.1 INTRODUCTION

Thornstein Veblen (1902–1983) said, "Necessity is the mother of invention." In biomedical science, this can be paraphrased as "an unmet medical need is the mother of invention." There is no question that early implantable pacemakers met a significant medical need, but they required substantial innovation to become practical and widely used. A pacemaker is composed of two primary devices, a pulse generator and at least one lead. The leads appear to be deceptively simple, yet they provide major contributions to pulse generator size, longevity, and features. They play a major role in device implantability and patient safety. Because of unmet medical needs, early pacemaker leads evolved from simple wires susceptible to numerous complications, to the sophisticated, far more reliable modern devices that serve patients today. Steroid-eluting electrodes were a breakthrough, but their efficacy is dependent on interactions with other design innovations.

5.2 BACKGROUND

5.2.1 Early Pacemakers, Unmet Medical Needs, and How They Were Resolved

Compared to modern pulse generators (Figure 5.1b), early versions were relatively simple, very large (bigger than hockey pucks), single-chamber (ventricular only)

Drug–Device Combinations for Chronic Diseases, First Edition. Edited by SuPing Lyu and Ronald A. Siegel.

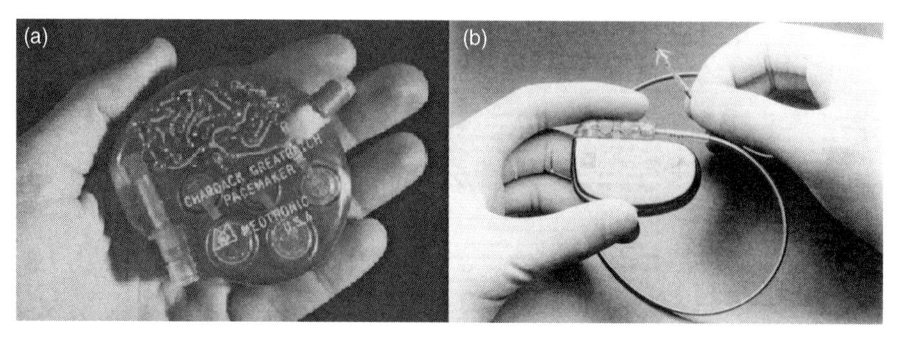

FIGURE 5.1 (a) An early "cordwood" circuit pacemaker with HgZn batteries, encapsulated in epoxy. The cylindrical structure near the subject's index finger is a 10-turn potentiometer. This was designed to receive a transcutaneous Keith needle, which when rotated, changed the pulse generator's pacing rate. (b) A modern single chamber pacemaker with a hermetically sealed hybrid circuit powered by a LiI cell. A ventricular, transvenous tied lead is attached. The device is noninvasively programmed, and uses telemetry to update the physician as to the device and patient's status. (Reproduced with permission from Medtronic plc.)

devices. They were epoxy-encapsulated "cordwood" circuitry, powered by HgZn batteries, with only one or two functions (initially stimulation only, then later also sensing (Figure 5.1a)). The earliest leads were silicone rubber-insulated wires (Figure 5.2a) sewn directly on the epicardial surface of the heart, requiring a thoracotomy with general anesthesia. This was supplanted by the much easier and safer to implant transvenous lead, requiring only local anesthesia.

Early lead complications included conductor fracture, dislodgment,[1] microdislodgment,[2] myocardial perforation,[3] and penetration.[4] Conductor fracture was dramatically reduced initially through the use of coiled multifilar MP35N[5] wires, coaxial designs, composite structures, and eventually multifilar microcables (Figure 5.2) [1,2]. Dislodgment- and perforation-related complications were resolved primarily by electrode fixation and managing lead stiffness [1]. The two fixation designs that eventually dominated the industry were passive (tined) and active (helical screw-in) electrodes (Figure 5.3). Electrode performance, however, was still in need of improvement.

A substantial, but elusive (at the time), unmet medical need was the reduction of the high and variable thresholds provided by the leads of the day.[6] Thresholds at implant were typically low, less than 1 V. During the subacute period, typically 1–2 weeks, but as long as 6 weeks (depending on the design), thresholds would rise to a peak, averaging around 1.5–3 V at 0.8–0.5 ms. Then, over time they would either

[1] Retraction of the stimulating electrode from the endocardium visible on X-ray.

[2] Slight retraction of the stimulating electrode from the endocardium not visible on X-ray.

[3] Perforation of the stimulating electrode through the myocardium and epicardium visible on X-ray.

[4] Penetration of the stimulating electrode into the myocardium, but not the epicardium, not usually visible on X-ray.

[5] The first super alloy composed of Ni, Co, Cr, and Mo, a trademark of Dupont.

[6] The minimum stimulus (voltage) required to consistently depolarize the heart outside the refractory period.

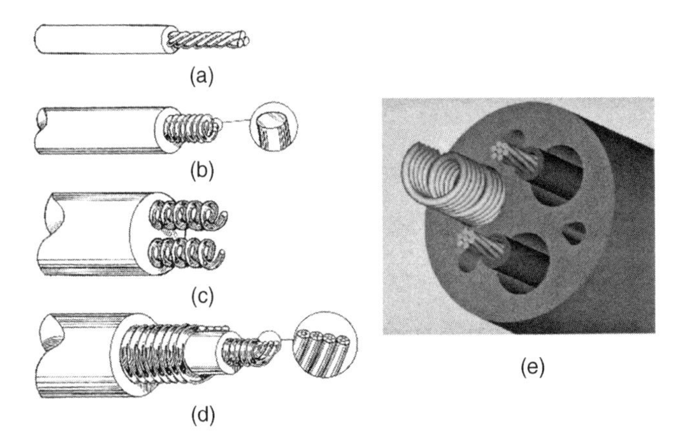

FIGURE 5.2 Lead conductors. (a) Early three-wire twist. (b) Unipolar coil. (c) Parallel bipolar coils. (d) Coaxial bipolar coil. (e), Multiconductor with one coil and two redundantly insulated microcables. (Reproduced with permission from Medtronic plc.)

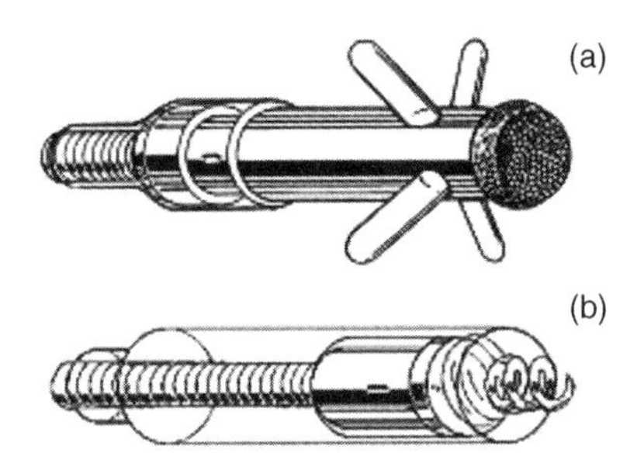

FIGURE 5.3 Distal tips of transvenous (a), passively fixed (tined), and (b) actively fixed (screw-in) pacemaker leads. (Reproduced with permission from Medtronic plc.)

remain stable or decline to an intermediate value (Figure 5.4) [3]. However, the goal was not to stimulate just the average patient. The goal was to safely pace at least 95% of patients. Looking at Figure 5.4, the 98th percentile (mean + 2 standard deviations) in the atrium at peak was about 3.2 V. In the ventricle, the 98th percentile at 8 weeks was also about 3.2 V. Because of circadian variations and other factors, a safety factor, preferably two to three times the threshold, was needed to ensure safe and consistent capture of the heart. By the early 1980s, pacemaker outputs were factory present at 5 V, 0.5 ms, which meant that in these subjects, safety factors were only about 1.6× threshold at these monitor periods. Thus, lower, stable thresholds were desired to improve patient safety. In addition, high thresholds impacted battery longevity, pulse generator size, and patient safety.

5.2.1.1 Leads and Battery Longevity The longevity of a pulse generator battery is determined by its charge capacity and the rate charge is drained from it. By definition,

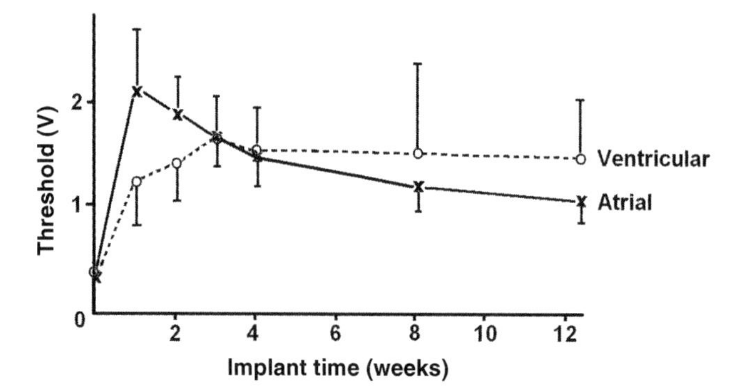

FIGURE 5.4 Canine cardiac thresholds (V) at 0.5 ms as a function of time. (Reproduced with permission from Medtronic plc.)

the rate of charge transfer is current. Thus, the longevity (L, in years) of a pulse generator is the deliverable battery capacity (C_b) in microampere-hours divided by the current used [4,5]:

$$L = \frac{C_b}{8760(I_C + I_l)} \tag{5.1}$$

where I_c is the "overhead" or "static" current (μA) required to operate the generator's circuitry, I_l is the "continuous" current over time (μA, not the current per pulse) used to pace the patient, and 8760 is the number of hours in a year. Since the battery capacity and circuit current drain are fixed for any given pulse generator design, the only parameter that can be controlled clinically is the lead current (I_l). The "continuous" current drained by the lead is

$$I_l = \frac{V_o^2 t}{V_b Z_p \text{CL} \times 10^{-6}} \tag{5.2}$$

where V_o is the output voltage, t is the pulse width (ms), V_b is the battery voltage, Z_p is the pacing impedance (Ω), CL is the cycle length (ms), and 10^{-6} is the number of μA per A. If Z_p, V_b, and CL are fixed, then only V_o and/or t can be adjusted to reduce current drain and increase longevity. Clearly, if stimulation safety factors permit, reduction in V_o is the preferred step.

Now assume that the factory-preset output in the early 1980s was used with a typical lead impedance of 500 Ω. A LiI cell may be rated at over 1 A/h, but it can only deliver about 0.750 A/h. Its voltage (V_b) is about 2.7 V, but some is consumed in the lost circuitry. Thus, V_b is about 2.6 V. If the patient is paced at 72 bpm (833 ms) 100% of the time, a 5 V, 0.5 ms output drains about 11.1 μA on a continuous basis. Thus, a LiI power source and a 1980s pacemaker circuit with about 4 μA current drain will pace one cardiac chamber for about 5.7 years according to Eq. (5.1). A dual chamber pacemaker would last only 2.8 years. The goal was to achieve 10+ years for single-chamber pacing and 5+ years for dual-chamber devices. If the output voltage could be safely reduced 50% to 2.5 V, then, under similar conditions, the current drain used by the lead would be only about 2.8 μA and the same pulse generator would last about 12.6 years for a single-chamber device and 6.3 years for dual-chamber pacing. This would require stable 98th percentile thresholds of ≤2.5 V (usually at 0.5 ms).

5.2.1.2 Leads, Pulse Generator Size, and Features If the current used for pacing could be reduced sufficiently, the battery capacity could also be reduced without sacrificing longevity. Thus, pulse generators could become smaller. Alternatively or in addition, some of the additional current saved could be used to drive new features.

5.2.1.3 Leads and Patient Safety Unfortunately, some patient's thresholds continued to increase after the peak period to the point where the pulse generator could no longer stimulate the heart (exit block) [6–8]. By the mid-1980s, reliable electrode fixation and controlled lead stiffness greatly reduced exit block in adults, but did not eliminate it. Children were particularly susceptible to exit block, experiencing up to a 4% incidence per year; 40% in 10 years [9]. For some of these patients, repeated exit block, even with frequent lead replacement, faced a life-threatening situation. The ability to eliminate such exit block and provide consistent high patient safety factors remained an unmet medical need.

5.3 THE QUEST FOR LOW-THRESHOLD LEADS

5.3.1 Mechanical Design

5.3.1.1 Electrode Size First, it must be recognized that in the 1970s and early 1980s, the mechanisms for device encapsulation were largely unknown. It was generally accepted that foreign objects, including cardiac electrodes, encapsulated due to the action of fibroblasts, but it was unknown just what was happening at electrode sites. It was known that the development of a fibrotic capsule surrounding the electrode separated it from viable myocardium, causing a decrease in the voltage delivered to myocyte membranes (Figure 5.5). Empirical studies established that threshold varied as an inverse function of electrode surface area [10–12]. A model developed by Irnich said that the electric field strength necessary to stimulate (ε) is a function of the voltage applied to the electrode (V) [13,14], the spherical electrode's radius (r_0), and the thickness of the fibrotic capsule (d) satisfy

$$\epsilon = \frac{V}{r_0}\left(\frac{r_0}{r_0 + d}\right)^2 \tag{5.3}$$

Since d was found empirically to be about 1 mm, the optimum size of a polished spherical electrode was determined to be about 1 mm in radius (5 mm^2 surface area). Anything smaller (where $d > r_0$) or larger (where $d < r_0$) would cause thresholds to increase. Thus, a relationship between the size of electrodes and chronic thresholds

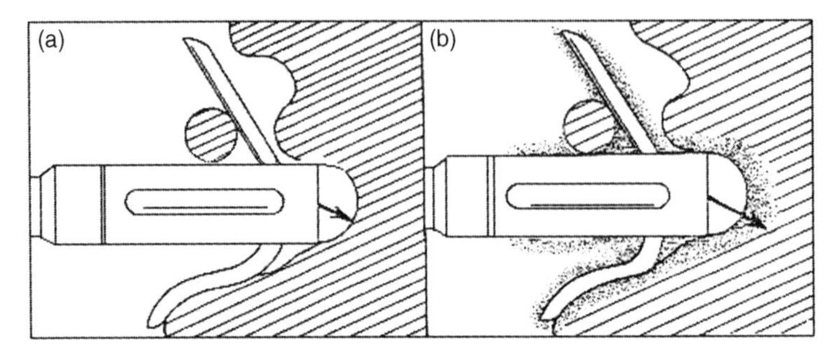

FIGURE 5.5 A transvenous electrode acutely positioned at the endocardial surface (a). The arrow indicates the electrode's radius. Part (b) shows the same electrode weeks later with a fibrotic capsule separating it from the viable cardiac tissue. The arrow now shows the virtual electrode, including the nonstimulatable collagenous capsule. (Reproduced with permission from Medtonic plc.)

was established. However, the electrodes in question developed sensing complications due to impedance mismatch with certain pulse generator's amplifiers [15,16]. The industry settled on a surface area of 8 mm^2 as the *de facto* standard, which avoided these sensing complications.

5.3.1.2 Electrode Surface Structure Some thought that electrode instability, even in the presence of fixation mechanisms, was responsible for threshold rise. Thus, totally porous and porous surface electrodes were developed with the intent that fibrous ingrowth would provide mechanical stability [17,18]. Clinical studies produced mixed results, although the trend was for improved performance [19–22]. In our own canine studies (Figure 5.6), the thresholds of a porous electrode were

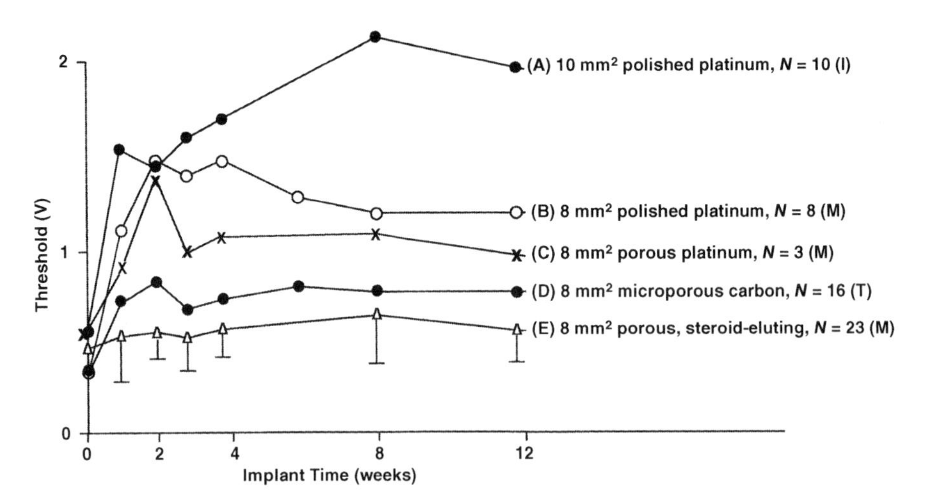

FIGURE 5.6 Canine thresholds at 0.5 ms as a function of time from several different manufacturers (I, M, and T). (Reproduced with permission from Medtonic plc.)

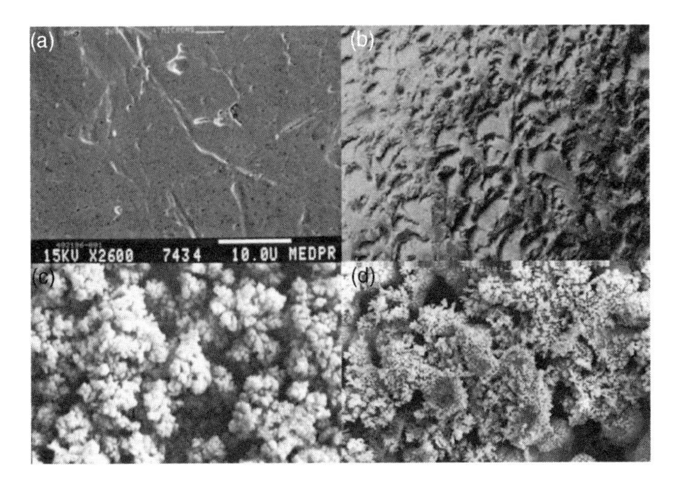

FIGURE 5.7 Electron micrographs of electrode surfaces. (a) Polished. (b) "Activated" vitreous carbon. (c) Platinized. (d) Fractal. (Reproduced with permission from Medtonic plc.)

marginally lower on average than the state-of-the-art polished tip of the same size. This finding was confirmed by human clinical studies [23].

About the same time, the microstructured vitreous carbon electrode surface was introduced (Figure 5.7b) [24]. As shown in Figure 5.6 (manufacturer C), thresholds over time were much improved over polished and porous electrodes. It was known that a thinner fibrotic capsule formed between the electrode and the viable tissue. The reasons for this, however, were still not well understood. Subsequent studies verified that microtextured electrode surfaces that did not allow tissue ingrowth (Figure 5.7b–d) were a major improvement [25–27]. It was subsequently discovered that these structures evoked a preferred orientation of collagen and fibroblasts on the device surface [28,29].

5.4 THE EVOLUTION OF STEROID-ELUTING ELECTRODES

5.4.1 Passive Fixation

It was theorized that if we could understand exactly what was happening at the electrode–tissue interface, we could possibly find means to control it and improve electrode performance even more. However, the means to directly study the *in vivo* events occurring over time at the electrode–tissue interface were not available to us. Thus, we designed a screening experiment using various drugs with anti-inflammatory, antirejection, anticlotting, and other properties that might indirectly point to a mechanism. The goal was to deliver these drugs directly to the electrode–tissue interface so that their effects would not be dispersed systemically. We used two different modifications of a marketed transvenous unipolar lead with an 8 mm^2 polished Pt-Ir ring–tip electrode (Figure 5.8a). One design employing a positive

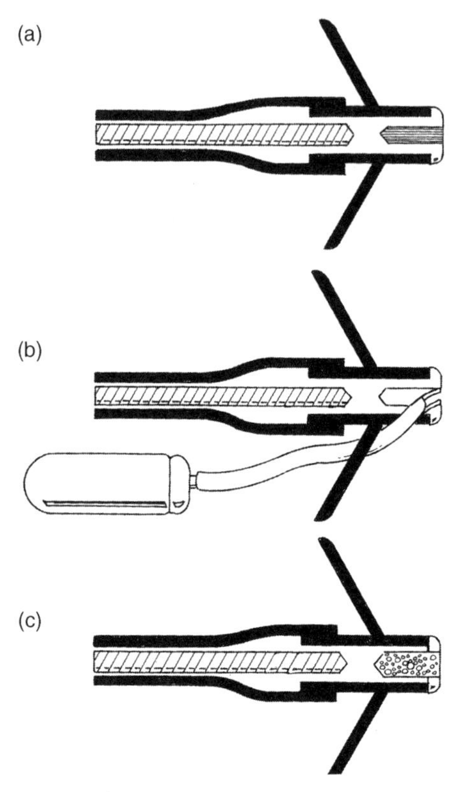

FIGURE 5.8 A polished 8 mm^2 ring–tip electrode (a), modified with an osmotic mini pump for liquid drug delivery and/or (b) modified with a solid water-soluble drug-compounded MCRD in the center of the ring (c). (Reproduced with permission from Medtonic plc.)

flow rate had a hole drilled through the side of the electrode. This facilitated insertion and fixation of a small catheter. The orifice of the catheter was located directly in the center of the ring electrode's annulus, in tissue contact (Figure 5.8b). The catheter was attached to the side of the lead, terminating at the proximal end in an osmotic mini pump (Alza). Once implanted, the pump delivered a constant 1 ml of fluid/h for 170 h. Another modification used sustained drug release. In this design, the silicone rubber in the center of the ring–tip electrode was replaced by silicone rubber compounded with solid particles of the water-soluble drug (Figure 5.8c). This design releases drug at a log rhythmically decreasing rate over time. Once implanted in canines, cardiac thresholds, pacing impedance, sensing, and so on were monitored for up to 12 weeks. Due to cost and logistical limitations, no more than three dogs with each design/drug were tested. Of course, such screening studies are by nature not statistically reliable, but they do point the direction of more detailed, future experiments. The drugs used and the maximum thresholds measured are listed in Table 5.1. The results pointed us toward the glucocorticosteroids, especially dexamethasone and its derivatives.

TABLE 5.1 Effects of Various Drugs Delivered Directly to the Electrode–Tissue Interface During a Three-Week Experiment

Drug	Concentration (meq/l) (mg/cc)	Delivery	Threshold (V) at 0.5 ms
None	NA	NA	1.5 ± 0.5
None	NA	NA	1.9 ± 0.9
Saline	0.9%	Pump	2.1 ± 0.9
KCl	3.3	Pump	1.4 ± 0.1
Propylene Glycol	—	Pump	1.8 ± 0.5
Dilantin	0.08	Pump	1.5 ± 0.07
Sulfinpyrazone	0.1	Pump	1.2 ± 0.4
Epinephrine	0.1	Pump	1.9 ± 0.7
Azathioprine	1.0	Pump	1.9 ± 0.8
Ibuprofen	0.1	Pump	1.7 ± 0.4
Ibuprofen	100	Pump	1.4 ± 0.4
Ibuprofen	(200)	MCRD	1.3 ± 0.2
Dexamethasone	2	Pump	0.9 ± 0.3
Dexamethasone	(200)	MCRD	1.1 ± 0.2
Prednisolone	1	Pump	1.6 ± 0.5
Heparin	(200)	MCRD	2.7 ± 3.0
Heparin	1000 U/ml	Pump	1.4 ± 0.06
Albumin	(200)	MCRD	1.5 ± 0.6
Dexamethasone phosphate	(100)	MCRD	0.8 ± 0.2
Dexamethasone phosphate	(200)	MCRD	0.8 ± 0.0
Dexamethasone phosphate	(400)	MCRD	1.0 ± 0.6
Cyclosporine		Pump	No improvement

Our earliest experimental canine design used the monolithic "controlled" release device (MCRD) concept containing dexamethasone sodium phosphate powder compounded in the ring–tip's central silicone rubber plug. While performance was improved over nonsteroid-eluting designs, the central silicone plug swelled and extruded from the electrode like a plunger. Since that raised questions about possible untoward (and unknown) complications, the design was abandoned.

As already noted, other researchers had published data suggesting that porous electrodes were superior in performance compared to polished surfaces [25–28]. Thus, we developed a porous titanium electrode with a central internal well containing the MCRD, with space to compensate for swelling (Figures 5.9 and 5.3a). In order to ensure immediate penetration of the pores with blood during implant and to prevent air blockage inhibiting steroid release, dexamethasone sodium phosphate was applied to the surface as a wetting agent. The performance of this design provided significantly lower and consistent thresholds in canines, as shown in Figure 5.6 (curve E). Note that the highest mean threshold + 2 standard deviations is ≤ 1 V, facilitating a safe 2.5:1 safety factor at 2.5 V, 0.5 ms. Pacing impedance and sensing were not adversely affected. Thus, this was the first steroid-eluting pacemaker lead evaluated clinically, then market released in 1983 [30]. The average adult human's thresholds

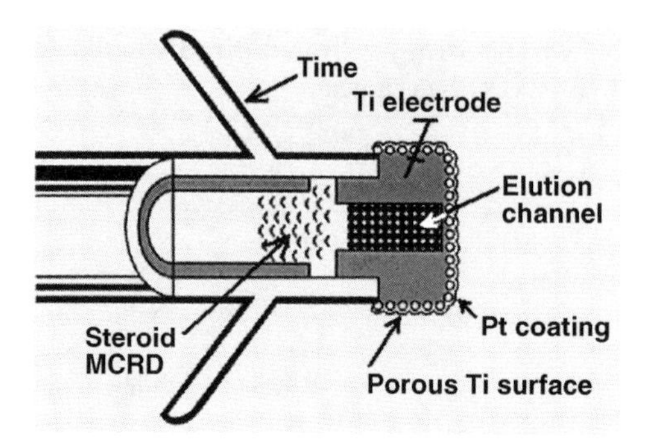

FIGURE 5.9 First-generation steroid-eluting electrode. (Reproduced with permission from Medtonic plc.)

remained low and consistent. Adult and pediatric patients with prior serious (and sometimes life-threatening) exit block maintained low and stable thresholds, effectively resolving this complication [31–33].

Subsequent designs were aimed at retaining the benefits of steroid elution while improving other characteristics, such as sensing and current drain. The second-generation steroid-eluting lead used a smaller fractal platinum electrode. This electrode was a platinum hemisphere, with spherical platinum particles sintered on its surface. This was "platinized" with platinum black.[7] The result was a smaller 5.8 mm^2 electrode as measured by its radius, but with an exceptionally high electrolyte-contacting surface area. This produced even lower stable thresholds, increased impedance (thus reduced current drain), and very high capacitance. The latter resulted in improved sensing and efficiency [34–36]. A third generation used an even smaller 1.5 mm^2 steroid-eluting, totally porous fractal platinum design to increase pacing impedance even more and further improve sensing without sacrificing threshold performance [37–39]. The first three generations of steroid-eluting electrodes are compared in Figure 5.10. Designs from other manufacturers included external MCRD rings at the proximal edge of the electrode.

5.4.2 Active Fixation

Active fixation transvenous steroid-eluting leads typically have evolved into three design categories. The "Bisping" type has a retracted distal helical corkscrew electrode during venous insertion. When the tip is in position, the terminal pin is rotated with a clamp-on tool (Figure 5.11). This causes the conductor coil to rotate within the insulation, extending the helical electrode. The sharpened helix engages and penetrates the myocardium to provide secure fixation. The "Dutcher"-type lead uses a stylet with a flat screwdriver blade at the distal tip. This is inserted, with the blade engaging a slot at the proximal end of the helix, inside the lead. When the stylet

[7]Electrodeposition of small particles of platinum, smaller than the wavelength of visible light. Light is absorbed making the surface appear black. The particles are in turn agglomerates of smaller particles, in the tens of ångström size range.

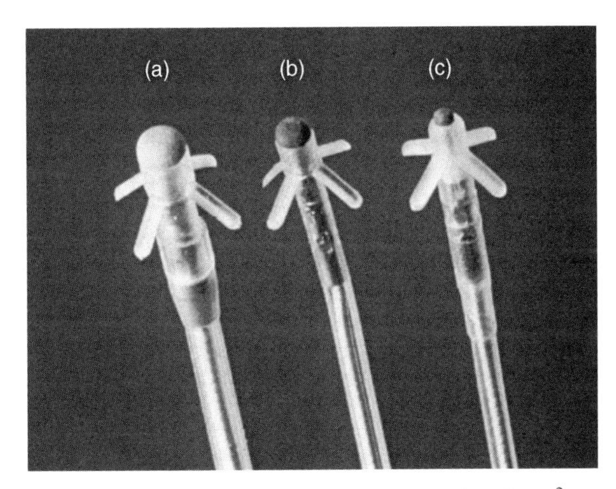

FIGURE 5.10 Three generations of steroid-eluting electrodes: 8 mm^2 porous surface (a), 5.8 mm^2 fractal (b), and 1.5 mm^2 totally porous (c). (Reproduced with permission from Medtonic plc.)

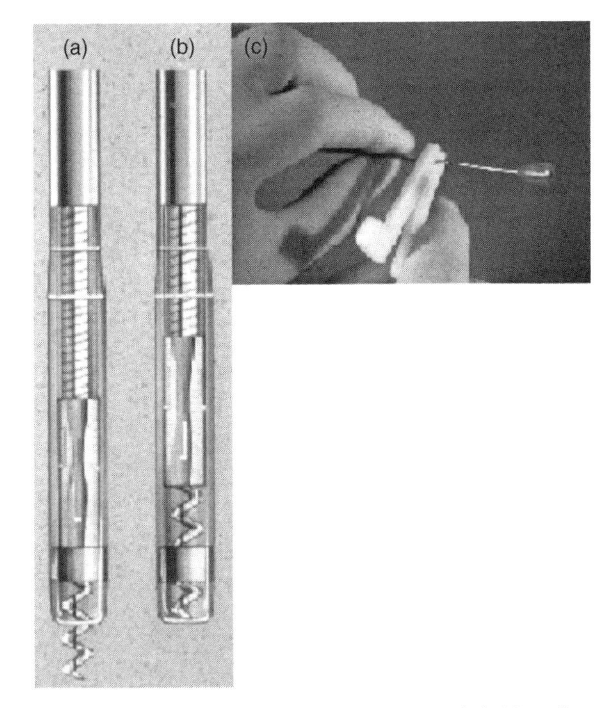

FIGURE 5.11 "Bisping' distal electrode with helix extended (a) and retracted (b). The rotation tool is shown on the terminal pin in part (c). (Reproduced with permission from Medtonic plc.)

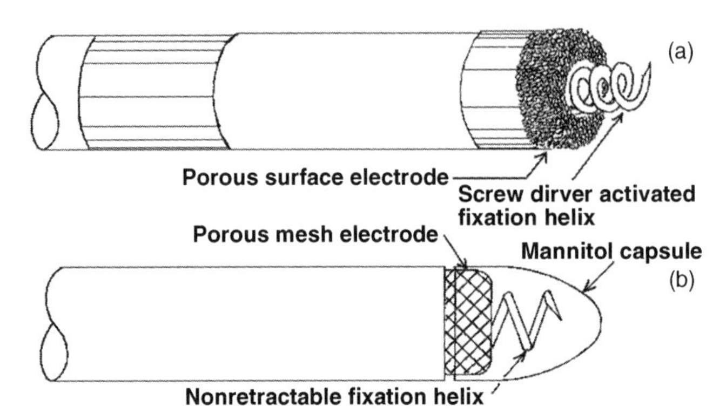

FIGURE 5.12 The "Dutcher" lead using a bladed stylet to extend the fixation helix (a) and a fixed helix protected with a mannitol capsule (b). (Reproduced with permission from Medtonic plc.)

is rotated, the helix is extended into myocardium (Figure 5.12a). The third design has a fixed helix. Some leads with unprotected tips are counter-rotated during venous insertion to avoid prematurely piercing the cardiovasculature, or snagging on structures such as valves. Others use a protective cap over the helix that dissolves within a short time after insertion (Figure 5.12b). Some of these designs use silicone–dexamethasone sodium phosphate or acetate MCRDs either as annular structures within or adjacent to the electrode or as a plug within the helix. The helices may additionally have a steroid coating.

5.5 INFLAMMATION MECHANISMS AND THE EFFECTS OF STEROID ELUTION

Breakthroughs in the pathology of inflammation were just being published in 1980 [40]. Soon after market release of the first-generation steroid-eluting lead, the physiological mechanisms became more evident [41,42]. The electrode encapsulation mechanism and its effect on cardiac thresholds as we know it today is shown schematically in Figure 5.13.

5.5.1 Threshold Changes in the Absence of Steroid Elution

When a transvenous electrode is positioned, the endocardial cell membranes become more porous resulting in edema. The lower resistivity edematous fluid causes a decrease in pacing impedance. Impedance continues to decrease for a minimum 1 week postimplant (regardless of design), and then increases to about the initial value as edema is resorbed (Figure 5.14). While impedance is a measure of the resistivity of tissue, threshold is a measure of cell membrane health. Within the first 15 min postimplant, thresholds actually drop (up to 30% for active fixation, less for passive designs). These lower thresholds remain stable for the first 3–4 days. This typically goes unnoticed in humans because the patients do not usually need to be monitored

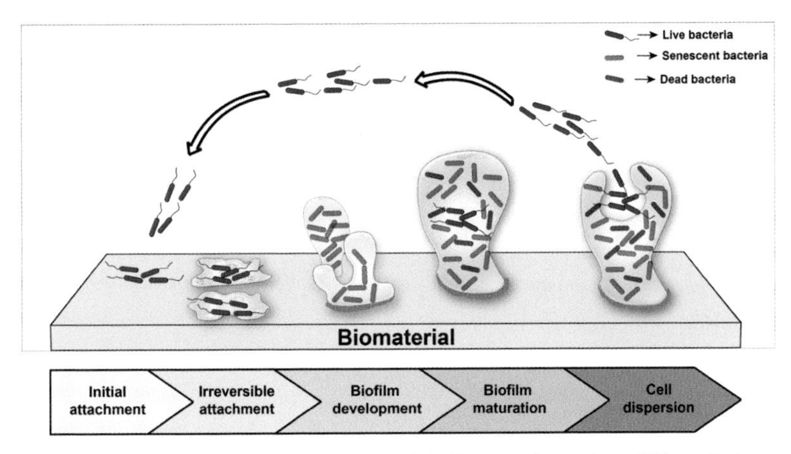

FIGURE 1.4 Bacterial seeding, colonization, biofilm transformation, differentiation, maturation, and further dissemination producing following biomaterial-associated contamination and infection.

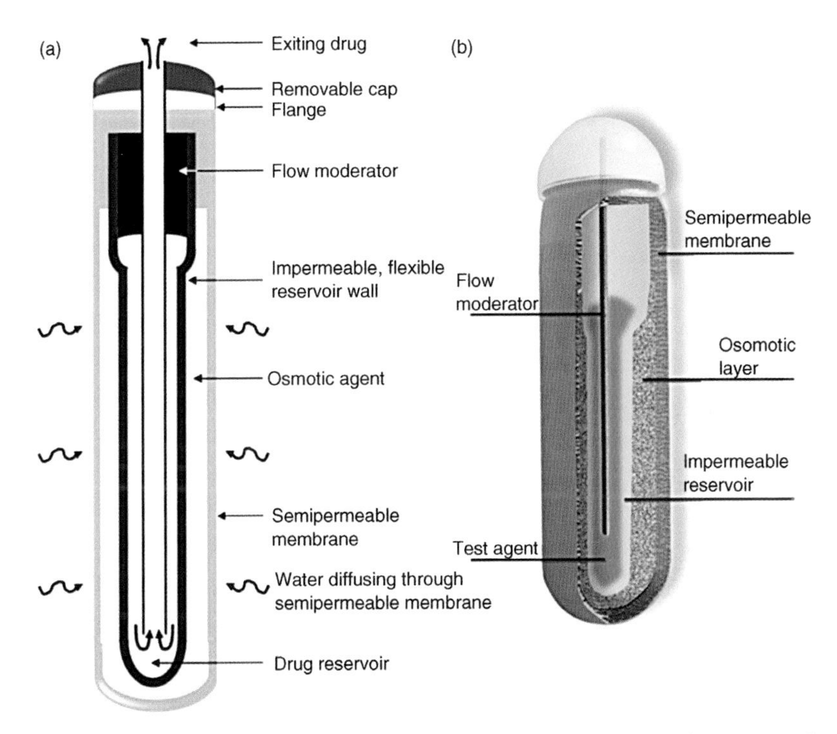

FIGURE 2.4 (a) Cross-sectional schematic of an osmotic pump. Adopted from schematics in an Alzet® catalog. (b) Alzet Osmotic Pump for drug delivery. (Obtained from Alzet catalog.).

Drug–Device Combinations for Chronic Diseases, First Edition. Edited by SuPing Lyu and Ronald A. Siegel.

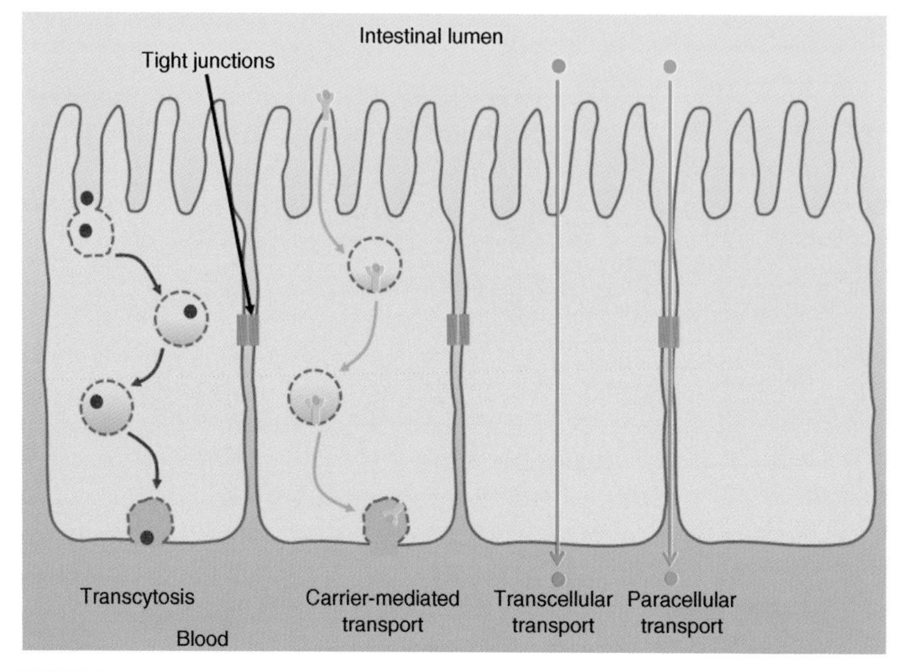

FIGURE 2.7 The four main pathways of drug/particle transport across the intestinal lumen: transcytosis, carrier-mediated transport, transcellular transport, and paracellular transport.

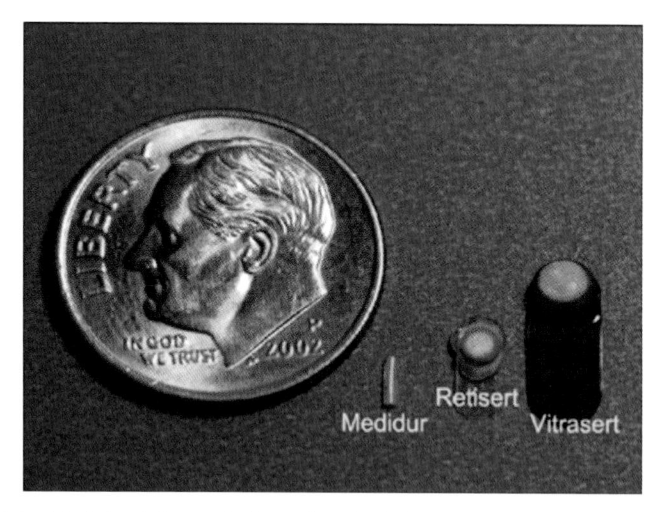

FIGURE 3.1 Medidur delivery platform showing Vitrasert, Retisert, and Medidur. (Reproduced with permission from pSivida, Ltd.)

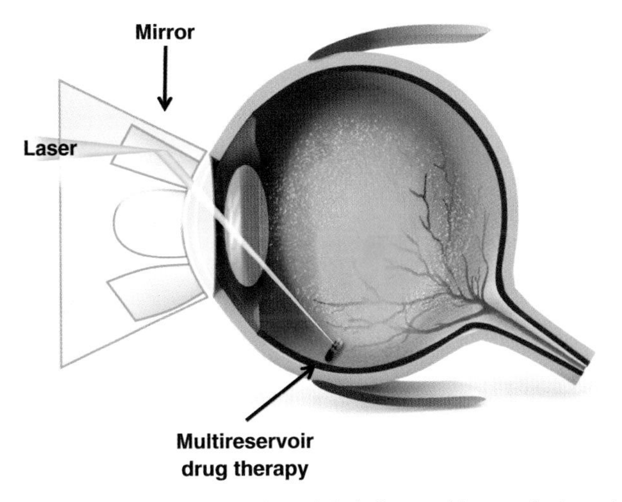

FIGURE 3.9 A laser activated system for ophthalmic a multireservoir, intravitreal implant. (Reproduced with permission from On Demand Therapeutics.)

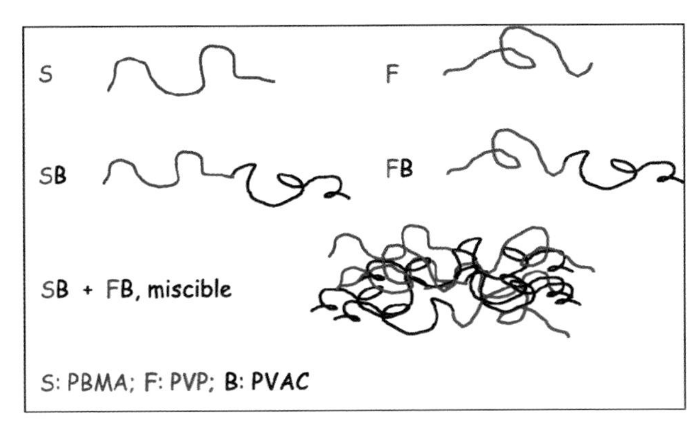

FIGURE 4.3 Miscible polymer blends made of SB and FB copolymers. This schematic is for demonstrating the concept, and not necessarily represents chain sequence or configuration of copolymers. (Reproduced with permission from Medtronic plc.)

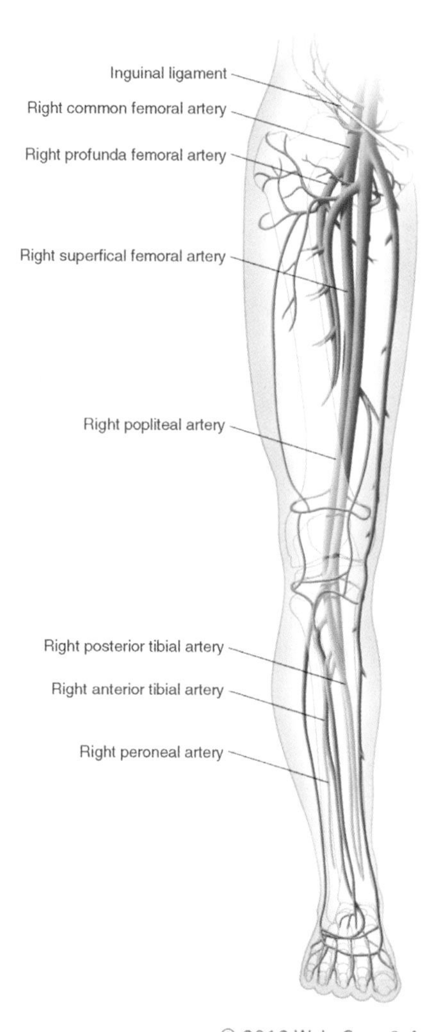

FIGURE 6.2 Lower limb vascular anatomy.

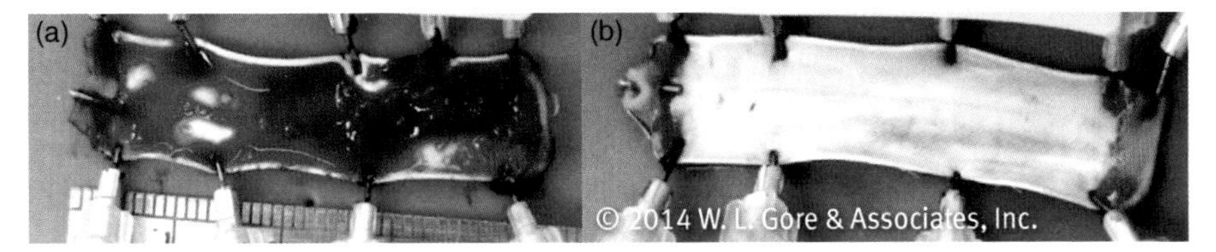

FIGURE 6.5 Representative photograph of CBAS–Heparin ePTFE vascular grafts after removal from the canine carotid interpositional model. The control vascular graft is occluded with thrombus (a), while the CBAS–ePTFE vascular graft is thrombus-free (b).

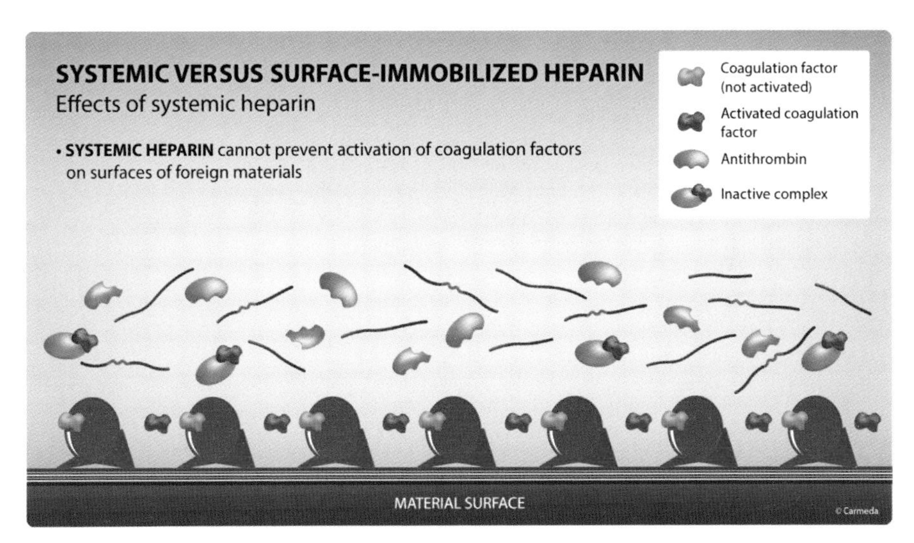

FIGURE 6.6 Systemic heparin attenuates unwanted and uncontrolled coagulation of the circulating blood and is often given during vascular graft implant procedures. During a vascular procedure, blood activation is a result of surgical trauma and is routinely controlled by systemic heparin.

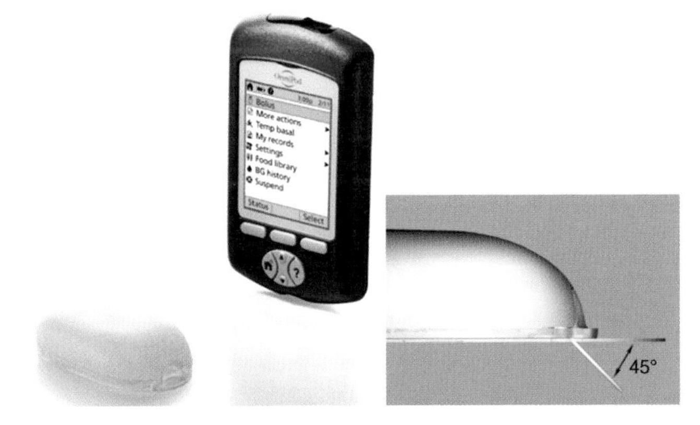

FIGURE 7.3 Omnipod "patch pump" system for SQ insulin infusion, including handheld remote patient programmer. Right panel shows SQ cannula. (Reproduced with permission from Lazar Partners ltd.) (*Source:* www.myomnipod.com.)

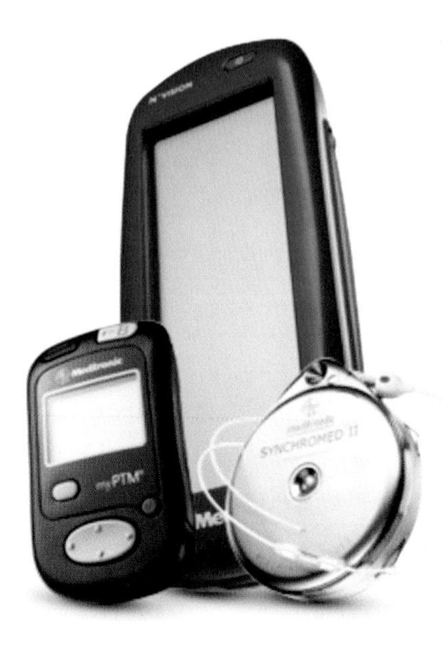

FIGURE 7.4 Medtronic implantable, programmable infusion system showing physician (blue) and patient (black) programmers. (Reproduced with permission from Medtronic, Inc.)

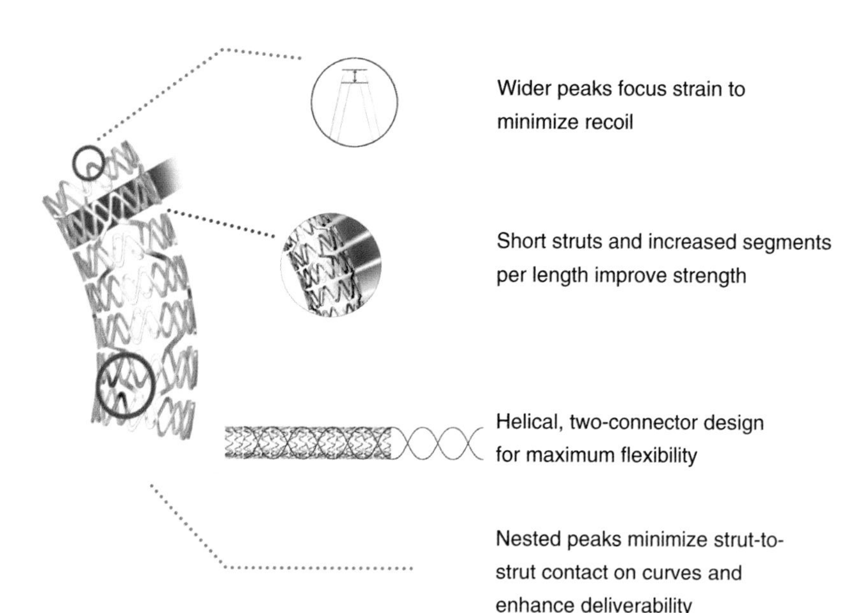

FIGURE 8.2 PROMUS Element workhorse stent. (Reproduced with permission from Boston Scientific Inc.)

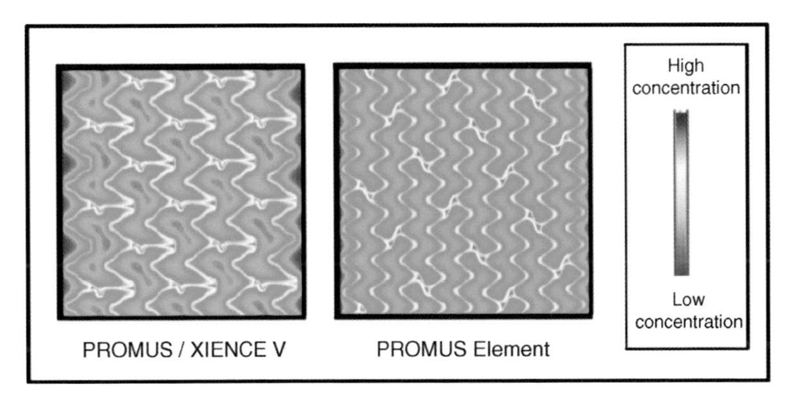

FIGURE 8.3 Computer simulation of drug distribution in vascular tissue. *Note:* Modeling results are for demonstration purposes only and may not necessarily be indicative of clinical performance. (Reproduced with permission from Boston Scientific Inc.)

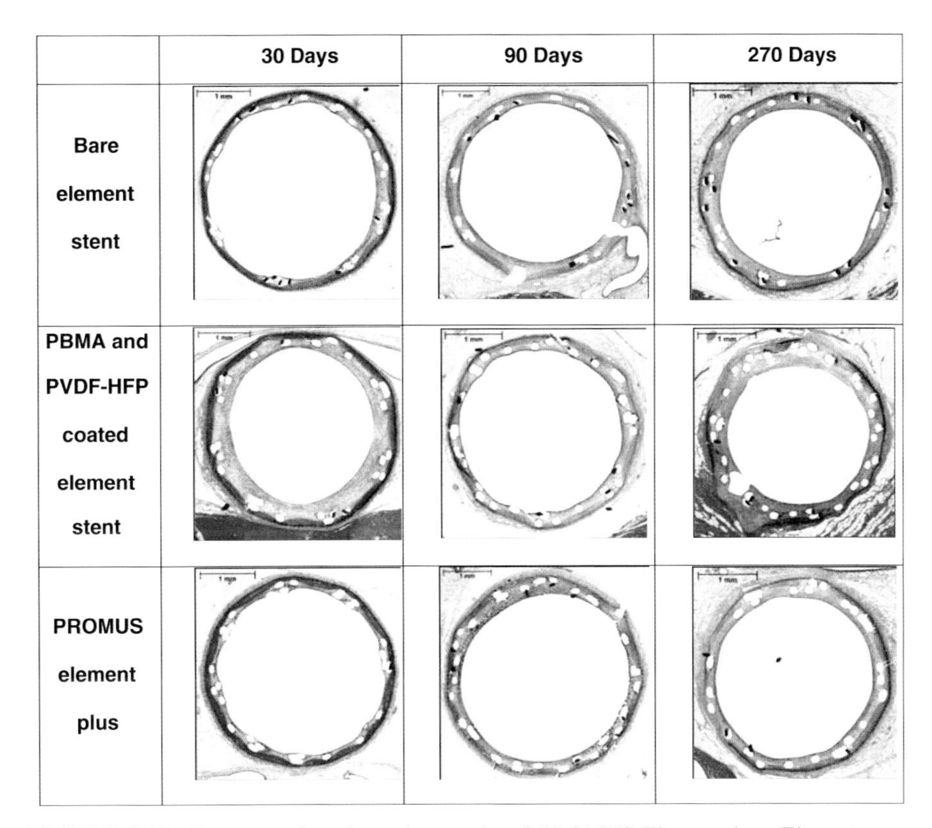

FIGURE 8.23 Representative photomicrographs of PROMUS Element, bare Element stent, and PBMA/PVDF-HFP-coated Element stent elastic–trichrome-stained sections of porcine coronary arteries implanted in an overlap configuration for 30, 90, and 270 days. (Reproduced with permission from Boston Scientific Inc.)

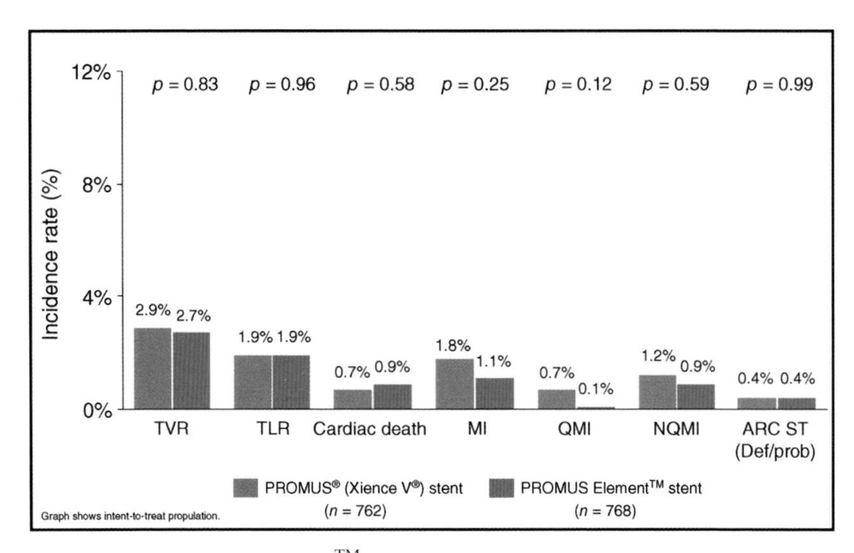

FIGURE 8.24 PROMUS Element™ Stent PLATINUM clinical trial 12 months data. TVR: target vessel revascularization; TLR: target lesion revascularization; MI: myocardial infarction; QMI: q-wave-elevated myocardial infarction; NQMI: non-q-wave-elevated myocardial infarction; ARC ST: definite and probable stent thrombosis as defined by the Academic Research Consortium. Presented by Gregg W. Stone, MD, ACC 2011. Xience V is a trademark of Abbott Laboratories group of companies. (Reproduced with permission from Boston Scientific Inc.)

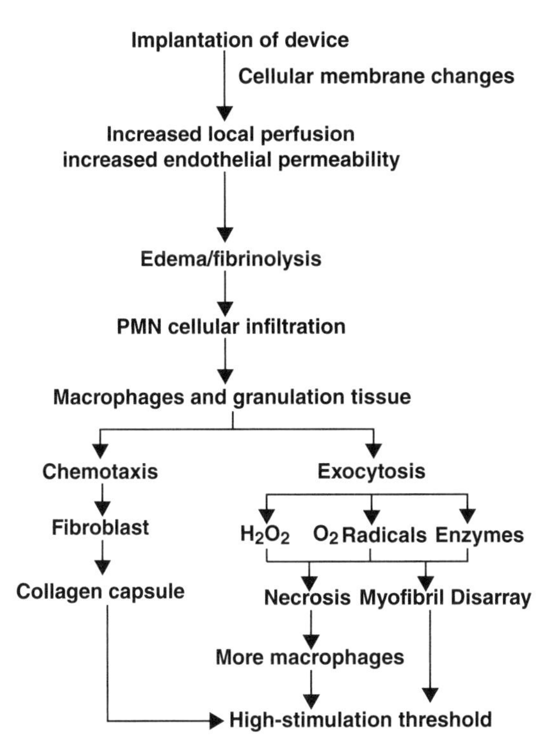

FIGURE 5.13 Inflammatory response to an implanted pacemaker electrode over time. (Reproduced with permission from Medtonic plc.)

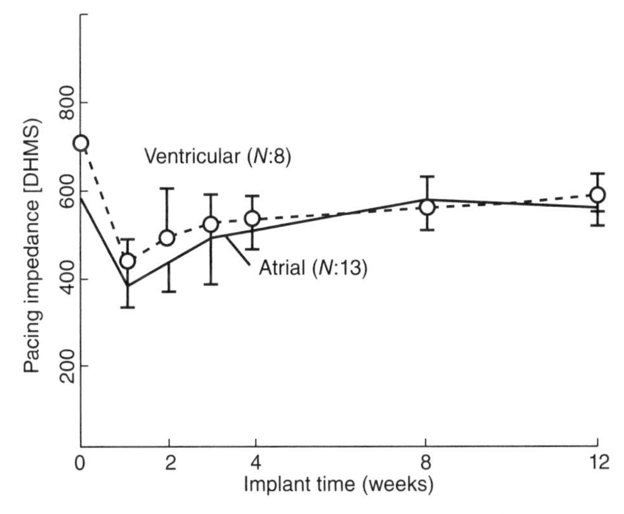

FIGURE 5.14 Pacing impedance in canines as a function of time. (Reproduced with permission from Medtonic plc.)

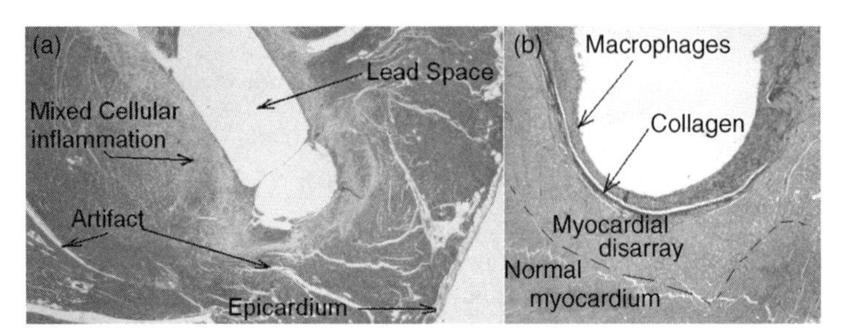

FIGURE 5.15 Inflamed tissue surrounding a transvenous lead 1 week postimplant at peak threshold (a) and a chronic steroid-eluting version with low, stable thresholds (b). (Reproduced with permission from Medtonic plc.)

that often just postimplant. Electrode–tissue interface sections taken from animals 3 days postimplant show no untoward pathology. That is, there are no inflammatory cells present at all! About 4 days postimplant, thresholds begin to rise, reaching a peak value, typically 1–3 weeks postimplant (Figure 5.4). This is caused by an intense mixed inflammatory response in the tissues surrounding the electrode, including monocytes, neutrophils, basophils, eosinophils, lymphocytes, and so on (Figure 5.15a). Because these cells are cytotoxic due to the release of inflammatory mediators, injury to myocyte membranes can occur. After the peak period, thresholds may remain stable, or decline to a stable intermediate value. During this post peak period, the mixed cellular response, with the exception of the monocytes, resolves and the electrode is encapsulated with a layer of collagen (Figure 5.15b). But there is much more to this story. The monocytes have migrated to the electrode's surface (Figure 5.16). They adhere, activate, and differentiate into macrophages. The latter spread out over the surface, fuse their membranes, and form foreign body giant cells (FBGC). At the same time, lysosomal release of mediators, including H_2O_2, super oxide anion, and hydroxyl radical, occurs. This "oxidative burst" declines substantially within about 1 week, but does not stop, continuing at a "trickle" for years. These mediators are highly cytotoxic causing the necrosis of adjacent myocytes. More monocytes settle over the FBGC and become activated, in part to phagocytose necrotic myocardium. Oxidative mediators diffuse into the surrounding tissue damaging membranes of cells farther "out" without necrosis. The collagenous beams and struts holding the myocardium in an orderly array are disrupted [43]. Thus, a zone of myocardial disarray is found between the collagenous capsule and an undamaged, healthy myocardium (Figure 5.15b). Meanwhile, as the macrophages settle on the electrode surface, they signal and activate fibroblasts to lay down a collagenous capsule separating the surface cells from viable tissue. However, viable does not necessarily mean healthy. It is believed that the thresholds of the disarrayed cell membranes are negatively affected. Supporting evidence for this is the approximately circular "holes" sometimes found in the disarrayed myocardium. These are "ghosts," fatty deposits remaining after phagocytosis of myocyte bundles by FBGCs during the acute inflammatory assault. It is believed that the thresholds of the disarrayed and damaged but viable cells are highest adjacent to the collagenous capsule, decreasing until they reach "normal" values where naturally ordered myocardium exists. Thus, it is

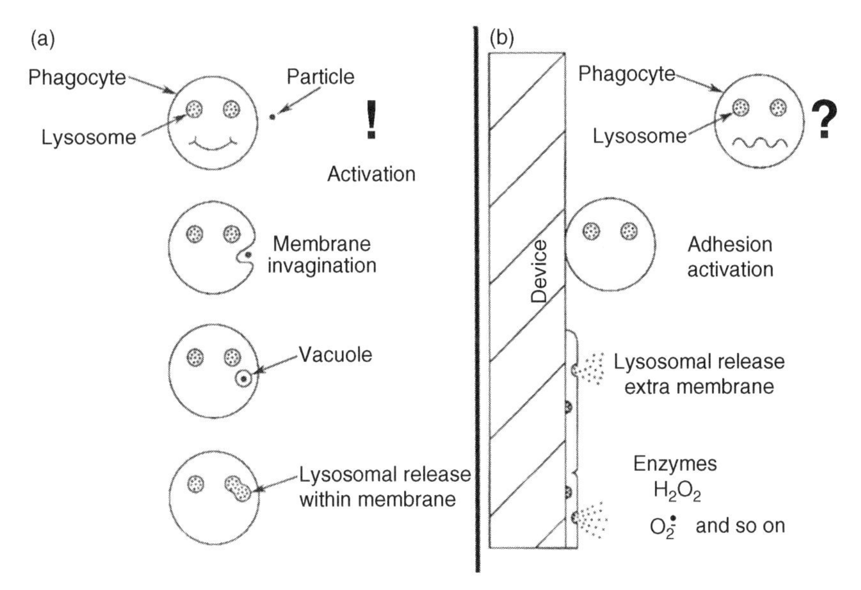

FIGURE 5.16 Part (a) shows schematically how phagocytosis works by ingesting a particle for destruction in a lysosome. Part (b) shows how a monocyte migrates to a surface, attaches, activates, and spreads forming a foreign body giant cell with lysosomal release onto the device and into the surrounding tissue. (Reproduced with permission from Medtonic plc.)

apparent that the known inflammatory processes over time correlate very well with electrical monitoring.

5.5.2 Threshold Changes in the Presence of Steroid Elution

Based on the already discussed mechanism, the secret to preventing cellular damage at the electrode–tissue interface is to control lysosomal release or leakage, both during the "peaking" phase and chronically. Fortunately, monocyte/macrophages have steroid receptors on their membranes (Figure 5.17) [44]. When these receptors are satisfied with glucocorticosteroid, the cell's membrane is stabilized, it cannot spread, and lysosomal release is mitigated. Figure 5.17b shows an electron micrograph of a monocyte in the absence of steroid, spread on a surface and activated. Figure 5.17c shows a similar macrophage on a surface in the presence of glucocorticosteroid. In the latter case, the macrophage retains its spherical shape due to membrane stabilization. Thus, steroid elution can prevent or minimize the myocellular damage and disarray that causes threshold increase.

5.5.3 Chronic Versus Acute Steroid Elution

There is occasionally some controversy as to whether or not chronic steroid elution is necessary. After all, just dipping the electrode in steroid solution is easier than manufacturing a device designed for chronic drug delivery. Often this controversy is

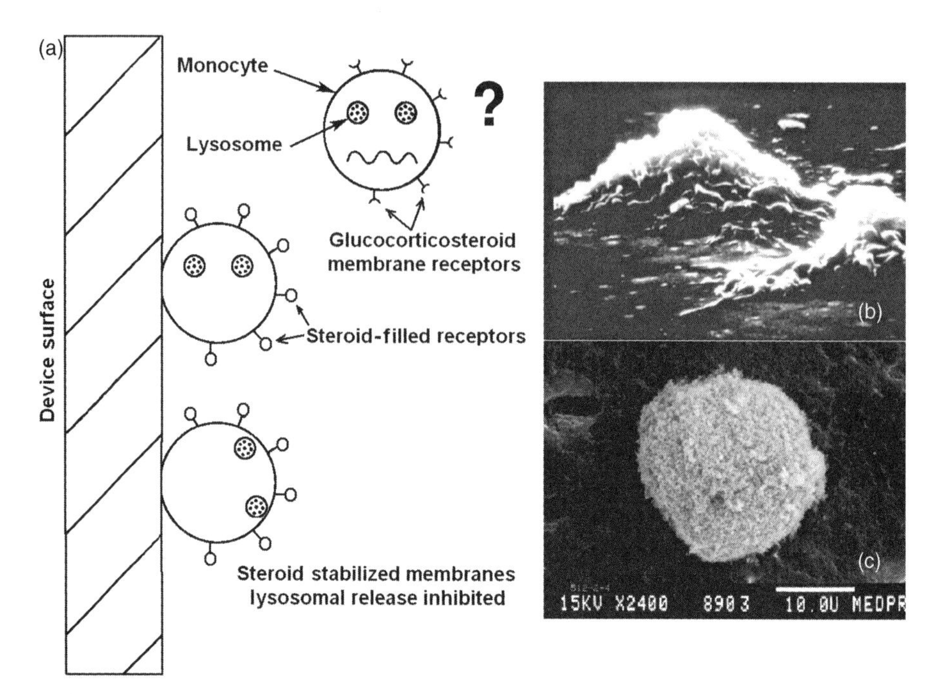

FIGURE 5.17 The process in the Figure 5.16b cartoon frustrated by steroid stabilized monocyte membranes (a). An activated macrophage on a device surface in the absence of steroid is shown in part (b). In the presence of steroid elution (c), the stabilized membrane cannot spread and the cell remains spherical. (Reproduced with permission from Medtonic plc.)

centered around the idea that acute drug delivery is all that is needed to suppress the acute threshold peak (Figure 5.4). The peaking phenomenon, however, has never been a significant complication. Prior to modern threshold tracking pulse generators, physicians would implant at factory-preset values, then adjust outputs accordingly after 6 weeks when thresholds declined. The loss of battery longevity in this case is negligible. With modern capture detection devices, the pacemaker automatically adjusts itself to provide the appropriate safety factors for the thresholds available. Acute exit block caused by dislodgment/microdislodgment and perforation/penetration was resolved by providing secure electrode fixation with appropriate lead body stiffness. Other complications, such as chronic exit block caused by exuberant inflammation (as in growing children), will not be mitigated after the steroid runs out. In addition, there are other chronic factors that can cause untoward threshold rise. Cardiac pacing patients typically have co morbidities requiring the use of certain drugs, especially certain antiarrhythmic agents such as propafenone or amiodarone [45]. Human clinical studies have demonstrated that thresholds can rise unacceptably for patients undergoing propafenone therapy paced with steroid-free screw-in leads [46]. Steroid-free electrodes with passive fixation (tines) produce less spectacular threshold increases, but high enough for concern. Passively fixed steroid-eluting electrodes show a relatively slight threshold increase, which does not typically require reprogramming. Mineralocorticosteroids are also known to increase thresholds. For example, a man with testicular cancer was receiving estrogen therapy. Titration of his estrogen levels reduced his thresholds sufficiently so that replacement of his lead with a steroid-eluting model was not required (K. Stokes, Personal communication). A pacemaker patient also on lithium therapy had her recurrent exit block resolved through implant of a steroid-eluting lead (K. Stokes, Personal communication). Electrolyte imbalances can cause threshold rise, and again steroid-eluting electrodes can mitigate the effect [47]. Children experiencing sudden chronic threshold rise due to illness such as summer colds would not be protected with only acute, short-term steroid delivery [8]. Thus, it is clear that chronic steroid elution is needed, does increase patient safety, and is highly reliable.

5.6 THE MECHANISM OF CHRONIC STEROID ELUTION: DOES THE STEROID RUN OUT?

5.6.1 Elution Rates

The question sometimes arises: What happens when the steroid runs out? This question can arise from *in vitro* tests that show effectively complete elution within a few weeks [48]. *In vitro* elution tests are misleading at best. Elution in a beaker occurs in a well-stirred medium with no barriers. The story *in vivo* is much different. The electrode is positioned either deep within trabeculi or in an endocardial pit (Figure 5.14). In a very short time, the electrode and its elution mechanism is encapsulated and separated from viable myocardium by macrophages and a collagen capsule. That these are substantial barriers to elution is demonstrated by a study of explanted leads with 8 and 5.8 mm^2 electrodes containing internal MCRDs (Figure 5.9). The leads were explanted from animals and humans over many years. The electrodes were cut open, the MCRDs removed, and submitted for steroid assay. The *in vivo* elution rate for this design is shown in Figure 5.18. The oldest implant time was 7.7 years. The extrapolated steroid retention at 20 years is about 15%. The elution rate declines at a log–log rate. Thus, steroid will not run out within the average patient's lifetime!

5.6.2 The Elution Mechanism

An interesting finding is that electrodes opened within a few hours of explant had wet MCRDs, but the compounded drug powder appeared to be "dry." This indicates that the steroid *in vivo* was a saturated solution, clearly a concentration impossible to deliver systemically. Thus, we believe that as the lead is inserted through the vasculature, the steroid coating in the pores allows blood to enter quickly. By the time the electrode reaches its final position, steroids within the pores and at the tissue facing surface of the MCRD are already dissolved and migrating toward the endocardium. The MCRD is a closed pore system. That is, every particle of dry steroid is separated from others by a thin silicone rubber membrane. Thus, serum must

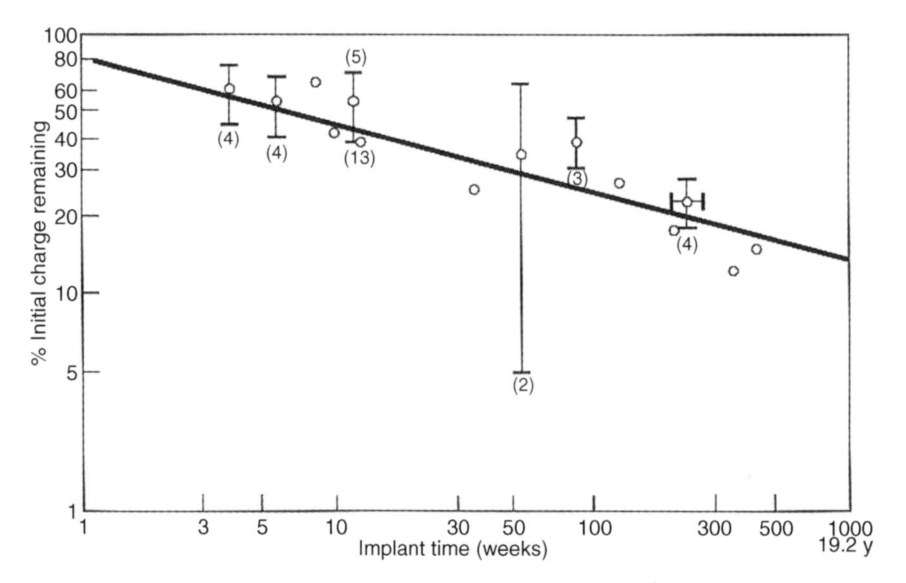

FIGURE 5.18 *In vivo* elution rate of steroid eluting 8 and 5.8 mm^2 porous electrodes. Note the log rhythmic axes. The amount of steroid remaining at 20 years is extrapolated to be about 15% of the initial charge. (Reproduced with permission from Medtonic plc.)

permeate the membrane, and dissolve the drug. Then, the dissolved drug must permeate the membrane in response to its concentration gradient to reach the endocardium. The next layer of drug must follow the same procedure, but now has to permeate two silicone rubber membranes, and so on. Initially, the concentration gradient is high, so elution is relatively rapid. As encapsulation proceeds and the drug solution becomes saturated, the elution rate must effectively stop. However, it is believed that macrophages take up steroid quickly and then release it slowly. Thus, this appears chronically to be a "demand" system where as one molecule of drug is taken up by a cell, another dissolves in the MCRD to begin its elution. The result is a log rhythmically decreasing elution rate as barriers (encapsulation) form, and macrophages on the device surface adsorb and release steroid molecules.

Here are more explanation with Higuchi' theory for your reference. Higuchi theory has been well accepted in correlating drug release kinetics with drug concentration and other parameters.

In a drug-loaded matrix system (e.g., polymer depot such as MCRD), let the initial drug loading (nominal concentration in the system) be C_0 and the solubility of drug in the matrix material be C_s. If $C_0 \gg C_s$, the drug stays in the matrix as dispersed particles and it elutes from the matrix into the eluting medium in two steps: first, the drug dissolves from the particles into the matrix at molecular level and second the dissolved drug molecules elute into the medium. There is a "skin" in the surface of the device. Inside the skin, the drug particles disappear because the drug in the particles dissolves in the matrix and elutes into the medium. There is drug concentration gradient in the skin that drives the drug to diffuse into the medium. The drug in the bulk needs to pass through this skin to reach the medium. When this skin is very thin compared the device size, the eluting can be approximated as a one-dimensional process and can be described with Higuchi's equation:

$$M(t) = A\sqrt{DC_s(2C_0 - C_s)t} \tag{5.4}$$

where $M(t)$ is the accumulative drug release, A is the surface area of the device where drug elutes, D is the diffusion coefficient of the drug in the matrix material, and t is the time. Total amount of drug release increases with time, but as a function of its square root. Then the elution rate decreases with time. Dexamethasone sodium phosphate is a salt and its solubility in the matrix (silicone) is low and diffusion coefficient is low as well. Therefore, the elution rate is low. However, the most significant factor is the drug concentration in the medium. In the original Higuchi equation, the medium is considered as a drug sink and the concentration is zero. In the implanted lead–tissue interface, the drug concentration near the medium (the MCRD–tissue interface) is controlled by the rate at which the drug diffuses away from the interface into the bulk tissue and rate at which the inflammatory cells consume the drug. Neither of these two rates is infinitely fast, actually both are slow. Therefore, the sink condition is not valid. Therefore, the drug concentration in the medium is not zero, but a finite value C_m. For this case, an approximation of the drug eluting can be

$$M(t) = A\sqrt{D(C_s - C_m)(2C_0 - C_s)t} \tag{5.5}$$

The rate is related to $(C_s - C_m)$ instead of C_s as in the original Higuchi equation. The higher the drug concentration at the MCRD–tissue interface, the slower the drug releases. It can reach zero if the concentration at the interface reaches the solubility limit. This may qualitatively explain why the drug in the MCRD lasts for years.

We believe that other closed-pore MCRD systems, such as external collars adjacent to the electrode, will perform similarly, because while the elution path is shorter, the barriers are similar. Open pore systems, however, will elute much more rapidly because the MCRD membrane barriers have been removed. Surface coatings cannot be expected to last more than a few weeks.

5.7 OTHER POSSIBLE DRUG ELUTION SYSTEMS

So far we have discussed only cardiac pacing and drug release for the suppression of localized inflammatory processes. Other uses in electrodes are also possible. For example, glucocorticosteroid elution from neuron electrodes has shown some limited promise [49,50]. Implants on rabbit's peripheral nerves were difficult to assess because of our inability to measure an improved response. Histopathology, however, demonstrated very significant improvements in nerve bundle preservation in the presence of steroid elution. Other possibilities might include the suppression of inflammation and encapsulation at implanted sensing electrodes such as those intended to measure blood glucose, pressure, pH, and a host of specialized molecules.

The delivery of genetic materials under certain conditions might allow one to correct cardiac defects, for example, by forming new nodal tissue, or repairing conduction system defects [51]. Since electrical monitoring of impedance relates to the presence of edema, is this a means to manage implantable drug delivery in heart failure patients? So far, it appears that the delivery of drugs from implanted electrodes for localized therapy or sensing has not been as thoroughly explored as it could be.

5.8 SUMMARY

The development of the glucocorticosteroid eluting pacemaker electrodes has been an important part of the drive to satisfy unmet medical needs. This, in combination with electrode fixation and optimization of size and surface structure, more biocompatible mechanical design is partially responsible for the evolution of modern, sophisticated, relatively small, multifunction, pulse generators. It has been very important in improving patient safety. Steroid elution in pacemakers serves to suppress the lysosomal release of inflammatory mediators from monocyte/macrophages at the device–tissue interface. This mitigates myocardial damage and threshold rise that would otherwise occur, sometimes to excessive degrees. It also helps to protect pacemaker patients from the threshold increasing effects of certain disease states and drug regimens. In the preferred electrode designs, elution continued over a period of many years, providing long-term patient safety.

REFERENCES

1. K.B. Stokes and N.L. Stephenson, The implantable cardiac pacing lead: just a simple wire? in S.S. Barold and J. Mugica (Eds.), *The Third Decade of Cardiac Pacing*, Futura Publishing Company Inc., Mount Kisco, NY, 1982, pp. 365–416.
2. H.G. Mond, Pacing leads for the new millennium. *Pacing Clin. Electrophysiol.* 24 (Part II) (2001) S92.
3. P. Doenecke, R. Flothner, G. Rettig, et al., Studies of short- and long-term threshold changes, *Engineering in Medicine 1: Advances in Pacemaker Technology*, Springer, New York, 1975, pp. 283–296.
4. K.B. Stokes, R. Taepke, and J. Gates, All impedances are not created equal: expediency vs efficiency. *Medtronic Sci. Technol.* 3 (1) (1994) 2–12.
5. K. Stokes, Estimating pulse generator longevity. *Medtronic Sci. Technol.* 3 (2) (1995) 26–28.
6. J.D. Morris, R.D. Judge, B.J. Leininger, and F.K. Vontz, Clinical experience and problems encountered with an implantable pacemaker. *J. Thorasc. Cardiovasc. Surg.* 50 (1965) 849.
7. F.M. Mowry, R.D. Judge, T.A. Preston, and J.D. Morris, Identification and management of exit block in patients with implanted pacemakers. *Circulation* 32 (II) (1965) 159.
8. R.B. Shepard, J. Kim, E.C. Colvin, et al., Pacing threshold spikes months and years after implant. *Pacing Clin. Electrophysiol.* 14 (Part II) (1991) 1835–1841.
9. W.G. Williams, P.S. Hesslein, and R. Kormos, Exit block in children with pacemakers. *Clin. Prog. Electrophysiol. Pacing* 4 (5) (1986) 478–489.

10. S.S. Barold, L.S. Ong, and R.A. Heinle, Stimulation and sensing thresholds for cardiac pacing: electrophysicologic and technical aspects. *Prog. Cardiovasc. Dis.* 24 (1981) 1.
11. S. Furman, B. Parker, and D. Escher, Decreasing electrode size and increasing efficiency of cardiac stimulation. *J. Surg. Res.* 11 (1971) 105.
12. N.P.D. Smyth, P.P. Tarjan, E. Chernoff, et al., The significance of electrode surface area and stimulating thresholds in permanent cardiac pacing. *J. Thorac. Cardiovasc. Surg.* 71 (1976) 559.
13. W. Irnich, Considerations in electrode design for permanent pacing, in H.J. Thalen (Ed.), *Cardiac Pacing; Proceedings of the Fourth International Symposium on Cardiac Pacing*, Van Gorcum & Co., Assen, The Netherlands, 1973, pp. 268–274.
14. W. Irnich, Engineering concepts of pacemaker electrodes, in M. Schaldach and S. Furman (Eds.), *Advances in Pacemaker Technology*, Springer, New York, 1975, p. 241.
15. H.C. Hughes, Jr., R.R. Brownlee, and G.F Tyers, Failure of demand pacing with small surface area electrodes. *Circulation* 54 (1) (1076) 128–132.
16. S.S. Barold, L.S. Ong, and R.A. Heinle, Matching characteristics of pulse generator and electrode. A clinician's concept of input and source impedance and their effect on demand function, in C. Meere (Ed.), Proceedings of the VIIth World Symposium on Cardiac Pacing, Montreal, PACESYMP, 1979. p. 34-3.
17. D.C. Amundson, W. McArthur, and M. Mosharrafa, The porous endocardial electrode. *Pacing Clin. Electrophysiol.* 2 (1) (1979) 40–50.
18. M. Hirshorn, L. Holley, J. Hales, et al., *In vivo* electrophysiological evaluation of porous titanium alloy cardiac pacemaker electrode tips. *Pacing Clin. Electrophysiol.* 3 (1980) 374.
19. D. MacGregor, G Wilson, P. Klement, and R.M. Pilliar, Improved electrophysiological performance using porous surfaced electrodes for atrial epicardial pacing. *Pacing Clin. Electrophysiol.* 6 (3 Part II) (1983) A-62.
20. C.L. Byrd, S.J. Schwartz, E. Wettenstein, et al., Two year analysis of polyurethane leads with porous-surfaced electrodes in a stable population. *Pacing Clin. Electrophysiol.* 8 (1985) 296.
21. L. Gould, C. Patel, and W. Becker, Long-term threshold stability with porous tip electrodes. *Pacing Clin. Electrophysiol.* 9 (6 Part II) (1986) 1202–1205.
22. N.D. Berman, S.E. Dickson, and I.H. Lipton, Acute and chronic clinical performance comparison of a porous and a solid electrode design. *Pacing Clin. Electrophysiol.* 5 (1982) 67–71.
23. F. Heinemann, R. Schallhorn, and J. Helland, Clinical comparison of available "low threshold" leads. *Eur. Heart J.* 9 (Suppl. 1) (1988) 268.
24. G.J. Richter, E. Weeidlich, F.V. Sturm, et al., Non-polarizable vitreous carbon pacing electrodes in animal experiments, in C. Meere (Ed.), Proceedings of the VIth World Symposium on Cardiac Pacing, PACESYMP, Montreal, 1979, Chapter 29-13.
25. G.A. Bornzin, K.B. Stokes, and W.A. Wiebusch, A low-threshold, low-polarization platinized endocardial electrode. *Pacing Clin. Electrophysiol.* 6 (A70) (1983) Abstract 260.
26. M. Djordjevic, P. Stojanov, D. Velimirovic, and D. Kocovic, Target lead-low threshold electrode. *Pacing Clin. Electrophysiol.* 9 (6 Part II) (1986) 1206–1210.
27. M. Schaldach, M. Hubmann, R. Hardt, and A. Weikl, Pacemaker electrodes made of titanium nitride. *Biomed. Technik* 34 (1989) 185–190.

28. E.T. den Braber, J.E. de Ruijter, L.A. Ginsel, et al., Orientation of ECM protein deposition, fibroblast cytoskeleton, and attachment complex components on silicone microgrooved surfaces. *J. Biomed. Mater. Res.* 40 (1998) 291–300.
29. J. Meyle, K. Gultig, H. Wolburg, and A.F. von Recum, Fibroblast anchorage to microtextured surfaces. *J. Biomed. Mater. Res.* 27 (12) (1993) 1553–1557.
30. K.B. Stokes, G.A. Bornzin, and W.A. Wiebusch, A steroid-eluting, low-threshold, low-polarizing electrode, in K. Steinbach (Ed.), *Cardiac Pacing*, Steinkopff Verlag, Darmstadt, 1983, pp. 369–376.
31. D.H. King, P.C. Gillette, C. Shannon, et al., Steroid-eluting endocardial pacing lead for treatment of exit block. *Am. Heart J.* 106 (6) (1983) 1438–1440.
32. G.C Timmis, V. Parsonnet, D.C. Westveer, et al., Late effects of a steroid-eluting porous titanium pacemaker lead electrode in man. *Pacing Clin. Electrophysiol.* 7 (Part 1) (1984) 479.
33. K. Stokes and T. Church, The elimination of exit block as a pacing complication using a transvenous steroid eluting lead, Pacing Clin. Electrophysiol., 10 (3 Part II) (1987) 748 (Abstract 475).
34. P.I. Hoff, K. Breivik, A. Tronstad, et al., A new steroid-eluting electrode for low threshold pacing, in F.P. Gomes (Ed.), *Cardiac Pacing, Electrophysiology, Tachyarrhythmias*, Futura Publishing Co., Mount Kisco NY, 1985, p. 1014.
35. R. Schallhorn and K. Oleson, Multi-center clinical experience with an improved steroid-eluting pacemaker lead. *Pacing Clin. Electrophysiol.* 11 (1988) 11 (Abstract).
36. H. Mond, P. Hunt, and D. Hunt, A second generation steroid eluting electrode. *RBM* 12 (1990) 1 (Abstract).
37. K. Stokes and T. Bird, A new efficient NanoTip lead. *Pacing Clin. Electrophysiol.* 13 (12 Part II) (1990) 1901–1919.
38. R. Fogel, F. Pirzada, J. Boone, John, et al., Initial clinical experience with high impedance, steroid eluting pacing electrodes. *Pacing Clin. Electrophysiol.* 15 (4) (1992) 571.
39. J. Champagne, F. Philippon, and G. O'Hara, Long term performance of high impedance pacing leads. *Can. J. Cardiol.* 12 (Suppl. E) (1886) 131E.
40. P.M. Henson, Mechanisms of exocytosis in phagocytic inflammatory cells. *Am. J. Pathol.* 101 (1980) 494–514.
41. J.M. Anderson, Inflammatory response to implants. *ASAIO Trans.* 11 (2) (1900) 101–107.
42. J.M. Anderson, Inflammation, wound healing and foreign body response, in *Biomaterial Science: An Introductory Text*, Society for Biomaterials, San Diego, CA, 1990.
43. T.F. Robinson, L. Cohen-Gould, and S.M. Factor, Skeletal framework of mammalian heart muscle: arrangement of inter- and pericellular connective tissue structures. *Lab. Invest.* 29 (4) (1983) 482–498.
44. Y. Sibille and H.Y. Reynolds, Macrophages and polymorphonuclear neutrophils in lung defense and injury. *Am. Rev. Respir. Dis.* 141 (1990) 471–502.
45. S.S. Barold, R. McVenes, and K. Stokes, Effects of drugs on pacing threshold in man and canines: old and new facts, in S.S. Barold and J. Mugica (Eds.), *New Perspectives in Cardiac Pacing*, Vol. 3, Futura Publishing, Mt. Kisco, NY, 1993, pp. 57–83.
46. D. Cornacchia, M. Fabbri, A. Maresta, et al., Effect of steroid eluting versus conventional electrodes on propafenone induced rise in chronic ventricular pacing threshold. *Pacing Clin. Electrophysiol.* 16 (12) (1993) 2279–2284.

47. R. McVenes, N. Hansen, S.P. Lathinen, and K. Stokes, The salty dog; serum sodium and potassium effects on modern pacing electrodes. *Pacing Clin. Electrophysiol.* 30 (1) (2007) 4–11.
48. G.C. Timmis, S. Gordon, D.C. Westveer, et al., A new steroid-eluting low threshold pacemaker lead, in K. Steinbach (Ed.), *Cardiac Pacing* Steinkopff Verlag, Darmstadt Germany, 1983, pp. 361–367.
49. P. Lee, K. Stokes, J. Gates, and G. Johnson, Steroid eluting cuff electrode for peripheral nerve stimulation (to Medtronic, Inc). U.S. Patent 5,092,332, March 3, 1992.
50. P. Lee, K. Stokes, and M. Colson, Steroid eluting electrode for peripheral nerve stimulation (to Medtronic, Inc). U.S. Patent 5,265,608, November 30, 1993.
51. J. Morissette and K. Stokes, System and method for enhancing cardiac sensing by cardiac pacemakers through genetic treatment (to Medtronic, Inc). U.S. Patent 7,337,011, February 26, 2008.

6

THROMBORESISTANT VASCULAR GRAFT

Paul C. Begovac,[1] Daniel Pond,[1] Jennifer Recknor,[1] and Elisabeth Scholander[2]

[1]*W. L. Gore & Associates, Inc., Flagstaff, AZ 86003-2400, USA*
[2]*Carmeda AB, Upplands Väsby, Sweden*

6.1 INTRODUCTION

Medical devices continue to undergo innovation by way of moving away from chemically inert systems to devices that are drug–device combination products, which may be suitable for more demanding clinical applications. The potential for enhancing existing device performance has now been demonstrated with devices such as drug-eluting stents and vascular grafts. While some products have shown proven patient benefit, the challenges of developing and commercializing drug–device combination products should not be underestimated.

This chapter seeks to provide an understanding of the development process leading to advancement of commercially available polymeric expanded polytetrafluorethylene (ePTFE) vascular grafts by the endpoint immobilization of heparin using the CARMEDA® BioActive Surface (CBAS® Surface) technology. The purpose of this endeavor has been to improve clinical outcomes for patients undergoing cardiovascular disease treatment and particularly to address the thrombotic failure of vascular grafts. We will describe the basic background of ePTFE vascular grafts, the history of early heparin approaches, and the basic science of the CBAS surface leading to prototyping. With initial prototyping success, the product development process through clinical commercialization will be explored.

Drug–Device Combinations for Chronic Diseases, First Edition. Edited by SuPing Lyu and Ronald A. Siegel.

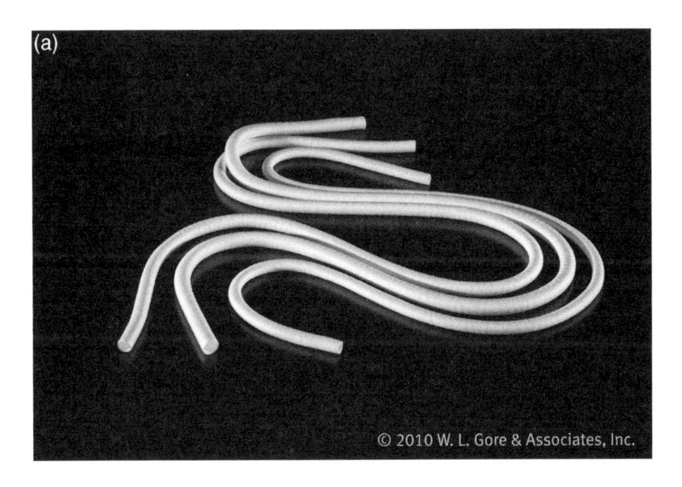

FIGURE 6.1 (a) ePTFE vascular graft. (b) PROPATEN Vascular Graft surface microstructure and immobilized heparin.

The continued commercial application of the CBAS surface is based on a decades-long clinical track record of proven usefulness of this technology for improving the hemocompatibility of this drug–device combination device. Today, the GORE ePTFE vascular graft with the CBAS surface, the GORE® PROPATEN® Vascular Graft (Figure 6.1), has improved the clinical performance of prosthetic small caliber vascular graft bypasses and has an important role in the management of lower extremity occlusive disease, with up to 4 year primary patency and limb salvage rates approaching historical results achieved with autologous vein conduits [1–4]. With several hundred thousands of these PROPATEN Vascular Grafts implanted worldwide, this graft has been noted as being widely used in contemporary practice [5]. The available worldwide experimental evidence and published clinical results point to significant durable clinical benefits of the CBAS heparin immobilization on the PROPATEN Vascular Graft imparting improved thromboresistance to the graft surface.

6.2 VASCULAR GRAFTING FOR PERIPHERAL ARTERIAL RECONSTRUCTION

For decades, vascular grafts have been used to treat cardiovascular disease [6]. The focus of peripheral vascular grafting is on arterial reconstruction when a patient's limbs are ischemic and/or potentially at risk of amputation due to occlusion of one or more infrainguinal blood vessels. Additionally, vascular grafts are used for angioaccess in patients receiving hemodialysis when arteriovenous fistulas are not feasible or have failed. Many vascular graft choices are available to the surgeon, each having advantages and disadvantages. Among these are synthetic vascular grafts, such as ePTFE, polyester, and grafts that are biologic in origin, such as autogenous vein, human umbilical vein, or cryopreserved allografts.

The bypass conduit of choice is the autologous saphenous vein, often referred to as the "gold standard." An ideal vascular graft would have characteristics that mimic the autologous saphenous vein, including minimizing the thromboreactive state, thromboresistance, biocompatibility, durability, resistance to infection, impermeability to blood, compliance, easy to sterilize, easy to implant, and readily available off the shelf [7].

Patency, limb salvage, and wound healing are standard outcome measures of vascular graft treatment as a result of critical or acute limb ischemia. Patency is a measure of blood flow through the conduit and reflects the success or failure of the prosthetic graft. If a graft has lost its patency, it has essentially thrombosed, has failed, is failing, or the outflow tract has thrombosed preventing blood flow through the graft. A graft is considered to have "primary" patency if it has had uninterrupted patency with either no procedure performed on it or a procedure to deal with disease progression in the adjacent native vessel. If a graft has "secondary" patency, blood flow is restored after occlusion by thrombectomy, thrombolysis, and transluminal angioplasty, or the graft or one of its anastomoses require revision or reconstruction [8].

Infrainguinal bypass graft procedures involve using a vascular graft to bypass an obstructed or highly diseased vessel below the inguinal ligament. These include bypasses from the femoral artery to vessels further distal to the femoral artery, such as the popliteal artery located above and below the knee or distal run-off vessels below the knee. Above-knee bypasses extend to a segment of the popliteal artery above the knee. Below-knee bypasses extend to the infrapopliteal, posterior tibial, anterior tibial, or peroneal arteries (Figure 6.2). Generally, the greater the number of patent distal run-off vessels in the patient's leg, the higher the likelihood of a patent prosthetic graft resulting in better graft patency and clinical outcomes. The same can also be said of a graft having a more proximal outflow vessel. For example, typically graft patencies for above-knee bypasses are higher than for below-knee bypasses over the same follow-up period [9].

The common femoral artery has generally been used as the site of origin for all bypasses to the popliteal and more distal arteries. However, when the superficial femoral, popliteal, or tibial arteries are relatively healthy or the vein length is limited, these vessels are used as inflow sites [10,11]. Furthermore, popliteal to pedal foot

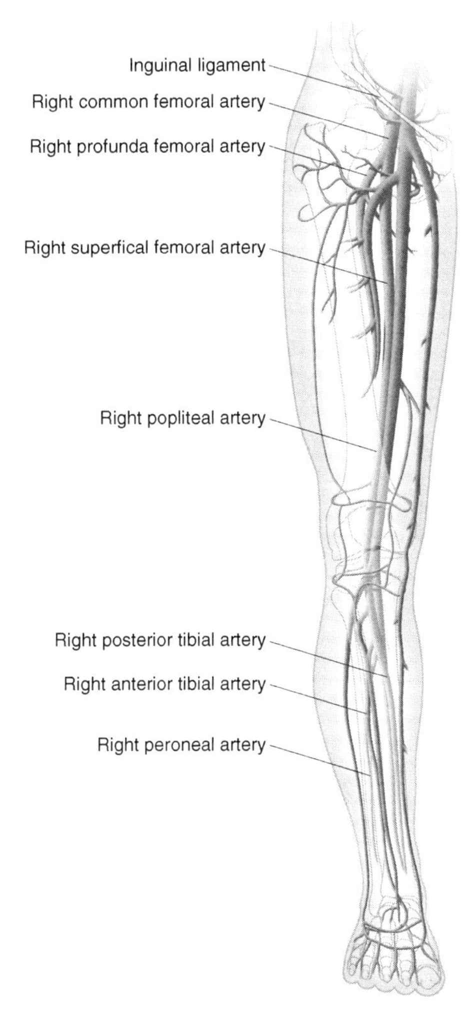

FIGURE 6.2 Lower limb vascular anatomy. (See colour plate section.)

bypass is restricted to a minority of patients with critical limb ischemia in whom the popliteal pulse provides adequate flow for bypass to an ischemic foot or nonhealing wound. Limb salvage is a measure of the "salvageability" of the limb without major amputation. It is commonly used to indicate salvage of the foot, not the limb [8].

Infrainguinal bypass grafts using autologous saphenous vein historically yield the highest reported patency rates. For femoropopliteal bypass, 5 year primary patency rates of 72–82% have been reported [12–14]. For infrapopliteal bypass, 5 year patency rates of 59–69% and limb salvage rates of 90% have been reported [4]. The greater saphenous vein is the graft conduit of choice for small vessel reconstruction as it most closely approximates the characteristics of the ideal vascular graft. However, at least 20% of current candidates for infrainguinal reconstructions have insufficient or unsuitable autologous veins available [12].

Although autologous saphenous veins are considered the gold standard conduit, most vascular surgeons deem synthetic vascular grafts to be appropriate for use in infrainguinal reconstruction for selected patients and procedural conditions. Synthetic vascular grafts offer the advantages of shorter operation times and smaller incisions. For above-knee femoropopliteal bypasses using ePTFE grafts, the patency rates at 5 years range from 39 to 65% with similar rates for polyester grafts [15,16]. Below-knee patency rates are generally poor, but adjunctive techniques, such as the addition of an interposition vein cuff or distal vein patch at the distal anastomosis [17,18], have led to improved results in bypasses below the knee.

When determining the most appropriate graft for an operation, graft performance for a particular reconstruction is evaluated along with a variety of additional considerations. These include the hemodynamic environment (inflow, outflow, and blood pressure), the presence of systemic or graft bed infection, the patient's ability to tolerate lengthy surgery, the need for and ability to tolerate chronic anticoagulation, overall health status, and projected life span [19]. These considerations must be balanced against the performance of the grafts available to determine the best course of action for treating the patient.

6.3 ePTFE VASCULAR GRAFTS

Ever since the introduction of prosthetic vascular grafts in the 1950s, the materials used have not been fully adequate for all clinical applications. Vascular grafts fail in clinical use primarily due to surgical technical error, thrombosis, intimal hyperplasia, and progression of cardiovascular disease. ePTFE and Dacron® are two of the most widely used graft materials. Clinically however, both of these materials have exhibited lower than desired patency rates in small-diameter applications (<6 mm). As a consequence, many experimental approaches have been used to improve the patency of these smaller diameter vascular grafts. Attempts made to improve synthetic vascular grafts during the 1970s through the mid-1990s, however, did not result in appreciable gains in clinical patency performance [20,21]. Bioinert surface modifications such as changing microstructures, using new polymers, and composite materials [22] were attempted. More complex bioactive approaches, including endothelial cell seeding, use of immobilized ligands such as cell adhesion peptides, and heparin [23,24] were also attempted without real success. While both preclinical and clinical studies did not demonstrate the outcomes desired, the search for the right combination of technologies continued into the 1990s [21].

ePTFE vascular grafts were first introduced in 1975 as a synthetic polymer option for treating vascular disease [25]. The early clinical results showed favorable outcomes with this newly applied polymer in many applications larger than 6 mm diameter [26]. Several advantages of ePTFE grafts included good biocompatibility, mechanical strength, and a simplified implant procedure with no requirement for preclotting, a requisite for Dacron grafts. The early burst and suture pullout force properties demonstrated long mechanical durability [27].

The widespread use of ePTFE grafts for treating peripheral vascular disease continued and the outcomes of prospective clinical studies showed that ePTFE grafts were inferior to suitable autologous saphenous vein. Even though ePTFE did not perform as well, it was regarded as the best prosthetic when saphenous vein was not available [28]. The overall 5-year primary patency rates of vein and PTFE above-knee femoropopliteal bypasses have been reported to be 74 and 39%, respectively [16]. Below-knee prosthetic bypasses have shown 1-year cumulative patency rates of 65%, declining to 29% by 2 years [17]. By comparison, historical ePTFE performance is suboptimal, and has borderline acceptability in small-diameter applications such as below-knee tibial or peroneal bypasses. Thrombogenicity and intimal hyperplasia appear to be the principal mechanisms of failure when standard ePTFE grafts are used as arterial substitutes in low-flow, high-resistance vascular beds [29].

With the positive clinical results of ePTFE grafts combined with the clinical need for continued improvement of synthetic graft materials for patients, efforts continued to be made to improve graft materials [30], particularly for below-knee applications. As a consequence, many experimental approaches have been used to improve the patency of these smaller diameter vascular grafts. Due to its clinical pharmacological effect and importance in preventing clot formation, there was an inherent attractiveness in using heparin.

6.4 EARLY HEPARIN SURFACES

The maintenance of homeostasis in our vascular system is a constant, active, ongoing process in which the healthy vascular endothelium plays an important role. To mimic this process for implanted cardiovascular devices during surgical or interventional vascular treatments in the absence of an endothelial cell surface, the device surfaces ideally should provide similar bioactive modes of action. Native vascular endothelium contains heparan sulfate, a compound similar in structure to heparin. Heparin is a naturally occurring substance found concentrated in mast cells, a component of the immune system abundant in the mucosa. Chemically, heparin is a polysaccharide glycosaminoglycan, heavily substituted with sulfate groups that make it the most densely negatively charged compound found in living tissue. Heparin, mainly isolated from pig intestinal mucosa, has a long history of clinical use as a systemic anticoagulant and its use enabled the early clinical work with extracorporeal heart–lung bypass and hemodialysis treatments [31,32]. It has multiple effects on the coagulation cascade, most notably functioning at the molecular level by binding the physiologic coagulation inhibitor antithrombin and inhibiting the thrombin conversion of fibrinogen to fibrin [33]. Coating a vascular graft with functionally active heparin could thus provide a surface partly mimicking the vascular endothelium that would, in a similar way, maintain homeostasis of the coagulation and anticoagulation systems during treatment. Furthermore, because of its known systemic anticoagulant effect, it was hypothesized that an immobilized heparinized surface would reduce or prevent thrombus formation on the vascular graft surface.

The use of heparin coatings to overcome the problems of thrombus formation on blood-contacting devices was first introduced by Gott et al. after a rather serendipitous finding [34,35]. When investigating the effect of negative electric charges on implanted graphite-coated tube samples in canine models, they disinfected the samples first with a solution of benzalkonium chloride and then rinsed with heparin to avoid thrombus formation during the insertion. Unintentionally, they had produced the first heparin coating consisting of heparin ionically bound to an adsorbed layer of benzalkonium chloride. The coating was shown to prevent thrombus formation up to 2 weeks in the canine model. These findings started the development of heparin coatings as a means of improving blood compatibility of materials, specifically thromboresistance.

The early heparin coatings were based on ionic immobilization of the highly negatively charged heparin molecule to material surfaces that had been prepared with positive charges, most often using surfactants carrying quaternary amino groups. Benzalkonium chloride and trimethyl dodecylammonium chloride (TDMAC) were widely used in combination with heparin [36] and later as water-insoluble ionic complexes, to prepare benzalkonium chloride heparin (HBAC) and TDMAC heparin coatings. One example of this type of coating is the DURAFLO® HBAC heparin coating [37]. Some of these ionically bound heparin coatings are still used today, mainly for short-term therapeutic treatments. For permanent implants, however, these ionic coatings did not provide the needed long-term protection against thrombus formation. The ionically bound heparin was rapidly eluted from the surface, usually within hours of implant. Initial attempts to stabilize the ionically bound coatings or to bind the heparin with more stable covalent bonds failed to give the required antithrombotic activity [38]. Therefore, it was assumed that the effect of the immobilized heparin to slowly leach off the material surface was needed to provide local inhibition of thrombus formation.

The results of the early attempts to achieve more permanently bound heparin coatings were discouraging. It was assumed that these permanent heparin coatings were nonfunctional and, therefore, would not provide any performance benefit. With the increased understanding of the molecular mechanisms of action of heparin in the late 1970s and early 1980s, the development of heparin coatings made rapid progress. A unique pentasaccharide sequence was discovered, present in approximately one third of unfractionated heparin molecules derived from commercial preparations. This sequence had a specific binding domain for the serine protease inhibitor antithrombin [39–43]. Heparin catalyzes the inhibition of clotting factors by antithrombin. When antithrombin is bound to these unique pentasaccharide sites of the heparin molecule, the inhibition rate of coagulation factors by antithrombin increased several orders of magnitude compared to the inhibitory capacity of antithrombin alone.

This important finding of the specific antithrombin binding domain on heparin enabled the development of the endpoint attachment technique introduced by Larm et al. 1983 [44,45]. By this endpoint binding of heparin to a base-coated material surface, the important pentasaccharide sequence of heparin is not compromised by the coupling chemistry, but instead, was preserved. The endpoint heparin maintained the ability to

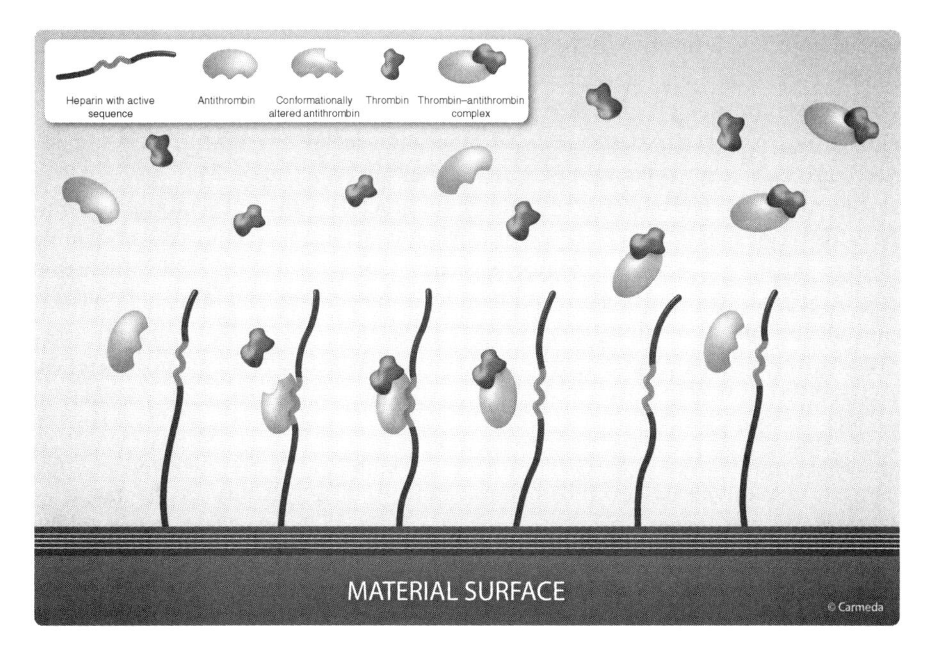

FIGURE 6.3 Illustration of CBAS surface showing material surface, base coating, and endpoint-immobilized heparin. Also shown are the reactants antithrombin, conformationally altered antithrombin, thrombin, and the inactive thrombin–antithrombin complex.

bind to the coagulation inhibitor antithrombin, even when immobilized with stable covalent bonds to the base-coated materials surface (Figure 6.3). Thus, the pivotal discovery was that the surface-bound heparin could thereby exert its anticoagulant activity similar to heparin in solution except that it was a local effect versus a systemic effect.

Another important aspect of the use of heparin in functionally active coatings is that when heparin is carefully bound with retained bioactivity, it is catalytic in its mode of action. It can repeatedly provide antithrombotic activity without being consumed by the reaction, just as is the case when heparin freely circulates in blood. This retention of catalytic activity provides the potential for long-term immobilized heparin functionality *in vivo*.

6.5 CARMEDA BIOACTIVE SURFACE TECHNOLOGY

The Carmeda BioActive Surface technology has been in clinical use for nearly 25 years. It is the most widely published of all commercially available technologies of its type providing evidence of the CBAS surface hemocompatibility and biocompatibility benefits for short-term and permanent product applications with very few adverse events reported. Over 400 publications and studies have examined the hemocompatible properties of the CBAS surface in controlled *in vitro* blood contact models or *in vivo* animal models and clinical studies. Furthermore, some of the key clinical performance benefits of endpoint immobilization of heparin on the CBAS surface have been demonstrated in broad application, including extracorporeal circuits, vascular stents, ventricular assist devices, and ePTFE vascular grafts.

The CBAS surface consists of endpoint-attached heparin covalently bound to a base matrix of alternating layers of polyethylenimine and an anionic polysaccharide dextran sulfate [44] applied in a layer-by-layer process. Heparin, modified by partial nitrous acid cleavage, is bound via a terminal aldehyde to amino groups to the last polyethylenimine layer of the base matrix [45]. The functional activity of the surface-bound heparin on the CBAS surface is quality controlled by a proprietary analytical test method based on the interaction of the surface-bound heparin with the coagulation inhibitor antithrombin.

The CBAS surface can be applied to most medical device materials (http://carmeda.com/upl/files/31627.pdf). The coating is thin, typically in the range of hundreds of nanometers. Since the heparin is bound by covalent bonds to the base coat, it does not leach off in contact with blood, but is permanently bound. If stored properly, CBAS-coated devices have an acceptable shelf life, typically at least 5 years. The CBAS coating can be sterilized by ethylene oxide, one of the common methods of device sterilization, without losing its mode of action. CBAS-coated devices also pass biocompatibility testing according to ISO 10993 [46].

While there are a wide variety of approaches for binding heparin to devices, it must be kept in mind that different immobilization techniques will affect the functional activity of the immobilized heparin as well as other coating properties to different extents. The performance of other heparin coatings with regard to thromboresistance does not necessarily resemble that of the CBAS surface. Heparin can be bound by covalent attachment to material surfaces in different ways that adversely affect heparin's functional properties. In contrast to endpoint covalent attachment as employed in the CBAS surface, heparin can be covalently bound by multipoint attachments along the heparin molecule, and the covalent bonds can involve the critical pentasaccharide sequence in heparin known to be essential for its anticoagulant activity. In such coatings, the surface-bound heparin cannot interact with coagulation inhibitors in the blood. Some heparin coatings may even activate rather than prevent clotting. Even endpoint attachment of heparin can be performed in different ways and will not necessarily result in functional properties [47] similar to CBAS surface. Each heparin coating technology must, therefore, be judged by its own merits. A more comprehensive review of heparin and alternative technologies for improving biocompatibility of device materials has been provided by Tanzi [22] and Jordan and Chaikof [23].

6.6 UNDERSTANDING CBAS SURFACE HEMOCOMPATIBILITY

6.6.1 Coagulation

Coagulation is the complex mechanism by which clots are formed upon injury in order to prevent blood loss. The coagulation cascade consists of a number of circulating plasma components called coagulation factors that play a key role in the formation of thrombus. After a triggering event, such as an injury or exposure to a foreign material, a complex series of reactions involving the activation of coagulation factors to functional enzymes are initiated. The reactions culminate in the activation of the coagulation factor thrombin, which, in turn, converts the soluble plasma protein fibrinogen into an insoluble fibrin clot. Activation of coagulation can occur through two different pathways: contact activation and tissue factor pathways, also known as the intrinsic and extrinsic pathways. Coagulation in response to exposure of blood to foreign materials is believed to occur by the contact activation (intrinsic) pathway [33].

Characteristic of this kind of defense mechanism, contact activation involves dramatic amplification in order to bring about an effective response even to a relatively mild trigger. To avoid unintentional clot formation, the coagulation system is under strict control of numerous inhibitors. Most important among these clotting inhibitors is antithrombin that circulates in blood at relatively high concentration. Antithrombin acts by binding to the activated coagulation factor thrombin, leading to the formation of a stable complex that is completely devoid of enzymatic activity. By a similar mechanism, antithrombin exerts effective control against unwanted activation of essentially all coagulation factors (Figure 6.4).

Heparin exerts its anticoagulant activity by enhancing the rate at which antithrombin neutralizes the target coagulation factor. A specific structural unit in the heparin molecule, the antithrombin binding sequence, binds to antithrombin resulting in a dramatic acceleration of the inhibitory capacity toward coagulation factors [33]. Once the inactive thrombin–antithrombin complex has been established, the binding between antithrombin and heparin dissociates and heparin is available for further

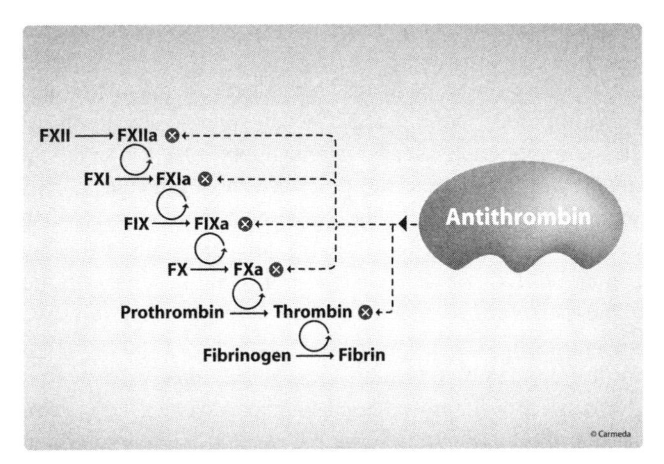

FIGURE 6.4 Contact activation coagulation cascade pathway. Representation of the intrinsic coagulation cascade and the various coagulation factor reactions that lead to fibrin formation as the final step. Reactions that are influenced by antithrombin are also shown (×) highlighting the broad effect of antithrombin on the coagulation cascade.

inhibitory cycles. Heparin thus serves as a catalyst in this reaction. It is not degraded or depleted while expressing its functional activity *in vivo*; therefore, it remains catalytically active on the graft surface.

The thromboresistant properties of the CBAS surface coating and its role in the inhibition of coagulation factors have been extensively studied and the key role of antithrombin demonstrated in numerous experiments. Substantial quantities of antithrombin are taken up by the CBAS surface from purified solutions of antithrombin as well as from plasma. Early studies showed the activated coagulation factors thrombin [48] and factor Xa (FXa) [49] to be neutralized by CBAS endpoint-immobilized heparin. In the absence of antithrombin, inhibition did not occur, confirming that the local mechanism of action of the CBAS-immobilized heparin closely resembles that of heparin administered systemically. Further studies revealed that activated coagulation factor XII (FXIIa), which is considered to be the trigger point of the clotting mechanism in response to contact with a foreign surface, was also inhibited by the CBAS surface [50]. Similar to thrombin and FXa, the inhibition of FXIIa was antithrombin dependent. In the absence of antithrombin (depleted plasma) or on a surface coated with heparin lacking the antithrombin binding sequence, extensive activation of FXII occurred, leading to immediate clotting [51]. These findings may have a wider implication for the blood compatibility of the CBAS surface, since FXII is part of the contact activation system, which is considered to be involved in the triggering mechanism of fibrinolytic and kallikrein–kinin systems, and in part of the complement system in addition to the contact activation of coagulation. It has also been reported that in blood exposed to CBAS surfaces *in vitro*, soluble levels of clinically relevant coagulation markers, including fibrinopeptide A, thrombin–antithrombin complex, and prothrombin fragments 1 and 2, are reduced compared to control [52–55].

6.6.2 Platelets

Platelets are circulating blood elements that adhere to injured tissue and foreign surfaces exposed to blood. The adhered platelets undergo activation, which triggers further platelet adhesion and platelet aggregation. Platelet activation is closely associated with the clotting mechanisms described above, and the aggregated platelets along with the fibrin network form the blood clot. Activation and aggregation of platelets often lead to loss of platelets from the circulating blood depending upon the extent of the injury. CBAS surface coating of artificial materials drastically reduces the adhesion and activation of platelets, as shown in an early *in vitro* study of the biocompatibility of cardiopulmonary bypass systems by Videm et al. [56]. The loss of platelets from fresh human blood in this large surface area system was significantly less when CBAS surface systems were used compared to uncoated systems. The results were confirmed by an *in vitro* study using a Chandler-like blood loop model [57].

In a further effort to get closer to the *in vivo* setting, *ex vivo* studies have also been performed. In contrast to uncoated devices, virtually no platelet adhesion was observed when CBAS-coated coronary stents or vascular grafts were tested in an *ex vivo* baboon shunt model [58,59]. It was later demonstrated that the antithrombin binding properties of the immobilized heparin were critical for the inhibition of platelet adhesion to the surface of the heparin-coated stents [60]. Heparin that was devoid of the antithrombin-binding pentasaccharide sequence did not differ from the uncoated control with respect to platelet adhesion in this baboon model, while immobilized heparin containing the antithrombin binding sequence effectively reduced platelet adhesion.

In summary, CBAS endpoint-immobilized heparin plays an important functional role in improved blood compatibility, with respect to both direct effects on the coagulation cascade and in the prevention of platelet adhesion and platelet aggregation resulting in the preservation of circulating platelets. The fact that CBAS surface inhibits the coagulation sequence at both an early stage (FXII) and the final conversion of fibrinogen to fibrin, if activated, demonstrates the power of the thromboresistant activity of CBAS endpoint-immobilized heparin. The key role of antithrombin on the ability of the CBAS surface to prevent activation of coagulation as well as to prevent platelet adhesion and activation is in accordance with the suggested functional relationship between the two systems.

6.6.3 Inflammatory Response

The thromboresistant effects based on blood coagulation and platelet function provide the most compelling evidence of the improved blood compatible functionality of CBAS surface with endpoint-immobilized heparin. However, contact between blood and artificial surfaces can also induce a number of other inflammatory responses, including activation of complement cascade systems and leukocyte activation, which can produce more widespread systemic responses. This inflammatory response as part of the overall biocompatibility of a biomaterial surface can affect the graft surface performance.

The complement system, which is part of the body defense against microorganisms, has the inherent capacity to discriminate between "self" and "non-self." Similar to the coagulation system, complement activation occurs by a cascade reaction sequence that can be triggered via different pathways and normally leads to an inflammatory reaction with swelling and pain. Early *in vitro* studies with the CBAS surface regarding the complement system demonstrated that analysis of the blood fluid phase was not sufficient to get a complete understanding of the biomaterial surface characteristics since the activation products may bind to the biomaterial substrate itself [61]. For example, uncoated TECOFLEX® tubing appeared to be complement compatible from fluid-phase measurements, while measurements of complement markers (C3b, C5a, and TCC) on the material surface showed considerable activation. Uncoated silicone, on the other hand, caused significant activation of the fluid phase, but no detectable activation products on the material surface. However, when both test materials were coated with the CBAS surface, formation of complement activation products was negligible both on the solid surface and in the fluid phase.

Many of the early functional studies on the CBAS surface with respect to effects on complement activation were performed in connection with cardiopulmonary bypass systems. These systems present a unique challenge for biomaterial compatibility since blood is exposed to a large total biomaterial surface area under flow conditions. In one of the initial *in vitro* human studies, cardiopulmonary bypass systems consisting of hollow fiber membrane oxygenators and pumps were perfused with fresh human donor blood. CBAS surface-treated systems demonstrated significantly less generation of complement activation products compared to uncoated systems and resulted in lower platelet loss [56]. Numerous *in vitro*, *ex vivo*, and clinical studies indicated that the CBAS surface coating markedly improves the biocompatibility of the cardiopulmonary systems by increasing the binding of complement regulators to the surface of the system, thereby effectively reducing the complement activation. Additional studies demonstrated that the CBAS surface has a beneficial effect with respect to reducing complement activation, both in the early and late stages of the complement cascade [62,63]. Specifically, it decreases the release of complement activation products such as C3bc, C5a, and TCC.

White blood cells or leukocytes, such as granulocytes (including neutrophils), monocytes, and lymphocytes, are cells of the immune system that defend the body against both infectious disease and foreign materials. Blood or tissue contact with artificial materials can activate populations of white blood cells. Leukocytes can be activated by numerous stimuli dependent on or independent of the complement system. Inflammatory mediators released from activated leukocytes are known to contribute to tissue destruction. The leukocyte compatibility of the CBAS surface has been examined *in vitro* with fresh human blood [64]. Exposure of blood to uncoated polyvinyl chloride (PVC) tubing led to a marked increase in the expression of adhesion and activation markers on both granulocytes and monocytes, while exposure to CBAS-heparinized PVC tubing did not. Since this effect on leukocytes correlated closely with the complement activation, it was concluded that the leukocyte cell activation was secondary to triggering by the complement system.

Use of an *in vitro* model of cardiopulmonary bypass with fresh human blood showed that the heparin coating reduced the initial decrease in absolute cell counts of monocytes and granulocytes reflecting reduced adhesion to the heparin-coated material surfaces [65]. The increases in complement activation markers C3bc and TCC as well as increases in plasma myeloperoxidase and lactoferrin were lower in the heparin-coated group compared to the control group.

In summary, there is ample evidence in support of the CBAS surface improving the biocompatibility of artificial surfaces in contact with blood by reducing inflammation-related reactions. Both the complement system and leukocytes undergo less activation in blood exposed to CBAS-coated surfaces than to uncoated materials. Interestingly, certain mechanisms leading to leukocyte activation have been shown to depend on complement activation while others did not. Irrespective of mechanism, however, activation was prevented in blood exposed to CBAS-coated surfaces.

6.7 EARLY CLINICAL EXPERIENCE WITH CBAS SURFACE

An early potentially beneficial application of the CBAS surface coating considered was gas exchange devices (oxygenators or artificial lungs) used for extracorporeal lung assist (ECLA) to treat patients with severe respiratory distress syndrome [66,67]. In such treatment, the patients' blood is pumped outside the body through an oxygenator where carbon dioxide is replaced by oxygen, and the oxygenated blood returned to the patient, thereby allowing damaged lung to rest and heal. In order to avoid coagulation in the large extracorporeal circuit, systemic heparin is given in high doses. However, trauma patients with multiple injuries cannot be systemically heparinized due to the bleeding risk. Even for nontrauma lung assist patients, prolonged treatment with systemic heparin may be hazardous and even counteract pulmonary healing. The applicability of the CBAS surface on extracorporeal systems for lung assist, used without systemic heparin, was therefore tested in canines by Bindslev et al. [68] with encouraging results.

The first treatment of a patient with a CBAS surface-coated extracorporeal lung assist system and only minimal amounts of systemic heparin was reported by Bindslev et al. [69] in 1987. The patient suffered from respiratory failure caused by complications from viral pneumonia. Subsequently, several *in vivo* studies as well as case reports and clinical studies on long-term extracorporeal lung assist systems coated with CBAS surface have followed the pioneering work by Bindslev et al. [69]. The beneficial effect of the CBAS surface was later shown in numerous clinical studies on cardiopulmonary bypass systems for thoracic surgery [70,71] and on ventricular assist devices for bridging to heart transplant [72]. Early clinical studies have also been performed on CBAS-coated permanent implants such as vascular stents [73] and on vascular grafts with a CBAS surface, as will be described in detail in this chapter.

6.8 PRODUCT DEVELOPMENT

6.8.1 Prototyping

The collective experience with the CBAS surface provided promise that combining the drug heparin in a specific manner with a commercially available ePTFE vascular graft may offer an improved product option. To this end, a collaboration was initiated to evaluate the potential of combining the CBAS heparin and ePTFE technologies. Under the appropriate contractual agreements between Carmeda AB and W.L. Gore & Associates, Inc., early prototyping resulted in a proprietary process that created a CBAS surface on an ePTFE vascular graft. Foremost in mind during this early prototyping effort was the clinical goal of developing a vascular graft with improved thromboresistance. While there are many avenues of scientific investigation that can be pursued, a clinically useful product is the final goal.

The PROPATEN Vascular Graft was developed with the intention of improving the clinical patient outcomes for cardiovascular disease treatment and particularly addressing the problems of thrombotic failure of peripheral vascular grafts. The CBAS surface is bound to the luminal surface of the PROPATEN Vascular Graft imparting ideal characteristics of a heparinized vascular graft, including uniformity of heparinization, retention of heparin on the graft flow surface, and functional maintenance of its heparin bioactivity. The product concept that emerged from this effort was the following:

The only immobilized heparin containing synthetic* below-knee vascular graft with improved thromboresistance**.

*Stable, bioactive heparin-coated ePTFE.

**Reduced platelet adhesion in nonhuman primate and canine models.

The initial prototyping efforts were conducted with the design constraints of maintaining the properties of existing ePTFE vascular grafts while combining the CBAS surface to create acceptable prototypes with respect to the above requirements.

Early evaluations of the heparin-coated vascular grafts focused first on benchtop testing to demonstrate the uniformity, mechanical stability, and heparin bioactivity testing. These results were mixed, and hence process optimization work was conducted to improve these parameters. The CBAS surface is only on the order of hundreds of nanometers thick. Therefore, it provides a macromolecular coating of the ePTFE nodes and fibrils without completely blocking the nodes and fibril structure of the base graft surface. Coating durability is best evaluated by heparin retention studies in which both the surface-bound heparin mass and heparin bioactivity, the ability to bind antithrombin, were shown to be maintained. Finally, heparin bioactivity levels were assessed to ensure that the coated ePTFE grafts maintained sufficient heparin functional activity after sterilization.

Finally, and most importantly, *in vivo* preclinical performance testing was conducted to evaluate the short-term, long-term, and histological responses of the finished product. Early *in vivo* acute shunt testing [74] readily showed improved thromboresistance with a dramatic decrease in thrombus accumulation grossly compared to uncoated controls in an acute 2 h non-anticoagulated severe canine carotid interposition shunt model (Figure 6.5).

Following these screening tests, long-term patency studies were conducted that demonstrated a significant difference in patency with the control grafts occluding fairly quickly within days, while the CBAS grafts maintained prolonged patency out to 6 months [74], the longest time point examined. The histological response of the CBAS grafts was benign, and there was significantly less neointimal tissue deposition on the CBAS grafts. Reduction of platelet deposition and less neointimal hyperplasia

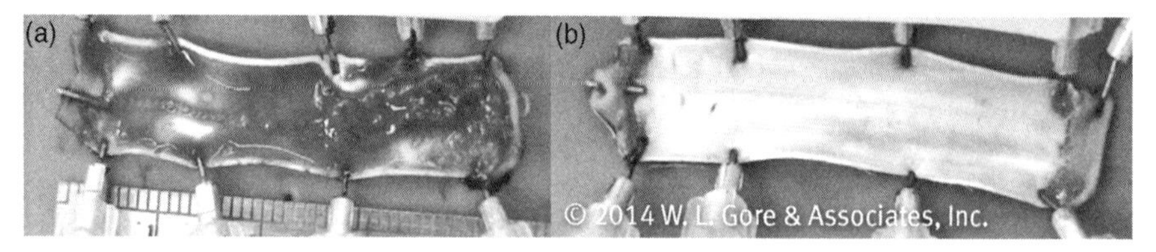

FIGURE 6.5 Representative photograph of CBAS–Heparin ePTFE vascular grafts after removal from the canine carotid interpositional model. The control vascular graft is occluded with thrombus (a), while the CBAS–ePTFE vascular graft is thrombus-free (b). (See colour plate section.)

were further demonstrated in animal models [59,75]. Having established the essential requirements for a CBAS heparin-coated ePTFE graft, the prototyping effort was poised to move into product development.

6.8.2 Regulatory Standards

A central requirement of any medical device product development program is the need to meet and comply with multiple domestic and international guidance documents and standards. As such, methods needed to evaluate a device–drug combination product have evolved over the years. When the PROPATEN Vascular Graft was first conceived, international standards such as ISO 7198 [76] and ISO 10993 [46] existed, but did not cover devices that included a medicinal substance or active pharmaceutical ingredient (API), such as heparin. As more device–drug products were developed, the need for additional guidance was identified. FDA developed and implemented the Office of Combination Products to help evaluate products combining devices with pharmaceuticals (e.g., drug-eluting coronary stents) spanning more than one branch of the FDA (e.g., CBER, CDER, and CDRH). AAMI and ISO also collaborated and specific standards were developed. Technical Information Report ANSI/AAMI/ISO TIR 12417 [77] was written specifically to help provide guidance to regulatory agencies and industry in this new area of product development. This standard along with the standards and guidance documents referenced therein are helpful to anyone developing a device–drug combination product. The types of questions and evaluation methods described in TIR12417 follow closely the process used in the development of the PROPATEN Vascular Graft.

In the past as well as in today's regulatory climate, the overriding factors that regulatory agencies use to approve a medical device weigh the relative risks and benefits to the patient. Regulatory agencies around the world are placing greater emphasis on risk management activities used throughout the development process to help ensure risks are mitigated appropriately and the benefits outweigh the remaining risks. This has become critical for combination devices since there are added risks associated with the medicinal substance on the device, whether the medicinal substance is the primary mode of action or is ancillary to the device. Therefore, a manufacturer must consider the potential risks and benefits during the entire device design and development process from conceptualization through commercialization and even postmarket surveillance. The development process for the PROPATEN Vascular Graft followed appropriate sections of ANSI/AAMI/ISO TIR12417:2011 [77].

6.8.3 Design Specifications

Early prototyping demonstrated product feasibility through a combination of *in vitro* and *in vivo* testing results. The foundation for PROPATEN Vascular Graft was the commercially available Gore Stretch Vascular graft that had been on the market worldwide for over 20 years. Product specifications and manufacturing processes were well developed, while a large body of published literature showing excellent clinical experience existed. The design intent was to take this well-established vascular graft and to apply the CBAS surface to the luminal surface of the vascular graft to improve thromboresistance over current grafts used in bypass vascular reconstructions. The essential functional requirements for a synthetic vascular graft include mechanical stability, biocompatibility, and acceptable host tissue response. All of these elements are largely met with a purely synthetic ePTFE vascular graft. In considering the addition of a heparin coating, many other elements come into play, including coating stability, coating uniformity, *in vivo* performance, and histology to demonstrate a durable, hemocompatible, and thromboresistant surface. Implied in this statement is the need for the heparinized surface to be uniform, stable, and durable plus the heparin must be functionally active to minimize clot formation at the graft surface–blood interface. The product concept, stated previously, remained the guiding design goal.

There are two important differences between this vascular device and many other drug-containing vascular devices. The product concept called for a stable and durable heparinized surface that is functionally active while maintaining the mechanical design attributes of the predicate ePTFE vascular graft. The immobilized heparin is intended to be permanent by design and the functional anticoagulative properties of the heparin are intended to be maintained on the surface of the graft. Achieving these two critical attributes has not only presented significant technical challenges but can also create confusion within regulatory authorities since it is *not* a drug-eluting device. Both of these considerations are discussed in more detail below.

In the present example, the mechanical design was well known and defined for the predicate ePTFE vascular graft that simplified the product development process. The primary mode of action of a vascular graft is to provide a conduit for blood flow (AAMI/ISO 7198 [76]). As a combination product, the application of endpoint immobilized heparin to the surface of the vascular graft provides a mode of action that is ancillary to the primary mode of action. Since the application of heparin to the surface to improve thromboresistance is ancillary to the device, new design and regulatory requirements were identified necessitating new design inputs and design outputs. Although it did not exist at the time, many of these and other applicable dosing requirements are now contained in ANSI/AAMI/ISO TIR12417:2011 [77] for the addition of a drug–device combination product. Additional guidance can be obtained from U.S. FDA Office of Combination Products [78] and the European Medical Device Directive (Essential Requirements: Annex I, 93/42/EEC as amended by Directive 2007/47/EC [79]) and other directives, physician input, risk analysis, and other sources. Table 6.1 provides a portion of a completed Essential Requirements List required for any European submission and Table 6.2 contains examples of product design inputs.

Product development requires a complete understanding of the fitness for use issues based on the device's clinical application. The development process of the basic vascular graft with a thromboresistant surface was accelerated by using an existing commercial vascular graft in the same application. The types of disease states in the femoropopliteal application were well known as were the treatment modalities. Clinical outcomes had been repeatedly published over decades, so targets could be set for determining improved patient outcomes. Recognition of many graft attributes,

TABLE 6.1 Example of an Essential Requirements List

Section	Requirements	External Standards	Reference Applicable Quality System Documentation	Addressed
1	The devices must be designed and manufactured such that, when used as intended, they will not compromise condition or the safety of patients	ISO 10555-1:2004—Sterile, Single-Use Intravascular Catheters—Part 1: General Requirements	Design History File	√
		ISO 14971:2001 A1 2003—Medical Devices: Application of Risk Management to Medical Devices	Risk Management Plan	
	Device-associated risks are acceptable compared to the benefits achieved—a high level of protection of health and safety	ISO 25539-1 2003—Cardiovascular Implants: Endovascular Devices—Part 1: Endovascular Prostheses	Risk Analysis and dFMEA	
2	Device design solutions should conform to safety standards and be state of the art	ASTM F2503-05—Standard Practice for Marking Medical Devices and Other Items for Safety in the Magnetic Resonance Environment	Design History File	√
	In selecting the most appropriate solutions the following principles apply:	EN 980:2003—Graphical Symbols for Use in the Labeling of Medical Devices	Risk Management Plan	
	Eliminate or reduce risks as far as possible	ISO 10555-1:2004—Sterile, Single-Use Intravascular Catheters—Part 1: General Requirements	Risk Analysis and dFMEA	
	Include adequate protection measures including alarms, in relation to minimize risks Inform users of the residual risks		Instructions for Use (IFU)	

TABLE 6.2 Sample Design Inputs

Intended Use	Peripheral Vascular Indication			
Input No.	Design Input	User Need/ Intended Use: Need (N), Use (U), Both (B)	Device Function: Critical (C), Noncritical (N)	Critical Technical Assumptions
1	Bioactive Surface (CBAS surface): device luminal surface is bioactive	U	C	Functional bioactive heparin
2	Packaging and sterilization are compatible with CBAS surface	U	C	Sterility assurance level 10^{-6}, not compromise CBAS surface
3	Package labels and IFU are compatible with CBAS surface	N	C	CBAS surface preserved
4	Durable CBAS surface: retained on the surface of the graft	U	C	Nonleaching heparin
5	Final assembly specification	B	C	Current vascular can be used w/ modification for CBAS surface
6	Finished goods specification	B	C	Current vascular graft can be used w/modification for CBAS surface

including resistance to blood or serous fluid leakage, suture retention, tensile strength, burst pressure, polymeric creep resistance, sterility assurance levels, and shelf life [76], helped define appropriate product specifications. Surgical procedures necessary to achieve good clinical outcomes had been refined and were well known, thereby creating an excellent platform from which to build upon. However, the vast body of clinical history indicated that there was room to improve the hemocompatability of the artificial surface of the vascular graft. Hence, the differentiating elements of the PROPATEN Vascular Graft product concept required CBAS surface heparin function, as illustrated and described in Figures 6.6 and 6.7.

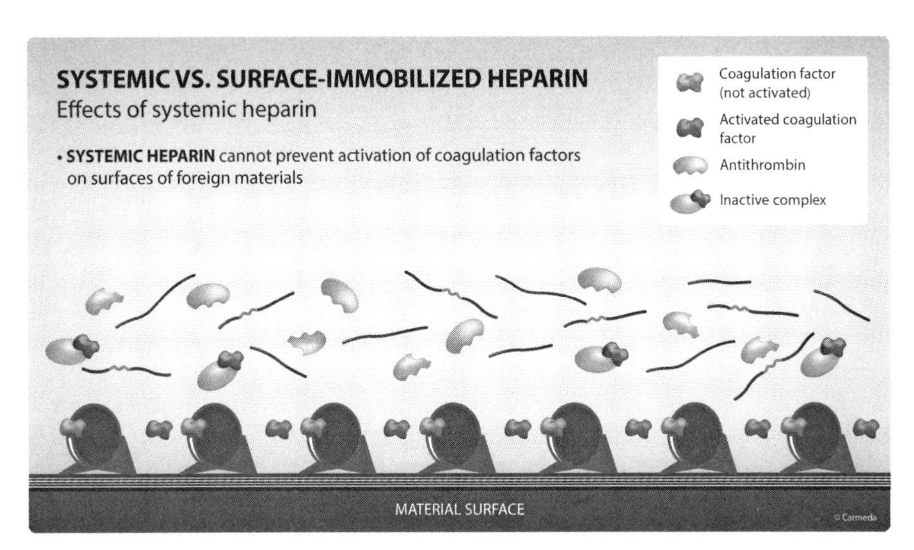

FIGURE 6.6 Systemic heparin attenuates unwanted and uncontrolled coagulation of the circulating blood and is often given during vascular graft implant procedures. During a vascular procedure, blood activation is a result of surgical trauma and is routinely controlled by systemic heparin. (See colour plate section.)

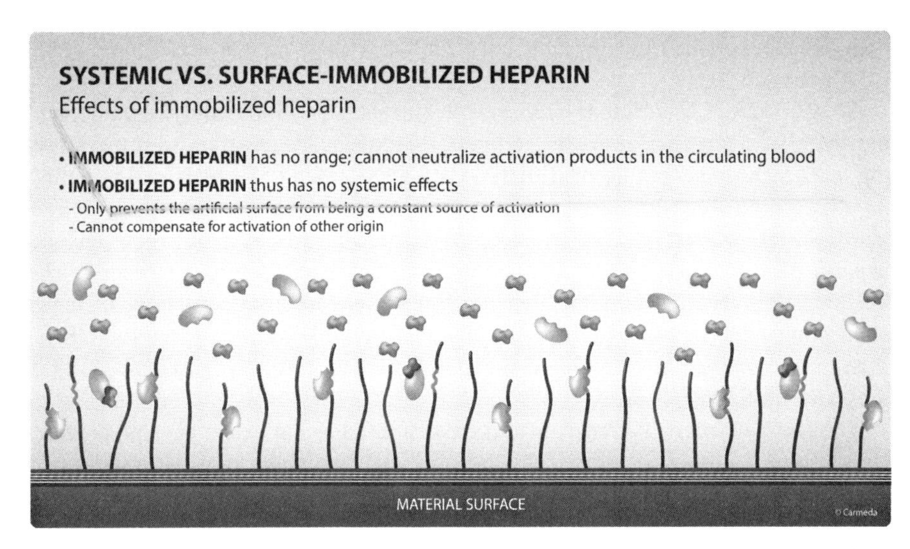

FIGURE 6.7 An immobilized heparin surface effectively suppresses activation by artificial surfaces of medical devices through inhibition of the coagulation cascade on the surface and reduction in inflammation. Postoperatively, a stable heparin surface of a vascular graft may reduce the need for sustained systemic anticoagulation. The long-term patency of vascular grafts with such a blood-contacting surface should, therefore, be improved.

While the clinical application, mechanical design attributes, failure modes, risks and surgical procedures were well understood, the specifications and performance parameters for the heparin surface of vascular grafts were not known, nor were methods for evaluating vascular graft performance. The complexity of creating a permanently stable and durable bond between heparin and a vascular graft made from inert ePTFE plus ultimate scale-up of such a manufacturing process was not trivial. Significant time was devoted to defining the design outputs in these key areas.

Another design choice was made that helped accelerate product development. Although it had never been bonded to ePTFE before, the heparin surface chosen was the CBAS surface that had already been in clinical use for nearly 5 years, with extensive publications of commercial heparin surface coating for medical devices, although not vascular grafts. At the time, there were over 140 peer-reviewed publications spanning in subject from the basic science of the surface, *in vitro* performance, *ex vivo* and *in vivo* animal studies, and clinical results. The application of the CBAS surface was based on the proven usefulness and safety of the technology for increasing the hemocompatibility of blood-contacting surfaces. The benefits of this approach included not only the demonstrated usefulness of the CBAS surface but also an initial manufacturing process, including an indication of evaluation techniques needed for the vascular graft application.

6.8.4 Manufacturing

The coating is performed in a multistep process where a base matrix is first applied followed by a final heparin immobilization step as described earlier. The heparin is bound to the base matrix with stable covalent bonds resulting in a stable and durable CBAS surface.

The PROPATEN Vascular Graft key chemicals heparin, polyethylenimine, and the anionic polysaccharide are tested to the component specifications. Furthermore, the manufacturing steps used in the PROPATEN Vascular Graft heparin coating process are also controlled throughout the manufacturing process. Most importantly, these devices are quality tested to ensure that mechanical properties and heparin functionality are maintained during the manufacturing process. The resulting product maintains the vascular graft microstructure consisting of interconnecting node and fibril structure without being able to visibly see the CBAS surface. Biocompatibility testing conducted using PROPATEN Vascular Grafts is presented in Section 8.5.

The heparin active pharmaceutical ingredient used as raw material to prepare the CBAS surface meets USP and EP pharmacopoeial monographs for heparin sodium of porcine origin from North America. The process used to create heparin for the CBAS surface from heparin sodium is similar to a process used to create low-molecular-weight heparins for injection. Both processes involve a partial, carefully controlled depolymerization of heparin and the basic chemical compositions of the heparin used to prepare the CBAS surface and low-molecular-weight heparins are similar. Key parameters for heparin API were identified and raw material specifications were written to ensure the conversion process is successful.

6.8.5 Quality Testing

It is important to note that the validated processes for immobilizing the heparin on the CBAS surface must yield robust surfaces with surface heparin bioactivity levels sufficient to demonstrate improved thromboresistance *in vivo* and in clinical studies. Therefore, qualitative and quantitative tests carried out to control the heparin on the graft surface were a necessity. The test methods used were fully documented and supported by appropriate validation data. Analytical data on multiple production scale batches were included in the validation. Quality control and lot release testing conducted on randomly selected samples from each production lot of finished product included various mechanical tests and heparin testing to ensure each lot of grafts was mechanically stable and CBAS surface remained functionally active.

All performance attributes of the predicate ePTFE graft from the product specification transferred directly to the new PROPATEN Vascular Graft product specification as part of the design outputs. The major remaining issue was to verify the relationship between CBAS surface bioactivity and thromboresistance. To examine the relationship between surface-bound heparin bioactivity and thromboresistance, CBAS surfaces with widely varying surface heparin bioactivity have been examined in several different blood contact models. These include an *in vitro* recirculating blood loop model (modified Chandler Loop), acute arteriovenous shunt animal models, as well as acute and chronic animal implant models.

Using an *in vitro* recirculating blood contact model (a modified Chandler Loop), polyvinyl chloride tubing with CBAS surfaces of widely varying surface heparin activities, as well as control substrates with no immobilized heparin, were tested by exposure to freshly drawn human blood. The blood was circulated in the tubing and then examined for the presence of thrombus. Blood flow in all control non-CBAS-treated surfaces rapidly became occluded by thrombus. In contrast, no thrombus formation was observed on CBAS surfaces.

In order to expand the understanding of CBAS surface thromboresistance in the cases where no thrombus was observed, quantitative methods were developed to probe the coagulation and platelet responses to CBAS surfaces in the presence of human blood. In practice, this experimental model is similar to those described for qualitative evaluation of thrombus in the blood loop model. Instead of reaching the clot endpoint, the blood is collected prior to visible clot formation and specific soluble markers of the coagulation process are measured, including platelet count and enzyme-linked immunosorbent assay (ELISA)-based measurements of commonly analyzed coagulation markers. This strategy for assessing material thrombogenicity is recommended in ISO 10993-4 [80] for hemocompatibility testing of medical devices and has been used to evaluate the performance of CBAS surfaces [52,53]. The results of these studies demonstrated appropriate CBAS surface functionality that helped establish the final device specification to ensure effectiveness.

As described earlier, a critical design requirement was that the CBAS surface be durable and not elute from the graft surface while maintaining functional activity. The purpose for evaluating the elution profile of the heparin on the device surface is to provide objective evidence that heparin does not come off the graft surface following processing, sterilization, and after shelf life stability testing. Heparin binding retention is evaluated through methods intended to measure the amount of heparin that is not bound to the surface. To ensure patient safety and mitigate the risk of excessive bleeding, studies have been performed to determine how much unbound heparin component remains on the surface and whether that amount is clinically relevant should it be released into the patient. Multiple studies were conducted to evaluate the residual amount of heparin that is released following processing. The cumulative amount of residual heparin released from a graft into solution over 7 days ranged from nondetectable to an amount two orders of magnitude below the typical amount of heparin sodium administered to patients intraoperatively (5000–15,000 IU) and was thus considered insignificant. Further *in vivo* testing was then required to demonstrate that the CBAS surface is stable in blood flow. As described earlier [74], a 90-day canine study incorporating ePTFE vascular grafts with and without the CBAS surface demonstrated improved thromboresistance of the CBAS devices over the 90-day period and proved the durability of the heparin as the CBAS surfaces still had measurable bioactivity.

In summary, several fundamental properties of CBAS surface performance have been consistently identified in the various experimental models:

- CBAS surfaces were consistently less thrombogenic than control non-CBAS-treated surfaces.
- CBAS surfaces exhibited thromboresistance as well as minimal activation of clinically relevant coagulation markers and attachment of platelets studied *in vitro*.
- Above minimum functional levels of heparin bioactivity, thromboresistance of CBAS surfaces was identical over a wide range of surface heparin activity represented by the manufacturing processes employed.
- Heparin is not eluted from the CBAS surface in amounts significant to the patient.

Based upon these collective studies, it was concluded that vascular grafts with the CBAS surface are effective with increased and durable thromboresistance.

6.8.6 Packaging

A sterilization process was needed that would not change the design attributes including no degradation of the thromboresistant properties of the graft while providing a sterility assurance level of 10^{-6}. Early work indicated that steam sterilization might impact the CBAS surface and it is well known that gamma radiation sterilization can impact the mechanical attributes of ePTFE. Therefore, ethylene oxide sterilization was chosen. The sterilization validation process utilized the "overkill" method wherein sterilization process parameters were chosen that would provide the necessary level of bacterial kill (based on bio-indicators and product challenge devices) and then doubling the process times. Once the process was identified, sterilized grafts were put through the battery of quality control tests and *in vitro* and *in vivo* testing to demonstrate thromboresistance was maintained after sterilization.

The graft mechanical characteristics and bioactivity must also be maintained over the course of its shelf life. This is accomplished through first packaging in a PVC tray with a Tyvek® lid, which acts as a sterile barrier. The inner layer of packaging is subsequently packaged in a foil pouch. The foil pouch serves as a sterile barrier and a moisture barrier. The foil pouch layer is subsequently packaged into a cardboard shelf carton. This packaging configuration is capable of maintaining adequate device protection when subjected to a variety of environmental conditions.

Proof that the CBAS surface maintains its desired function throughout the shelf life of the device, potential interactions with other materials, and potential degradation of the heparin or base matrix were obtained through several real-time and accelerated stability tests. Finished product samples were placed in storage under normal environmental conditions to detect changes during real-time testing as well as increased temperature and humidity designed to increase the kinetics of any changes that were occurring over time. These tests demonstrated that there was no degradation to the grafts or the CBAS surface through the quality control tests. Confirmation was obtained by taking poststorage samples and subjecting them to the blood loop model and determining there was no change in thromboresistance. These results culminated in the establishment of the labeled storage conditions and shelf life of the grafts.

6.8.7 Regulatory Approval Pathways

Two basic pathways were used to achieve market approval and subsequent commercialization. In the United States, because of the similarities between the graft design and clinical application of the PROPATEN Vascular Graft and its predicate graft, the 510(k) approval process was chosen. An Investigational Device Exemption (IDE) was required that was shortened due to the cumulative clinical success in the United States and Europe and the 510(k) was approved in 2006. In Europe, the PROPATEN Vascular Graft received CE approval based on the demonstration of graft usefulness outweighing new potential risks as demonstrated by the extensive preclinical studies. What remained to be seen was the long-term performance in the human patient population.

Virtually all regulatory pathways require a manufacturer to demonstrate that the usefulness of the new device outweighs the risks associated with that device. The CBAS surface has been in use for decades, and today it is the most extensively published commercial heparin surface coating for medical devices. The continued commercial application of the CBAS surface is based on the proven usefulness of the technology for increasing the hemocompatibility of blood-contacting surfaces and on its documented benefit for improving clinically meaningful outcomes in patients who receive CBAS-treated medical devices.

6.9 CLINICAL PERFORMANCE, CLINICAL USEFULNESS, AND CLINICAL CHALLENGES

6.9.1 Clinical Performance

Once the development challenges are complete and the regulatory pathways are followed, what remains is the final clinical evidence and proof of patient benefit. All indications were that the PROPATEN Vascular Graft could play an important role in the management of peripheral arterial obstructive disease. In animal models and clinical applications, evidence has indicated that PROPATEN Vascular Graft is superior to uncoated grafts with patency rates comparable to autologous veins in humans [11,15,59,74]. By substantially reducing acute graft thrombosis within weeks after implantation, PROPATEN Vascular Graft provides beneficial effects that untreated grafts do not [81].

Early product and clinical evaluations of PROPATEN Vascular Graft were conducted in Europe and the United States with the intent of characterizing the performance of PROPATEN Vascular Graft in above-knee peripheral bypass applications. As the above-knee bypass application treats less medically challenging patient populations and presents lower risk than below-knee peripheral bypass applications, these early evaluations were essentially designed to investigate how PROPATEN Vascular Graft compared to the current treatment paradigm at the time for the same patient population, standard ePTFE vascular grafts. As the positive clinical performance of PROPATEN Vascular Graft above-knee was demonstrated, more challenging patient populations treated with PROPATEN Vascular Graft continued to be studied worldwide.

Following the initial European product evaluation, clinical results for PROPATEN Vascular Graft bypasses in both above-knee (Table 6.3) and below-knee (Table 6.4)

TABLE 6.3 Clinical Results of the PROPATEN Vascular Graft and Autologous Saphenous Vein for Above-Knee Bypasses

		Primary Patency (%)		
Author (Year)	No. of patients	1 year	2 year	3 year
Autologous Saphenous Vein				
Veith et al. (1986) [4]	85	83	78	70
Mills in Rutherford RB (2005) [9]	Review: All series published since 1981	84	82	73
Daenens et al. (2009) [1]	12	91	80	NI
PROPATEN Vascular Graft				
Peeters et al. (2008) [84]	75	81	78	75
Daenens et al. (2009) [1]	86	92	83	NI
Pulli et al. (2010) [85]	101	80	72	72
Samson and Morales (2013) [3]	47	88	88	88

NI: not investigated.

TABLE 6.4 Clinical Results of the PROPATEN Vascular Graft and Autologous Saphenous Vein for Below-Knee Bypasses

			Primary Patency (%)			
Author (Year)	Indication	No. of Patients	1 year (FP/FC)	2 year (FP/FC)	3 year (FP/FC)	4 year (FP/FC)
Autologous Saphenous Vein						
Daenens et al. (2009) [1]	BK FP and FC	98	72/69	72/64	NI	NI
Dorigo et al. (2012) [2]	BK FP, FC, and Tibioperoneal Trunk	394	80	68	66	61
Neville et al. (2012) [82]	BK FC	50	86	NI	NI	NI
PROPATEN Vascular Graft						
Walluscheck et al. (2005) [83]	BK FP and FC	31	92	81	NI	NI
Peeters et al (2008) [84]	BK FP and FC	78	86/71	79/60	75/60	NI
Daenens et al. (2009) [1]	BK FP and FC	154	92/79	83/69	NI	NI
Pulli et al. (2010) [85]	BK FP and FC	324	75/66	67/57	61/52	NI
Dorigo et al. (2012) [2]	BK FP, FC and Tibioperoneal Trunk	556	73	62	52	45
Neville et al. (2012) [82]	BK FC	62	75	NI	NI	NI
Monaca et al. (2013) [86][a]	BK FP and Tibioperoneal Trunk	212	80	62	58	55

BK: below-knee; FP: femoro-popliteal; FC: femoro-crural; NI: not investigated.
[a]9-year follow-up reported.

applications began to be reported. Treatment with PROPATEN Vascular Graft resulted in excellent early- and mid-term patency results comparable with autologous saphenous vein in above-knee bypasses. When PROPATEN Vascular Graft was used for below-knee bypasses, graft thrombosis was significantly reduced, or the bypass was associated with less acute thrombosis compared to the use of standard ePTFE vascular grafts. It was soon discovered that even for one of the most challenging patient populations, those patients with critical limb ischemia and, typically, having femorodistal or femorocrural bypasses, comparative results to autologous saphenous vein were demonstrated.

In a retrospective case–control study with a mean follow-up of 19 months, below-knee bypasses were compared between PROPATEN Vascular Graft, autologous saphenous vein, and standard ePTFE with no significant differences among these groups [87]. Early graft thrombosis, fewer than 30 days postoperatively, occurred in 5, 3, and 10 cases, respectively. The 18-month primary patency rates were 53% (PROPATEN Vascular Graft), 75% (autologous saphenous vein), and 40% (standard ePTFE). Statistical analyses showed significantly better results for autologous saphenous vein and PROPATEN Vascular Graft compared with standard ePTFE and there was not a significant difference between autologous saphenous vein and PROPATEN Vascular Graft in early graft thrombosis. Longer term follow-up with an inclusion of additional below-knee bypass patients in this study revealed 2-year primary patency rates for PROPATEN Vascular Graft of 81% reported for below-knee bypasses with limb salvage of 84% at 2 years [88].

A further retrospective, nonrandomized study of PROPATEN Vascular Graft and autologous saphenous vein grafts implanted in below-knee femoropopliteal and infrapopliteal positions showed no differences in terms of graft occlusion and amputation between the PROPATEN Vascular Graft and autologous saphenous vein. The 1- and 2-year primary patency rates in PROPATEN Vascular Graft and autologous saphenous vein groups were 78/76% and 80/80%, respectively [89]. At 2 years, primary patency rates of 69% for femorocrural PROPATEN Vascular Graft bypasses and 64% for femorocrural autologous saphenous vein bypasses were reported in the largest single-center nonrandomized retrospective series of PROPATEN Vascular Graft implantations at the time (240 patients were treated with a PROPATEN Vascular Graft and 110 patients with autologous saphenous vein) [1]. The authors concluded that there was no significant difference between patency rates of PROPATEN Vascular Graft and autologous saphenous vein grafts. Adjunctive procedures were employed at the distal anastomotic site in 37% of these femorocrural bypasses. In the majority of cases, adjunctive techniques are applied at the distal anastomotic site of femorocrural bypasses due to the mismatch between a 6 mm prosthetic graft and a 2–3 mm often-calcified crural vessel [1,87]. It has been shown that results below the knee for PROPATEN Vascular Graft can be optimized by the stringent use of a patch of bovine pericardium, due to the specific material characteristics [83]. Promising results with regard to patency and limb salvage rates have also been demonstrated without the use of adjunctive procedures such as cuffs or patches at 1 year in a prospective, multicenter trial [90,91], at 2 years with all patients having critical limb ischemia [92], and in a longer study out to 3 years with nearly 50% of the patient population having critical limb ischemia [84]. These results suggest that the early patency results reported can be prolonged and are maintained over time in the clinical application [84].

The previous result was later replicated in a 5-year retrospective study with patients having limb salvage rates of 84% at 5 years, the majority of which having severe peripheral arterial disease treated with femorocrural bypasses [93]. It was further demonstrated in a prospective, multicenter trial, enrolling 142 patients with above-knee and below-knee PROPATEN Vascular Graft bypasses, that the primary patency rate decreased as the number of patent run-off vessels decreased [94].

In podium presentations and publication, it was reported that PROPATEN Vascular Grafts had improved clinical performance over standard ePTFE, especially in the most challenging patient populations [3,95]. Recent prospective and retrospective studies have led to the conclusion that peripheral arterial disease treatment using PROPATEN Vascular Graft is a clinically acceptable, safe alternative to treatment with native vein, especially disadvantaged vein [2]. The Scandinavian PROPATEN trial prospectively evaluated PROPATEN Vascular Graft across 11 centers in patients with chronic limb ischemia [95]. PROPATEN Vascular Graft was randomized against Stretch ePTFE Vascular Grafts in femoropopliteal (above-knee and below-knee) or femoral–femoral bypasses and demonstrated statistically significant improvement versus ePTFE in primary patency, secondary patency, and in patients with critical limb ischemia. It was determined that as severity of the disease increases, the benefit of PROPATEN Vascular Graft increases [95].

Single-center studies in the United States have demonstrated similar comparisons. Tibial artery bypass performance was studied retrospectively using PROPATEN Vascular Graft compared to great autologous saphenous vein in patients with critical limb ischemia finding that at 1 year patency and limb salvage rates were comparable to intact great saphenous vein [82]. A single-center, prospective U.S. study showed statistically better overall primary patency rates for PROPATEN Vascular Graft compared to historical standard ePTFE results at 1, 2, and 3 years for femoropopliteal (above-knee and below-knee) positions [3]. Over 9 years, 556 above-knee and below-knee patients with critical limb ischemia were treated with a PROPATEN Vascular Graft and retrospectively enrolled in an Italian multicenter registry [2]. Early and late follow-up results of below-knee bypasses performed using PROPATEN Vascular Graft in patients with peripheral arterial obstructive disease were compared with those obtained in patients operated on with autologous saphenous vein in the same centers over the same period of time. Secondary patency and limb salvage were not statistically different in the two groups at 4 years.

The longest PROPATEN Vascular Graft patency results to date are reported from the same large, multicenter Italian registry [86]. Nine year results were reported for below-knee patency and limb salvage in a subset of Italian Registry patients that had a low risk for thrombotic events and received PROPATEN Vascular Graft, saving the great saphenous vein for future distal revascularization. In these patients with low thrombotic risk, mid-and long-term patency of vein and PROPATEN Vascular Graft were demonstrated to be comparable.

It is important to consider some points with regard to patients receiving a readily available, off-the-shelf PROPATEN Vascular Graft even if vein is available. First, the patients for these procedures are increasingly older at the time of scheduling and present with more comorbid conditions. More high-risk patients with long occlusions that are not suitable for interventional therapy are presenting for surgery. These patients specifically benefit from anesthesiologic management in regional anesthesia with short operation times. Vein harvesting is time consuming and produces an expanded wound area compared with the use of a prosthetic graft. The actuarial survival rate of these high-risk patients with cardiovascular diseases is low, and a significant number of patients will die with patent bypass in a short-term period. The primary target is to achieve limb salvage and optimize quality of life for patients with severe critical limb ischemia. In addition to using the PROPATEN Vascular Graft for a patient who does not have a suitable vein, the PROPATEN Vascular Graft is considered to be a viable alternative for older patients with increased comorbidity independent of vein availability [96]. Furthermore, PROPATEN Vascular Graft is an advantageous treatment paradigm as it can be revised effectively, while it can be difficult to revise vein [2].

6.9.2 Clinical Usefulness

The proprietary CBAS endpoint immobilization of heparin preserves long-term heparin function by creating a stable, nonleaching heparin surface and allowing preservation of the heparin active sequence. Numerous publications and studies have examined the hemocompatible properties of the CBAS surface in controlled *in vitro* blood contact models or *in vivo* animal models. Many of these studies employed analytical methods similar or identical to those described in ISO 10993-4 [80]. Additionally, a number of clinical studies that provide data on the clinical utility of the CBAS surface are also presented. The CBAS surface has been shown to exhibit several hemocompatible properties that establish its clinical usefulness. These properties are divided into the following categories:

(1) *Inhibition of the coagulation cascade and the complement system.* The action of the immobilized CBAS surface to inhibit the activity of coagulation proteases, including FX, FXII, and thrombin, has been extensively described using *in vitro* blood contact models. Reduction of coagulation biomarkers was also observed in an *in vitro* study with human blood [97]. Similar inhibition of the complement system has also been described.

(2) *Reduced platelet attachment and activation.* The CBAS surface has also been shown to improve the platelet compatibility of medical device surfaces through reduction of platelet attachment and activation. This action appears to be dependent on the presence of the antithrombin binding domain of heparin, and, as a result, the reduced binding of platelets may be secondary to the direct anticoagulant action of the surface-bound heparin. The platelet compatibility of the CBAS surface has been observed using *in vitro* models, *in vivo* animal models, and in human clinical studies.

(3) *Reduced deposition of thrombus on device surfaces.* This effect is the result of inhibition of coagulation and platelet attachment. Reduced thrombus relative to untreated control devices has been observed using *in vitro* models, animal models, and in human clinical studies.

(4) *Improved patency and reduced subacute thrombosis of devices.* This property directly relates to device performance. Improvement in patency due to the CBAS surface has been observed in several animal models. Reduction in subacute thrombosis in CBAS-treated coronary stents was directly measured in an animal model [98] and has also been suggested in a clinical registry-based study [99].

(5) *Reduced intimal hyperplasia associated with coronary stents and ePTFE vascular grafts.* Indications of these effects have been observed in both canine and nonhuman primate animal models [59,75,100].

Many of these properties have been demonstrated employing *in vitro* and *in vivo* investigations. One of the most important clinical observations supporting clinical usefulness come from histological examination of a small number of PROPATEN Vascular Graft explant specimens. In order to resist thrombus buildup, it is essential that heparin remain on the surface and retain its potential for bioactive function. Controlled preclinical studies of the use of the PROPATEN Vascular Graft out to 3 months and anecdotal human explants at approximately 8 months, 3 years, and 8 years postimplantation have demonstrated surface-bound heparin bioactivity levels that were within the original manufacturing specifications for release of this product. This is in line with longevity reported with ventricular assist devices [101]. These observations of long-term *in vivo* heparin bioactivity illustrate the potential for sustained functionally active heparin on the graft surface. PROPATEN Vascular Graft has improved patency compared to nonheparinized ePTFE with the potential to improve patient quality of life with less revisions and greater limb salvage. This evidence supports the product concept and exhibits the usefulness of the proprietary CBAS endpoint covalent heparin bonding to achieve local thromboresistance at the graft surface while avoiding the systemic effects of heparin.

6.9.3 Clinical Challenges

The basis of the clinical benefit of PROPATEN Vascular Graft appears to be primarily due to its thromboresistant properties as well as antiplatelet effects. The body of clinical evidence published to date on PROPATEN Vascular Graft has shown patency improvement over traditional ePTFE grafts and competitive patency comparisons to the gold standard conduit, autologous saphenous vein. However, since its commercial availability, this graft has presented specific clinical considerations that are not routine among all surgeons using PROPATEN Vascular Graft as well as challenges to the clinical community for which a deeper understanding of the scientific mechanism of action behind PROPATEN Vascular Graft has been and continues to be investigated.

6.10 HEPARIN-INDUCED THROMBOCYTOPENIA

Heparin-induced thrombocytopenia (HIT) is a possible adverse effect of heparin therapy. It occurs in two forms, of which the milder (HIT or HIT I) is transient and reversible. The severe form (HITT or HIT II) is delayed in onset and is mediated by an immune mechanism. It manifests clinically as a severe decline in circulating platelets (thrombocytopenia) sometimes accompanied by thrombosis. When HIT II is diagnosed, heparin is usually withdrawn and replaced by an alternative anticoagulant therapy.

HIT II is caused by antibodies that form in response to complexes of heparin and platelet factor 4 (PF4) on platelet surfaces, resulting in aggregation and activation of the platelets. The rate of HIT is low but significant, ranging from <1 to 5% in surgical patients who receive heparin systemically [102,103]. The incidence of HIT differs depending on the patient population (women > men, surgical patients > medical patients) and on the kind of heparin administered (unfractionated > low molecular weight).

Almost exclusively, heparin used clinically takes the form of injectable heparin given to systemically anticoagulate blood and reduce the risk of thrombosis. Nearly all patients having a vascular procedure routinely get systemic heparin administration during the procedure. The risk of HIT II from systemic heparin and the molecular mechanisms responsible for the antibody-mediated reduction of platelets have been extensively described in the literature [103]. Heparin used in small amounts to flush indwelling catheters is another source of heparin exposure in the clinic, and HIT II related to heparin flushes has also been reported [104]. Medical devices containing immobilized heparin represent another mode of heparin exposure to patients, but relatively little is known about the interaction of HIT antibodies with this source of heparin. Complicating this evaluation is that a variety of different heparin bonding technologies for medical devices is in use commercially. The amount and type of heparin used, and the method by which the heparin is bonded to the device differ substantially between manufacturers [47]. Some heparin-bonded devices are designed to elute heparin over time and others, similar to the CBAS surface, contain a stable covalently bound heparin that is essentially noneluting into the blood.

As the antibodies against heparin in complex with platelet factor 4 are predominantly responsible for the development of HIT II, they are often used to diagnose the condition. In the prospective, randomized U.S. clinical study comparing PROPATEN Vascular Graft and the GORE® Stretch Vascular Graft (uncoated ePTFE graft), no reported cases of HIT in either group ($n = 200$) were found. In a small study involving 10 patients receiving PROPATEN Vascular Grafts and 10 patients receiving uncoated ePTFE grafts, no antibodies could be demonstrated in either group up to 6 weeks after implantation [55]. Dr. Theodore Warkentin, an international authority on HIT who has coauthored recognized diagnostic and treatment guidelines on this syndrome [105], reported not finding any convincing evidence that the CBAS surface has a role in causing or contributing to HIT [106].

In the published cases of HIT following implantation of PROPATEN Vascular Graft, in which symptoms persist or worsen after withdrawal of systemic heparin therapy, the available evidence points to the cause being a form of HIT called "delayed-onset HIT" in which the graft is likely an "innocent bystander." Resolution of thrombocytopenia has occurred even when the grafts remained *in situ* and in circulation demonstrating HIT antibody transience [107]. Thus, thrombocytopenia that has resolved upon explantation of the heparin-coated device is likely coincidental with the natural history of HIT. If heparin-bonded grafts were directly activating and maintaining HIT, platelet-rich graft thrombosis would be expected to be the most common feature of these cases. However, published data show that this is not the case.

In conclusion, from the vast number of cardiovascular interventions where CBAS-coated devices have been used, reports on heparin-induced thrombocytopenia are negligible [107]. Establishing a cause and effect relationship is particularly challenging because patients receiving CBAS-coated devices usually have had prior heparin exposure and/or receive heparin during the surgical procedure to implant the device. Finally, if noneluting heparin-bonded grafts were directly activating platelets and thereby maintaining clinically evident HIT, it is expected that the most prominent clinical feature would be graft thrombosis or associated thromboembolism. Furthermore, overall the published data demonstrate that the benefits of the PROPATEN Vascular Graft outweigh any risks associated with the implantation of this thromboresistant vascular graft.

6.11 WORKING HYPOTHESIS

When medical devices are exposed to the circulating blood, plasma proteins are instantaneously adsorbed to the polymeric or metallic artificial surface. The specific biochemistry of this plasma protein deposition likely varies depending upon the chemistry of the underlying artificial material surface. Protein adsorption, however, inevitably involves conformational changes that lead to activation of the coagulation system.

While not fully substantiated in every respect, a reasonable working hypothesis for how the CBAS surface improves blood compatibility and overall biocompatibility would be the following: The CBAS surface makes protein adsorption more selective and independent of the underlying material. By adsorbing inhibitors of coagulation, notably antithrombin, the endpoint-immobilized heparin has the capacity to suppress the initiation of the coagulation cascade by the artificial surface. As a consequence of the suppression of the coagulation, adhesion, activation, and aggregation of platelets are minimized. The dampening of the inflammatory response to the CBAS surface most likely occurs by similar mechanisms, that is, by preventing activation of components of the complement system. Moreover, as there appears to be a connection between remodeling of clot and neointimal hyperplasia, the long-term inhibition of thrombus formation may even to some degree affect this response to the vascular implant.

The molecular mechanisms of action underlying how the CBAS surface improves clinical performance of the PROPATEN Vascular Graft continue to be investigated using *in vitro* and *in vivo* experimentation. Continued scientific and clinical investigation will help us to understand more completely how the CBAS surface performs as well as how it can be improved upon.

6.12 SUMMARY

Development of a drug–device combination product is a daunting technical, product development, manufacturing, and commercial challenge. The increased complexities of trying to maintain the drug functionality on a device surface while effectively demonstrating patient benefit require a long-term view. These efforts can easily take a decade from conception to proven clinical performance. Critical to this process is the need to be rigorous and objective to a fault, with respect to both the technologies employed and the regulatory and clinical environment in which the drug–device combination product must be proven. In today's increasingly complex environment, drug–device combination products will continue to require deep understanding of the device, the drug, and the combination of the two to provide reasonable assurances that the device is safe and effective. Successfully delivering these technology advances with real understanding truly will advance patient care and outcomes in a highly meaningful way.

The synergy between two successful technologies to achieve improved patient outcomes is what drug–device combination products are intended to provide. The collective worldwide published scientific and clinical experience of the PROPATEN Vascular Graft with the CBAS surface demonstrate the significant long-term clinical benefits and improved patient outcomes due to the improved thromboresistance functionality of the graft surface. Ultimately, the desired patient quality of life improvement as a primary goal with any highly advanced drug–device combination product has been achieved with this thromboresistant vascular graft.

CARMEDA® and CBAS® are registered trademarks of Carmeda AB.
GORE® and PROPATEN® are registered trademarks of W. L. Gore & Assoc., Inc.
DACRON® is a registered trademark of Invista North America S. A. R. L. Corp.
DURAFLO® is a registered trademark of Edwards Lifesciences Corporation.
TECOFLEX® is a registered trademark of Thermedics, Inc.

REFERENCES

1. K. Daenens, S. Schepers, I. Fourneau, S. Houthoofd, and A. Nevelsteen, Heparin-bonded ePTFE grafts compared with vein grafts in femoropopliteal and femorocrural bypasses: 1- and 2-year results. *J. Vasc. Surg.* 49 (2009) 1210–1216.
2. W. Dorigo, P. Raffaele, and G. Piffaretti, Results from an Italian multicentric registry comparing heparin-bonded ePTFE graft and autologous saphenous vein in below-knee femoro-popliteal bypasses. *J. Cardiovasc. Surg.* 53 (2012) 187–193.
3. R.H. Samson and L. Morales, Improved three year patency rates for heparin-bonded ePTFE femoropopliteal bypass grafts vs. ePTFE grafts without heparin. Presented at the 35th International Charing Cross Symposium (CX 35); April 6–9, 2013; London, UK, in R.M. Greenhalgh (Ed.), *Vascular and Endovascular Challenges Update*, BIBA Publishing, London, UK, 2013, pp. 315–320.
4. F. J. Veith, S. K. Gupta, E. Ascer, S. White-Flores, R.H. Samson, L.A. Scher, J.B. Towne, V.M. Bernard, P. Bonier, W.R. Flinn, P. Astelford, J.S.T. Yao, and J.J. Bergan, Six year prospective multicenter randomized comparison of autologous saphenous vein and expanded PTFE in infrainguinal reconstruction. *J. Vasc. Surg.* 3 (1986) 104–114.
5. C.P. Twine and A.D. McLain, Graft type for femoro-popliteal bypass surgery (review). *Cochrane Database Syst. Rev.* 5 (2010) CD001487.
6. J.T. Scales, Tissue reactions to synthetic materials. *Proc. R. Soc. Med.* 46 (1953) 647.

7. J.J. Rocotta, Vascular conduits: an overview, in R.B. Rutherford (Ed.), *Vascular Surgery*, Sixth ed., W.B. Saunders, Philadelphia, 2005, pp. 688–695.
8. R.B. Rutherford, J.D. Baker, C. Ernst, K.W. Johnson, J.M Porter, S. Ahn, and D.N. Jones, Recommended standards for reports dealing with lower extremity ischemia: revised version. *J. Vasc. Surg.* 26 (1997) 517–538.
9. J.L. Mills, Infrainguinal bypass, in R.B. Rutherford (Ed.), *Vascular Surgery*, Sixth ed., W.B. Saunders, Philadelphia, 2005, pp. 1154–1174.
10. F.J. Veith, E. Ascer, S.K. Gupta, S. White-Flores, S. Sprayregan, and L.A. Scher, Tibiotibial vein bypass grafts: a new operation for limb salvage. *J. Vasc. Surg.* 2 (1995) 552–557.
11. F.J. Veith, K. Gupta, R.H. Samson, S. White-Flores, G. Janko, and L.A. Scher, Superficial femoral and popliteal arteries as inflow site for distal bypasses. *Surgery*, 90 (1981) 980–990.
12. L.M. Taylor, J.M. Edwards, and J.M. Porter, Present status of reversed vein bypass grafting: five year results of a modern series. *J. Vasc. Surg.* 11 (1990) 193–205.
13. G.L. Moneta and J.M. Porter, Arterial substitutes in peripheral vascular surgery: a review. *J. Long-Term Eff. Med. Implants* 5 (1995) 47–67.
14. M. Luther and M. Lapantalo, Infrainguinal reconstructions: influence of surgical experience on outcome. *Cardiovasc. Surg.* 6 (1998) 351–357.
15. C. Devine and C. McCollum, Heparin-bonded Dacron or polytetrafluorethylene for femoropopliteal bypass: five-year results of a prospective randomized multicenter clinical trial. *J. Vasc. Surg.* 40 (2004) 924–931.
16. P. Klinkert, P.N. Post, P.J. Breslau, and J.H. van Bockel, Saphenous vein versus PTFE for above-knee femoropopliteal bypass: a review of the literature. *Eur. J. Vasc. Endovasc. Surg.* 27 (2004) 357–362.
17. P.A. Stonebridge, R.J. Prescott, and C.V. Ruckley, Randomized trial comparing infrainguinal polytetrafluoroethylene bypass grafting with and without vein interposition cuff at the distal anastomosis. The Joint Vascular Group. *J. Vasc. Surg.* 26 (1997) 543–550.
18. R.F. Neville, B. Tempesta, and A.N. Sidway, Tibial bypass for limb salvage using polytetrafluoroethylene with a distal vein patch. *J. Vasc. Surg.* 33 (2001) 266–272.
19. M.C. Stoner and W.A. Abbott, Vascular grafts: characteristics and rational selection of prostheses, in W.S. Moore (Ed.), *Vascular and Endovascular Surgery: A Comprehensive Review*, 7th ed., Elsevier, Philadelphia, 2006, 452–468.
20. W.M. Abbott and M.E. Landis, Vascular grafts: characteristics and rational selection of prostheses, in *Vascular Surgery: A Comprehensive Review*, Sixth ed., W.B. Saunders, California, 2002, pp. 419–436.
21. R.J. Zdrahala, Small caliber vascular grafts: Part I. State of the art. *J. Biomater. Appl.* 10 (1996) 309–329.
22. M.C. Tanzi, Bioactive technologies for hemocompatibility. *Expert Rev. Med. Devices*, 2 (4) (2005) 473–492.
23. S.W. Jordan and E.L. Chaikof, Novel thromboresistant materials. *J. Vasc. Surg.*, 45 (6) (2007) Supplement 1:A104-A115.
24. C.O. Esquivel, C.G. Björck, S.E. Bergentz, D. Bergqvist, R. Larsson, S.N. Carson, P. Dougan, and B. Nilsson, Reduced thrombogenic characteristics of expanded polytetrafluoroethylene and polyurethane arterial grafts after heparin bonding. *Surgery* 95 (1) (1984) 102–107.

25. J.S.T. Yao. Dr. Ben Eiseman and the accidental discovery of Gore-Tex® graft, in M.K. Eskandari, M.D. Morasch, W.H. Pearce, J.S.T. Yao (Eds.), *Contemporary Vascular Surgery*, People's Medical Publishing House, Shelton, CT, 2011, pp. 1–7.
26. C.D. Campbell, D.H. Brooks, and H.T. Bahnson, Expanded microporous polytetrafluoroethylene (Gore-Tex) as a vascular conduit, in P.N. Sawyer, M.J. Kaplitt (Eds.), *Vascular Grafts*, Appleton-Century Crofts, New York, 1978, pp. 335–348.
27. M.E. McClurken, J.M. McHaney, and W.M. Colone, Physical properties and test methods for expanded polytetrafluoroethylene (PTFE) grafts, in H.E. Kambic, A. Kantrowitz, P. Sung (Eds.) *Vascular Graft Update: Safety and Performance, ASTM 898*, American Society for Testing and Materials, Philadelphia, 1986, pp. 82–94.
28. L.M. Graham and J.J. Bergan, Expanded polytetrafluoroethylene vascular grafts: clinical and experimental observations, in J.C. Stanley (Ed.), *Biologic and Synthetic Vascular Prostheses*, Grune & Stratton, New York, 1982, pp. 563–586.
29. A. Lumsden, A unique combination of ePTFE and proprietary endpoint covalent bonding of heparin for lower extremity revascularization: The GORE PROPATEN Vascular Graft. *Vasc. Dis. Manage.* (Suppl. B) (2007) 11B–14B.
30. S.J. Sheehan, S.P. Colgan, D.J. Moore, and G.D. Shank, The role of PTFE grafts in lower arterial reconstruction, in F.J. Veith (Ed.), *Current Critical Problems in Vascular Surgery*, Vol. 5, Quality Medical Publishing Inc., St. Louis, 1993, pp. 81–84.
31. D.A. Lane and U. Lindahl, Heparin: chemical and biological properties, in *Clinical Applications*, CRC Press, Inc., Boca Raton, FL, 1989, pp. 1–623.
32. U. Lindahl, 'Heparin': from anticoagulant drug into the new biology. *Glycoconjug. J.* 17 (7–9) (2000) 597–605.
33. I. Björk, S.T. Olson, and J.D. Shore, Molecular mechanisms of the accelerating effect of heparin on the reactions between antithrombin and clotting proteases, in D.A. Lane, U. Lindahl (Eds.), *Heparin: Chemical and Biological Properties, Clinical Applications*, R.C. Press, Florida, 1989, pp. 229–255.
34. V.L. Gott, J.D. Whiffen, and R.C. Dutton, Heparin bonding on colloidal graphite surfaces. *Science* 142 (1963) 1297–1298.
35. V.L. Gott, Wall-bonded heparin: historical background and current clinical applications. *Adv. Exp. Med. Biol.* 52 (1975) 351–363.
36. R.D. Falb, R.I. Leininger, G. Grode, and J. Crowley, Surface-bonded heparin. *Adv. Exp. Med. Biol.* 52 (1975) 365–374.
37. J.M. Toomasian, L.C. Hsu, R.B. Hirschl, K.F. Heiss, K.A. Hultquist, and R.H. Bartlett, Evaluation of Duraflo II heparin coating in prolonged extracorporeal membrane oxygenation. *Trans. Am. Soc. Artif. Intern. Organs* 34 (1988) 410–414.
38. R.L. Larsson, M.B. Hjelte, J.C. Eriksson, H.R. Lagergren, and P. Olsson, The stability of glutardialdehyde-stabilized 35-heparinized surfaces in contact with blood. *Thromb. Hemost.* 37 (1977) 262–273.
39. L.O. Andersson, T.W. Barrowcliffe, E. Holmer, E.A. Johnson, G.E. Sims, Anticoagulant properties of heparin fractionated by affinity chromatography on matrix-bound antithrombin III and by gel filtration. *Thromb. Res.* 9 (1976) 575–583.
40. M. Höök, I. Björk, J. Hopwood, and U. Lindahl, Anticoagulant activity of heparin: separation of high-activity and low-activity heparin species by affinity chromatography on immobilized antithrombin. *FEBS Lett.* 66 (1976) 90–93.

41. L.H. Lam, J.E. Silbert, and R.D. Rosenberg, The separation of active and inactive forms of heparin. *Biochem. Biophys. Res. Commun.* 69 (1976) 570–577.
42. U. Lindahl, G. Bäckström, L. Thunberg, and I.G. Leder, Evidence for a 3-*O*-sulfated D-glucosamine residue in the antithrombin-binding sequence of heparin. *Proc. Natl. Acad. Sci. USA* 77 (1980) 6551–6555.
43. L. Thunberg, G. Bäckström, and U. Lindahl, Further characterization of the antithrombin-binding sequence in heparin. *Carbohydr. Res.* 100 (1982) 393–410.
44. O. Larm, R. Larsson, and P. Olsson, A new non-thrombogenic surface prepared by selective covalent binding of heparin via a modified reducing terminal residue. *Biomater. Med. Devices Artif. Organs* 11 (1983) 161–173.
45. O. Larm, Process for covalent coupling for the production of conjugates, and polysaccharide containing products thereby obtained. U.S. Patent 4,613,665. September 23, 1986.
46. BS EN ISO 10993-1: October 2009, Biological Evaluation of Medical Devices. Part 1: Evaluation and Testing Within a Risk Management Process, ISO 10993-1:2009, British Standards, 1–32.
47. H.P. Wendel and G. Ziemer, Coating-techniques to improve the hemocompatibility of artificial devices used for extracorporeal circulation. *Euro. J. Cardiovasc. Surg.* 16 (3) (1999) 342–350.
48. B. Pasche, K. Kodama, O. Larm, P. Olsson, and J. Swedenborg, Thrombin inactivation on surfaces with covalently bonded heparin. *Thrombos. Res.* 4 (1986) 739–748.
49. K. Kodama, B. Pasche, P. Olsson, J. Swedenborg, L. Adolfsson, O. Larm, and J. Riesenfeld, Antithrombin III binding to surface immobilized heparin and its relation to F Xa inhibition. *Thromb. Hemost.* 58 (1987) 1064–1067.
50. J. Sanchez, G. Elgue, J. Riesenfeld, and P. Olsson, Studies of adsorption, activation, and inhibition of factor XII on immobilized heparin. *Thromb. Res.* 89 (1998) 41–50.
51. J. Sanchez, G. Elgue, J. Riesenfeld, and P. Olsson, Control of contact activation on end-point immobilized heparin: the role of antithrombin and the specific antithrombin-binding sequence. *J. Bio. Mater. Res.* 29 (1995) 655–661.
52. G. Elgue, M. Blombäck, P. Olsson, and J. Riesenfeld, On the mechanism of coagulation inhibition on surfaces with end point immobilized heparin. *Thromb. Hemost.* 70 (2) (1993) 289–293.
53. C. Arnander, P. Olsson, and O. Larm, Influence of blood flow and the effect of protamine on the thromboresistant properties of a covalently bonded heparin surface. *J. Biomed. Mater. Res.* 22 (1988) 859–868.
54. N. Weber, H.P. Wendel, and G. Ziemer, Hemocompatibility of heparin-coated surfaces and the role of selective plasma protein adsorption. *Biomaterials* 23 (2) (2002) 429–439.
55. J.M.M. Heyligers, T. Lisman, H.J. Verhagen, C. Weeterings, P.G. de Groot, and F.L. Moll, A heparin-bonded vascular graft generates no systemic effect on markers of hemostasis activation or detectable heparin-induced thrombocytopenia-associated antibodies in humans. *J. Vasc. Surg.* 47 (2008) 324–329.
56. V. Videm, T.E. Mollnes, P. Garred, and J.L. Svennevig, Biocompatibility of extracorporeal circulation: *in vitro* comparison of heparin-coated and uncoated oxygenator circuits. *J. Thorac. Cardiovasc. Surg.* 101 (1991) 654–660.

57. T.E. Mollnes, V. Videm, D. Christiansen, G. Bergseth, J. Riesenfeld, and T. Hovig, Platelet compatibility of an artificial surface modified with functionally active heparin. *Thromb. Hemost.* 82 (1999) 1132–1136.

58. J.F. Kocsis, G. Llanos, and E. Holmer, Heparin-coated stents. *J. Long Term Eff. Med. Implants* 10 (2000) 19–45.

59. P.H. Lin, C. Chen, R.L. Bush, Q. Yao, A.B. Lumsden, and S.R. Hanson, Small-caliber heparin-coated ePTFE grafts reduce platelet deposition and neointimal hyperplasia in a baboon model. *J. Vasc. Surg.* 39 (2004) 1322–1328.

60. E. Holmer, S. Hanson, N. Chronos, G. Llanos, and J.F. Kocsis, On the mechanism for the non-thrombogenic properties of the Carmeda BioActive Surface on coronary stents: results from a baboon AV-shunt study. Abstract presented at Surfaces in Biomaterials, September 2–4, 1999, Scottsdale, AZ.

61. T.E. Mollnes, J. Riesenfeld, P. Garred, E. Nordström, K. Høgåsen, E. Fosse, O. Gotze, and M. Harboe, A new model for evaluation of biocompatibility: combined determination of neoepitopes in blood and on artificial surfaces demonstrates reduced complement activation by immobilization of heparin. *Artif. Organs* 19 (1995) 909–917.

62. V. Videm, J.L. Svennevig, E. Fosse, G. Semb, A. Osterud, and T.E. Mollnes, Reduced complement activation with heparin-coated oxygenator and tubings in coronary bypass operations. *J. Thorac. Cardiovasc. Surg.* 103 (1992) 806–813.

63. E. Fosse, O. Moen, E. Johnson, G. Semb, V. Brockmeier, T.E. Mollnes, M.K. Fagerhol, and P. Venge, Reduced complement and granulocyte activation with heparin-coated cardiopulmonary bypass. *Ann. Thorac. Surg.* 58 (1994) 472–477.

64. P. Garred and T.E. Mollnes, Immobilized heparin inhibits the increase in leukocyte surface expression of adhesion molecules. *Artif. Organs* 21 (1997) 293–299.

65. H.E. Høgevold, O. Moen, E. Fosse, P. Venge, J. Bråten, C. Andersson, and T. Lyberg, Effects of heparin coating on the expression of CD11b, CD11c and CD62 L by leucocytes in extracorporeal circulation *in vitro*. *Perfusion* 12 (1997) 9–20.

66. J.D. Hill, T.G. O'Brien, J.J. Murray, L. Dontigny, M.L. Bramson, J.J. Osborn, and F. Gerbode, Prolonged extracorporeal oxygenation for acute post-traumatic respiratory failure (shock-lung syndrome): use of the Bramson membrane lung. New Engl. J. Med. 286 (1972) 629–634.

67. L. Gattinoni, T. Kolobow, G. Damia, A. Agostoni, and A. Pesenti, Extracorporeal carbon dioxide removal (ECCO2R): a new form of respiratory assistance. *Int. J. Artif. Organs* 2 (1979) 183–185.

68. L. Bindslev, I. Gouda, J. Inacio, K. Kodama, H. Lagergren, O. Larm, E. Nilsson, and P. Olsson, Extracorporeal elimination of carbon dioxide using a surface-heparinized veno-venous bypass system. *Trans. Am. Soc. Artif. Intern. Organs* 32 (1986) 530–533.

69. L. Bindslev, J. Eklund, O. Norlander, J. Swedenborg, P. Olsson, E. Nilsson, O. Larm, I. Gouda, A. Malmberg, and E. Scholander, Treatment of acute respiratory failure by extracorporeal carbon dioxide elimination performed with a surface heparinized artificial lung. *Anesthesiology* 67 (1987) 117–120.

70. J. Borowiec, S. Thelin, L. Bagge, J. Hultman, and H.E. Hansson, Decreased blood loss after cardiopulmonary bypass using heparin-coated circuit and 50% reduction of heparin dose. *Scand. J. Thorac. Cardiovasc. Surg.* 26 (3) (1992) 177–185.

71. G.S. Aldea, M. Doursounian, P. O'Gara, P. Treanor, O. M. Shapira, H. Lazar, and R.J. Shemin, Heparin-bonded circuits with a reduced anticoagulation protocol in primary CABG: a prospective, randomized study. *Ann. Thorac. Surg.* 62 (1996) 410–418.
72. F. Kaufmann, E. Henning, M. Loebe, and R. Hetzer, Improving the antithrombogenity of artificial surfaces through heparin coating: clinical experience with the pneumatic extracorporeal Berlin heart assist device. *Cardiovasc. Eng.* 10 (1996) 40–44.
73. P.W. Serruys, B. van Hout, H. Bonnier, V. Legrand, E. Garcia, C. Macaya, E. Sousa, W. van der Giessen, A. Colombo, R. Seabra-Gomes, F. Kiemeneij, P. Ruyrok, J. Ormiston, H. Emanuelsson, J. Fajadet, M. Haude, S. Klugmann, and M. Morel, Randomised comparison of implantation of heparin-coated stents with balloon angioplasty in selected patients with coronary artery disease (Benestent II). *Lancet* 352 (1998) 673–681.
74. P.C. Begovac, R.C. Thomson, J.L. Fisher, A. Hughson, and A. Gällhagen, Improvements in GORE-TEX® Vascular Graft performance by Carmeda® bioactive surface heparin immobilization. *Eur. J. Vasc. Endovasc. Surg.* 25 (2003) 432–437.
75. P.H. Lin, R.L. Bush, Q. Yao, A.B. Lumsden, and C. Chen, Evaluation of platelet deposition and neointimal hyperplasia of heparin-coated small-caliber ePTFE grafts in a canine femoral artery bypass model. *J. Surg. Res.* 118 (2004) 45–52.
76. BS ISO 7198:1998, Cardiovascular Implants: Tubular Vascular Prostheses, British Standards, 1–50.
77. ANSI/AAMI/ISO TIR 12417:2011, Cardiovascular implants and extracorporeal systems: vascular device–drug combination products, Association for the Advancement of Medical Instrumentation, pp. 1–66.
78. Guidance for industry and FDA staff: classification of products as drugs and devices and additional product classification issues, June 2011, pp. 1–12. Available at http://www.fda.gov/CombinationProducts/default.htm.
79. European Medical Device Directive, Essential Requirements Checklist, Essential Requirements: Annex I, 93/42/EEC, as amended by Directive 2007/47/EC, pp. 1–22.
80. ISO 10993-4-4:2009, Biological Evaluation of Medical Devices. Part 4: Selection of Tests for Interactions with the Blood, British Standard, 1–46.
81. J.M.M. Heyligers, H.J. Verhagen, J.I. Roumans, C. Weeterings, P.G. de Groot, F.L. Moll, and T. Lisman, Heparin immobilization reduces thrombogenicity of small-caliber expanded polytetrafluoroethylene grafts. *J. Vasc. Surg.* 43 (2006) 587–591.
82. R. F. Neville, A. Capone, R. Amdur, M. Lidsky, J. Babrowicz, and A.N. Sidway, A comparison of tibial artery bypass performed with heparin-bonded expanded polytetrafluoroethylene and great saphenous vein to treat critical limb ischemia. *J. Vasc. Surg.* 56 (4) (2012) 1008–1014.
83. K.P. Walluscheck, S. Bierkandt, M. Brandt, and J. Cremer, Infrainguinal ePTFE vascular graft with bioactive surface heparin bonding: first clinical results. *J. Cardiovasc. Surg. (Torino)* 46 (2005) 425–430.
84. P. Peeters, J. Verbist, K. Deloose, and M. Bosiers, Will heparin-bonded PTFE replace autologous venous conduits in infrapoliteal bypass? *J. Vasc. Endovasc. Surg.* 15 (2008) 143–148.
85. R. Pulli, W. Dorigo, P. Castelli, V. Dorrucci, F. Ferilli, et al., Midterm results from a multicenter registry on the treatment of infrainguinal critical limb ischemia using a heparin-bonded ePTFE graft. *J. Vasc. Surg.* 51 (2010) 1167–1177.

86. V. Monaca, G. Battaglia, S.A. Turiano, R. Tringale, and S. Catalfamo, Sub-popliteal revascularization. Criteria analysis for use of E-P.T.F.E. (Propaten ®) as first choice conduit. *Ital. J. Vasc. Endovasc. Surg.* 20 (2013) 165–169.
87. W. Dorigo, R. Pulli, and A.A. Innocenti, Lower limb below-knee revascularization with a new bioactive prosthetic graft: a case–control study. *Ital. J. Endovasc. Surg.* 12 (2005) 75–81.
88. W. Dorigo, F. Di Carlo, N. Troisi, G. Pratesi, A.A Innocenti, R. Pulli, and C. Pratesi, Lower limb revascularization with a new bioactive prosthetic graft: early and late results. *Ann. Vasc. Surg.* 22 (1) (2008), 79–87.
89. G. Battaglia, R. Tringale, and V. Monaca, Retrospective comparison of a heparin bonded ePTFE graft and saphenous vein for infragenicular bypass: implications for standard treatment protocol. *J. Cardiovasc. Surg. (Torino)* 47 (2006) 41–47.
90. M. Bosiers, K. Deloose, J. Verbist, H. Schroë, G. Lauwers, W. Lansink, and P. Peeters, Heparin-bonded expanded polytetrafluoroethylene vascular graft for femoropopliteal and femorocrural bypass grafting: 1-year results. *J. Vasc. Surg.* 43 (2) (2006) 313–319.
91. M. Bosiers, K. Deloose, J. Verbist, and P. Peeters, Two-year results with heparin-bonded PTFE (Gore Propaten) grafts for femoropopliteal and femorodistal bypasses are encouraging. Abstract presented at 32nd Annual Veith Symposium, November 17–20, 2005, New York.
92. V. Dorucci, F. Griselli, G. Petralia, L. Spinamano, and R. Adornetto, Heparin-bonded expanded polytetrafluoroethylene grafts for infragenicular bypass in patients with critical limb ischemia: 2 year results. *J. Cardiovasc. Surg.* 49 (2) (2008) 145–149.
93. H. Lösel-Sadée and C. Alefelder, Heparin-bonded expanded polytetrafluoroethylene graft for infragenicular bypass: five-year results. *J. Cardiovasc. Surg.* 50 (3) (2009) 339–343.
94. B. Hugl, A. Nevelsteen, K. Daenens, M.A. Perez, P. Heider, M. Railo, H. Schelzig, B. Gluecklich, K. Balzer, F. Vermassen, P. De Smit, and G. Fraedrich, PEPE II Study Group, PEPE II: a multicenter study with an end-point heparin-bonded expanded polytetrafluoroethylene vascular graft for above and below knee bypass surgery: determinants of patency. *J. Cardiovasc. Surg.* 50 (2009) 195–203.
95. J. S. Lindholt, B. Gottschalksen, N. Johannesen, D. Dueholm, H. Ravn, E.D. Christiansen, B. Viddal, T. Flørenes, G. Pedersen, M. Rasmussen, M. Carstensen, N. Grøndal, and H. Fasting, The Scandinavian Propaten® Trial: 1-year patency of PTFE vascular prostheses with heparin-bonded luminal surfaces compared to ordinary pure PTFE vascular prostheses: a randomised clinical controlled multi-centre trial. *Euro. J. Vasc. Endovasc. Surg.* 41 (5) (2011) 668–673.
96. K. P. Walluscheck, Ten years of experience with the heparin-bonded ePTFE Graft: the newest advancement in vascular surgery. *Vasc. Dis. Manage. Suppl.* 7 (9) (2010) 3–27.
97. S. Bannan, A. Danby, D. Cowan, S. Ashraf, M. Gesinde, and P. Martin, Cell activation and thrombin generation in heparin bonded cardiopulmonary bypass circuits using a novel *in vitro* model. *Eur. J. Cardiothorac. Surg.* 12 (2) (1997) 268–275.
98. P.A. Hårdhammar, H.M. van Beusekom, H.U. Emanuelsson, S.H. Hofma, P.A. Albertsson, P.D. Verdouw, E. Boersma, P.W. Serruys, and W.J. van der Giessen, Reduction in thrombotic events with heparin-coated Palmaz–Schatz stents in normal porcine coronary arteries. *Circulation* 93 (3) (1996) 423–430.
99. R. Mehran, E. Nikolsky, E. Camenzind, M. Zelizko, I. Kranjec, R. Seabra-Gomes, M. Negoita, S. Slack, and C. Lotan, An Internet-based registry examining the efficacy of

heparin coating in patients undergoing coronary stent implantation. *Am. Heart J.* 150 (2005) 1171–1176.

100. P.H. Lin, N.A. Chronos, M.M. Marijianowski, C. Chen, B. Conklin, R.L. Bush, A.B. Lumsden, and S.R. Hanson, Carotid stenting using heparin-coated balloon-expandable stent reduces intimal hyperplasia in a baboon model. *J. Surg. Res.* 112 (2003) 84–90.
101. J. Riesenfeld, D. Ries, and R. Hetzer, Analysis of the heparin coating of an EXCOR® Ventricular Assist Device after 855 days in a patient. Abstract 180 presented at the 32nd Society for Biomaterials Annual Meeting, April 18–21, 2007, Chicago, IL.
102. L. Linkins and D. Lee, Frequency of heparin induced thrombocytopenia, in T.E. Warkentin, A. Greinacher (Eds.), *Heparin-Induced Thrombocytopenia*, 5th ed., (4) Informa Healthcare, New York, 2012, 110–150.
103. J.G. Kelton and T.E. Warkentin, Heparin-induced thrombocytopenia: a historical perspective. *Blood* 112 (7) (2008) 2607–2616.
104. I. McNulty, E. Katz, and K.Y. Kim, Thrombocytopenia following heparin flush. *Prog. Cardiovasc. Nurs.* 20 (4) (2005) 143–147.
105. T. E. Warkentin, A. Greinacher, A. Koster and A.M. Lincoff, Treatment and prevention of heparin-induced thrombocytopenia: American College of Chest Physicians Evidence-Based Clinical Practice Guidelines (8th Edition). *Chest* 133 (2008) 340–380.
106. T.E. Warkentin, Heparin-coated intravascular devices and heparin-induced thrombocytopenia, in T.E. Warkentin, A. Greinacher (Eds.), *Heparin-Induced Thrombocytopenia*, 5th ed., Informa Healthcare, New York, 2013, 573–590.
107. K. Kasirajan, Outcomes after heparin-induced thrombocytopenia in patients with Propaten vascular grafts. *Ann. Vasc. Surg.* 26 (6) (2012) 802–808.

7

DEVICE-ENABLED DRUG INFUSION THERAPIES

Keith R. Hildebrand
Medtronic, Inc., Neuromodulation, Minneapolis, MN 55432, USA

7.1 INTRODUCTION

This chapter focuses on device-based infusion systems and therapies, namely, external and implantable drug delivery systems, that infuse therapeutic agents over sustained periods of time (months to years) to patients beyond the confines of a hospital or clinic for the management of chronic diseases. Technologies that deliver injectable drug formulations provide many advantages, one of which is liberating patients so that they are free to go about their activities of daily living. Two major categories of infusion devices will be discussed. The first is external delivery systems that are used most commonly for subcutaneous (SQ) infusion of drugs; cell phone-sized devices often worn at the waist to deliver insulin for the treatment of diabetes are the prototypical example. The other major device category that will be discussed is the fully implantable systems (Figure 7.1) that are currently used most commonly to manage two chronic diseases: (1) intractable pain and (2) severe spasticity. Because of the invasive nature and added cost of this device category, they are used for treating disease states that are resistant to standard medical management. This resistance may involve dose-limiting side effects associated with standard systemic (oral or transdermal) administration of therapeutic agents. A detailed analysis of the technical aspects of pump functions and materials is not the intent of this chapter, but rather a review of general concepts related to the delivery system requirements, the primary components comprising the delivery systems and

Drug–Device Combinations for Chronic Diseases, First Edition. Edited by SuPing Lyu and Ronald A. Siegel.

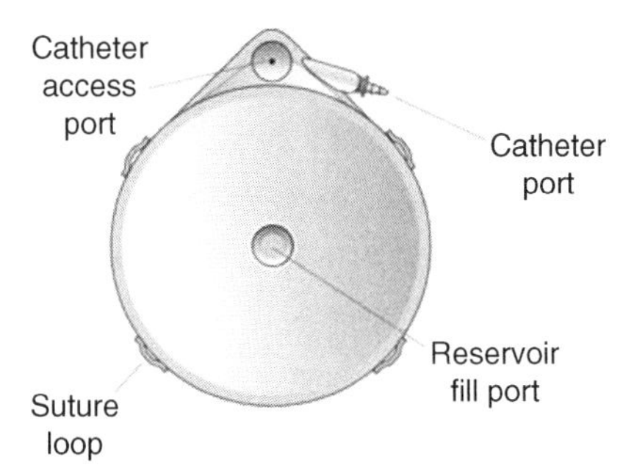

FIGURE 7.1 Implantable drug pump. (Reproduced with permission from Medtronic, Inc.)

their primary functions, clinical applications and associated routes of administration, advantages of infusion therapy over traditional drug delivery, interactions between the body and the delivery system and the drug and the device, challenges, considerations for the development of new therapies, new indications that are currently being investigated, and possibilities for future infusion therapy advancements.

7.2 BASIC DEVICE REQUIREMENTS AND ROUTES OF INFUSION

The critical function of any drug-infusion device is to accurately and reliably dispense injectable medications into the body using a defined route of administration for a prescribed duration of infusion. These basic requirements that apply to a simple syringe-hypodermic-needle-delivery system are essential for more sophisticated infusion systems designed for long-term treatment of chronic diseases. As an example, for insulin infusion pumps, accuracy of dosing and timing relative to blood glucose levels or activities, for example, eating or exercise, which may affect these levels, is critical for optimum treatment of diabetes and to prevent potentially life-threatening over- or underdosing of insulin. For implantable systems in which the pump or infusion catheter cannot be replaced without surgery, it is critical that the key dispensing features of the device remain functional for the anticipated duration of the device within the body. Specific infusion devices are designed based to a large extent on the route of administration used for the drug. Routes of administration for chronic infusion include SQ, intraintestinal, intravenous, intrathecal (into the subarachnoid space of the central nervous system, which contains cerebrospinal fluid), intra-arterial, and intraparenchymal, that is, directly into tissue such as specific tissue targets in the brain for treating neurodegenerative diseases. The route of administration selected for infusion is based on a number of considerations including the location of the therapeutic target within the body, the ability to maintain access to the administration route for chronic therapy, the pharmacokinetics associated with the chosen route of administration, and risks and complications that may be associated with a particular route of administration. Because of increased levels of invasiveness and costs associated with chronic infusion devices, these therapies must offer significant advantages over standard oral and transdermal drug delivery for the therapies to be adopted by patients, clinicians, providers, and payers. These advantages will be discussed in this chapter.

7.3 INFUSION THERAPY COMPONENTS AND FUNCTIONS

7.3.1 The Pumps

The main component of the infusion system is the pumping mechanism that provides the force needed for the medication to be transferred from a reservoir located in the pump to the body. Two pumping mechanisms commonly used are gear-driven, syringe-driver pumps and rotary peristaltic pumps. These mechanisms have been used commonly for many years in bench or bedside versions of infusion devices used in the clinic, hospital, or in-home palliative care [1]. Fixed-rate implantable infusion pumps are pressurized with gas and push the drug solution through a flow restrictor, which has been designed specifically to produce a given rate of delivery. Implantable peristaltic pumps push the drug solution through internal flexible tubing in small increments as mechanical rollers within the pump compress segments of the tubing and rotate the entrapped fluid toward the exit port of the pump. Peristaltic pumping means that the drug is infused in a series of repeated small pulses as opposed to a continuous stream of fluid provided by gear-driven syringe drivers or pressured gas pumps. Because of the low infusion rates, for example, 100–250 mcl/day, typical of implantable peristaltic pumps that target the intrathecal (IT) space and the dampening provided by the tubing and catheter that interface between the pump and the body, pulsations of fluid at the point of delivery are negligible.

A rotary, gear-driven motor required for implanted peristaltic pumps requires robust engineering to ensure long-term delivery performance. Continuously moving gears and repeated tubing compression not only consume battery power but may also be potential sources of device failure, which is eliminated with fixed-rate, constant-pressure pumps that use passive flow restrictors to determine dosage metering [2]. The significant clinical advantage of programmable implantable pumps, such as SynchroMed® II, manufactured by Medtronic Inc., is that the peristaltic mechanism can be easily programmed to change the rate of continuous infusion, administer periodic drug boluses, and to provide flexible delivery patterns that can be tailored to the needs of individual patients, which may change over the course of a day or week.

In syringe-driven, peristaltic, and pressurized fixed-rate devices, the pumping mechanism provides the pressure to propel the drug solution in one direction through the fluid path of the pump and generally an attached catheter or cannula that directs the medication to a specific site of administration within the body.

The reservoir of the infusion device stores and confines the drug solution within the pump. Keeping the drug solution confined within the reservoir and fluid path of the pump is important; leaking drug solution can corrode mechanical or electrical components of the pump and lead to malfunctions or pump failure [3]. The reservoir of a syringe pump is simply the barrel of the syringe typically composed of polypropylene. This is advantageous since most parenteral formulations are already administered acutely and often chronically via standard polypropylene syringes that are inexpensive, sterile, and easy to replace. The syringe of the infusion device may be custom configured to fit and load into the pump in a specific manner to conserve space and prevent syringe loading errors for small external pumps. Implantable pump reservoirs have a bellows made of titanium that pushes the drug solution through an inline 0.2 micron filter to the peristaltic tubing. The reservoir bellows is pressurized using a gas propellant that fills the space between the outside of the bellows and the outer case of the pump. The filter ensures that any bacteria or particles inadvertently introduced when the formulation is prepared and the reservoir is filled are prevented from entering the body. This filter does not remove endotoxins and drug product used for infusion must meet specific requirements for endotoxin levels depending on the route of administration.

Because the reservoirs of implanted pumps contain a fixed volume of drug, for example, 20 or 40 ml for SynchroMed II pumps, a mechanism is required to allow the pump to be refilled with drug solution after all or most of the medication has been dispensed. A polymer septum is located centrally on a raised port on one side of the pump (Figure 7.1) that can be pierced with a needle to allow percutaneous access to the reservoir. Needles used to fill the reservoir are noncoring in order to maximize the life span and integrity of the septum. The refill septum is generally located by palpation of the skin over the pump. The skin above the pump is aseptically prepared, a noncoring needle is used to cross the septum and new drug solution is passed through a 0.2 micron filter when the reservoir is refilled. Because refills are performed in the clinics by health care professionals, time periods between refills are generally extended by at least 1 month and sometimes 6 months with 3-month intervals being common. Refill intervals are often more frequent when infusion therapy is initiated in order to titrate the drug to an individual patient based on clinical efficacy and side effects. Refill intervals may also be adjusted over the course of therapy as the patient's clinical condition changes, for example, more IT pain therapy may be needed with advancing cancer or IT medication requirements may change as a result of changes in other non-IT medications the patient is prescribed.

Other factors that affect refill intervals are the properties of the agent infused, since the drug must be stable in the pump for the duration of the refill interval. For new drugs, refill intervals may be extended as longer term supplemental stability data in the pump is generated. Refill intervals may decrease over time if the drugs used in the pump induce tolerance, for example, morphine for chronic pain [4]. Means by which clinicians may extend refill intervals are (1) using a pump with the largest reservoir; (2) using a drug solution that is more concentrated; (3) using an analgesic that has greater pharmacologic potency; (4) adjusting the medications that the patient is receiving by non-IT routes; and (5) adding additional off-label medications to the IT infusate [5]. Using drug combinations in implantable pumps is not an FDA-approved practice and is associated with a number of risks. Physicians may determine that the clinical benefits of combination drug infusates outweigh the risks for particular patients, for example, in terminal cancer. Using multiple drugs to treat chronic pain has become a common clinical practice [5].

7.3.2 The Catheters

The other critical device component of the drug-infusion system, whether the pump is external or implanted, is the catheter. The catheter delivers the therapy from the exit port of the pump to the target tissue of the patient. It is critical that this function be maintained for the anticipated life of the infusion system without dislodgments, occlusions, or breaches in the integrity of the catheter wall. Although drug-infusion catheters may vary in length, material, flexibility, and other properties depending on the specific application, they are generally designed for a specific route of administration as well as the surgical methods that will be used to position or implant and secure the catheter within the body. For example, vascular catheters used for hepatic arterial infusion of cancer drugs [6] must address concerns of occlusion due to blood clotting, which is not an issue with catheters that infuse drugs into cerebrospinal fluid (CSF). An issue relevant to many infusion catheters with regard to catheter patency is the flow rate of the infusate through the catheter lumen. High flow rates, continuous infusions and intermittent bolus delivery can provide mechanical advantages in maintaining catheter patency. Some catheters have been designed with slit valves to minimize the influx of biological fluids [7] and with multiple exit ports to prevent cessation of therapy should one exit port become occluded.

Implanted catheters or catheter tips generally include a radio-opaque marker so that the position of the catheter can be detected with fluoroscopy during implant surgery and if needed for future diagnostic purposes. IT catheters are often made of silicone impregnated with barium for radio-opacity [8]. This design allows the entire length of the catheter body to be observed under fluoroscopy, whereas some newer designs have radiographic markers only at the catheter tip. Surgical implant of the infusion system introduces the risk of cutting, puncturing, or kinking of the catheter as well as connection (pump to catheter, IT catheter to SQ catheter) integrity issues. Small imperfections that may be introduced during implantation, for example, a needle puncture of a silicone catheter for IT infusion, can lead to failures over time as the forces of a moving body and infusing drug enlarge the defect [9].

An IT catheter is introduced through a spinal needle with the tip of the needle preplaced in the subarachnoid space; the catheter is advanced through the needle with the aid of a stylet placed in the lumen of the catheter to provide a means of advancing the flexible catheter toward the head of the patient. To date, the catheter has been the component of the implantable IT infusion system most prone to complications [8,10,11]. Medtronic has recently introduced a new IT catheter (Ascenda®) with a trilaminar design (Figure 7.2). This catheter continues to use the same drug-contacting material, silicone, as earlier generation catheters but includes two outer layers of different materials intended to increase the durability of the device and reduce the incidence of dislodgments, kinking, and integrity breaches that occur with silicone-only designs.

Because the pump is implanted below the skin some distance from the IT catheter, a connecting catheter or extension may be used to complete the fluid path from the pump to the catheter. Alternatively, one-piece catheters are available and preferred by some physicians. Such catheter systems must be of sufficient length and flexibility to

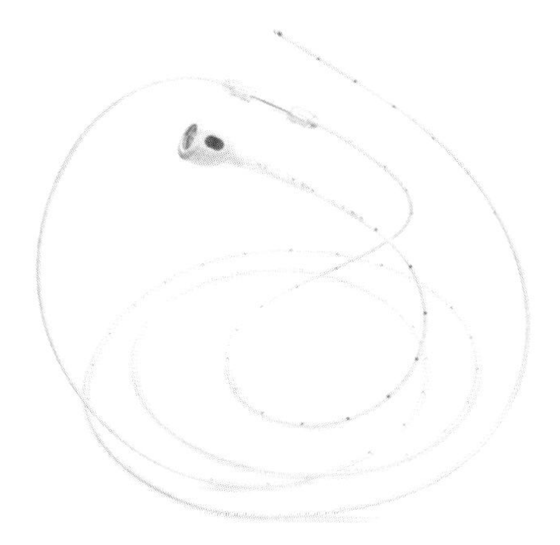

FIGURE 7.2 Ascenda three-layer intrathecal catheter by Medtronic with sutureless connectors for pump and between catheters. (Reproduced with permission from Medtronic, Inc.)

accommodate body movements in the long term so that tensile forces and friction are minimized and mechanical failures including catheter dislodgments are prevented. Connections between the pump and catheter(s) are also critical in maintaining integrity of the drug flow path over the life of the implanted system. Connections must be easy for the surgeon to make and assess for integrity at the time of implant. Sutureless connectors have been developed to reduce the risks of inadvertent nicks or cuts in the catheter, which may be made by suture needles, cutting instruments, or ligatures.

Unlike IT catheters that are introduced through a needle, catheters or cannulas used for SQ infusions are typically introduced over a metal needle that is then removed to leave behind a short 6–9 mm soft flexible plastic cannula with the tip remaining in the SQ space. A variety of cannula materials, configurations, and insertion tools are available to accommodate the needs and preferences of a wide variety of patients [12]. Depending on the indication, these cannulas remain in place for 8–36 h on average. For insulin delivery it is common to rotate the delivery site at least every 3 days to minimize local irritation and the risk of infection. For SQ infusion of apomorphine for treating Parkinson's disease, the site is generally changed every day. This is because (1) apomorphine induces significant local site reactions and (2) apomorphine infusion is generally discontinued over night to minimize psychotropic side effects that may be induced by activation of D_2-dopaminergic receptors [13,14]. SQ infusion cannulas are secured to the local infusion site with a circumferential patch that adheres to the skin. Many cannula designs allow the infusion tubing that interfaces between the pump and the cannula to be easily connected and reconnected to the cannula for planned interruptions in drug infusion, for example, therapy cessation during sleeping or showering.

This SQ pump-tubing-cannula design is often referred to as tethered pump, which is generally worn on a belt. Tethered pumps may lead to short-term connection issues that the patient may not always be aware of, especially if the infusion is occurring

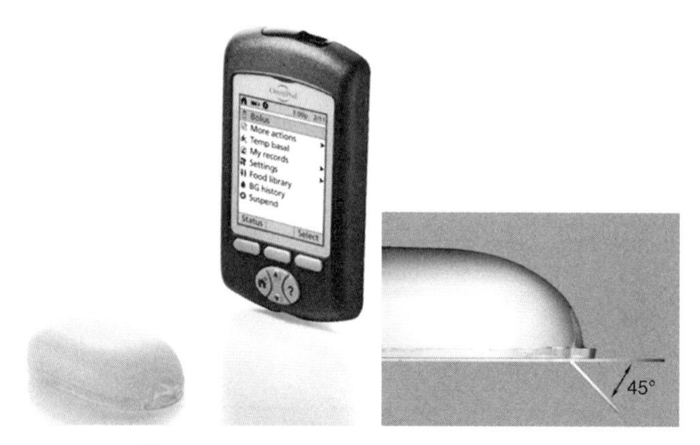

FIGURE 7.3 Omnipod® "patch pump" system for SQ insulin infusion, including handheld remote patient programmer. Right panel shows SQ cannula. (Reproduced with permission from Lazar Partners ltd.) (See colour plate section.)

while sleeping. This issue has spurred the development of the so-called patch pump designs in which the entire drug delivery path is integrated into the body of the device. These pumps generally have an integrated cannula that can be activated and inserted below the skin of the patient. An example of this type of pump is the Omnipod® system manufactured by Insulet [15] (Figure 7.3). An adhesive is used to secure the patch pump to the skin, for example, on the abdomen, thigh, or arm of the patient. A patch pump has similarities to a transdermal patch but has increased thickness to accommodate the pumping mechanism, drug reservoir, and electronics. Unlike a transdermal patch, patch pumps do not rely on surface contact area to increase drug delivery since the drug is infused directly into the subcutis and absorbed by the vasculature. Both types of patches are generally used to provide systemic delivery of medications, although some transdermal patches are placed at the local site of need, for example, at the transdermal lidocaine for the treatment of focal pain associated with postherpetic neuralgia [16].

7.3.3 Additional Device Components

The delivery parameters of an implantable infusion device can be monitored and changed noninvasively using radio telemetry and a device programmer (Figure 7.4). This is the same technology used to monitor and program other implantable devices such as cardiac pacemakers and neurostimulators. The programmer is a separate computer-like device that is maintained at the clinic or hospital and used by the health care professional to control the functions of the implanted delivery system at the time of initial pump implant and during the long-term course of therapy. At a minimum at each pump refill, the programmer is used to assess the historical and current status of the device and update the in-pump memory with regard to drug(s), drug concentration, and delivery rates and patterns.

Technology is also available with SynchroMed implantable infusion systems, which allow the patient to program the pump to deliver additional boluses of drug

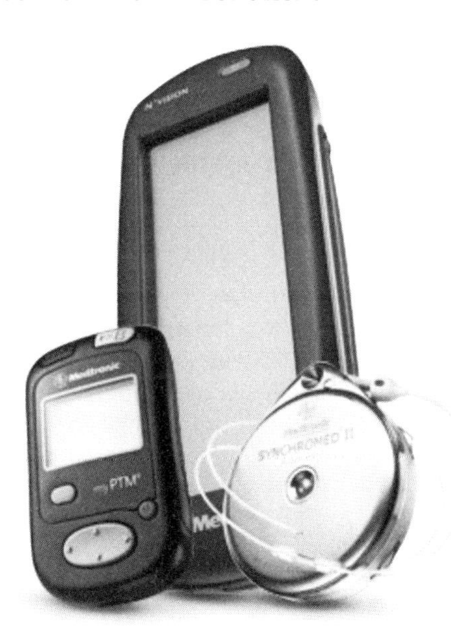

FIGURE 7.4 Medtronic implantable, programmable infusion system showing physician (blue) and patient (black) programmers. (Reproduced with permission from Medtronic, Inc.) (See colour plate section.)

during the day on top of their baseline continuous infusion rate [17,18]. Patient programmers are small handheld devices that communicate with the pump via telemetry (Figure 7.4). For safety, the physician maintains ultimate control of the patient programmer by limiting the size, number, and the rate at which boluses may be administered in a given day for a specific patient. This technology allows the patient to take a safe and active role in managing their medical condition. This feature is commonly used for patient-controlled analgesia, allowing the patient to administer a drug bolus to treat break-through pain. Because of the variability in chronic pain over the course of a single day and between days, patient-controlled activators allow dosing to be quickly tailored to changing symptoms. This feature allows a sophisticated implanted targeted delivery system to function with the same flexibility as a fast-acting oral medication that can be self-administered when needed.

Another important component of an implantable pump is the catheter access port (CAP). The CAP is located on the peripheral aspect of the pump body (Figure 7.1) and allows access by the physician directly to the fluid path of the catheter, that is, the reservoir and pump tubing are bypassed. Like the refill septum, the CAP is accessed percutaneously using a syringe and needle of a smaller gauge than the needle used to access the pump reservoir. The CAP allows diagnostic fluids such as radiocontrast agents to be introduced to assess for catheter patency, integrity, and dislodgment issues. Drugs may be introduced into the target tissue acutely using the CAP. For catheters located in a vascular space, intermittent heparin infusions may be used to maintain catheter patency [6]. Although not designed for this purpose (and often not feasible depending on the location of the tip of the catheter), clinicians occasionally use the CAP to withdraw intermittent samples of CSF for cytology, chemistry, or biomarkers. The CAP is not appropriate for collecting pharmacokinetic samples of the infused agent, since upon aspiration, the sample will be contaminated from drug solution in the fluid path of the pump proximal to the CAP.

7.3.4 The Drug Component of the Therapy System

The drug delivered by the infusion device provides the primary means of therapy for drug infusion systems. By interacting with biological substrates within the body, for example, receptors, ion channels, and enzymes, the drug produces biological effects that ultimately result in the clinical benefits of the infusion therapy. This is why in terms of regulatory classification of the infusion system as a combination product, the drug component represents the primary mode of action. This is in contrast to a drug-eluting stent, a combination product where the device component of the therapy system, that is, opening the lumen of a compromised artery, provides the primary mode of action. Because secondary biological effects may indeed be introduced by the placement or implantation of the device system and by administration of the solvent or vehicle that is used to carry the drug, vehicle-delivering control devices are assessed in animals and patients for safety and efficacy with results compared to systems infusing different doses of the active drug [19,20]. These data are collected and used to support regulatory approval of this type of combination product.

The agent used in any infusion system must be stable and compatible with the device components that it contacts (see interactions section) under conditions and for durations appropriate for the intended clinical use [21,22]. This ultimately dictates which specific drugs can be used in a specific infusion device, how drugs can be formulated since it is a formulated drug, that is, the drug product that is used in the device, and what specific materials can be used in the components of the device that contact the drug product. Implantable infusion systems that are refilled with drug intermittently require drugs that are stable when stored for prolonged periods at body temperature. Although external pumps allow the room-temperature drug product to be replenished frequently providing less demanding storage conditions, stability and compatibility are nonetheless critical to assess. Recombinant human insulin (Humalog®) and other proteins that are infused below the skin present unique challenges related to aggregation, which may decrease the amount of active protein infused and increase immunogenic responses against the protein. Development of neutralizing antibodies by the body can render the infused protein less effective and potentially lead to other medical complications [23].

When considering therapeutic candidates for use in implantable infusion systems, several properties of the active agent and formulation should be considered. Preservatives and organic solvents including antimicrobial agents and antioxidants should be avoided in formulations for IT infusion as they may produce neurotoxicity or affect the integrity of device components [24,25]. Drug solutions infused into the CNS should be approximately isosmotic and have a pH that ideally is as close as possible to 7, although a range of 4–8 may be acceptable depending on the specific formulation and route of administration (IT versus intraparenchymal). For IT infusion, drugs that are hydrophilic such as baclofen or peptides (ziconotide) and do not readily cross the blood–brain barrier (BBB) are advantageous. Once administered on the CNS side of the BBB, these drugs tend to stay and distribute via the CSF to spinal targets and are generally eliminated from the CNS compartment by bulk CSF clearance [26]. Drugs that induce significant development of tolerance, that is, loss of biological response with prolonged drug exposure or the need to increase dose over time to maintain the effect, are not ideal for long-term infusion therapies. Morphine, which is commonly infused IT for chronic pain, can induce tolerance and significant dose escalation if used aggressively with high doses and concentrations [4]. Drugs that have high water solubility and/or high pharmacologic potency may be advantageous as they allow for the option of smaller pumps (reservoir volumes) and/or longer refill intervals. On the contrary, depending on the drug, high-concentration formulations may produce local toxicity issues [27].

Currently, only four drugs are approved as monotherapies and manufactured specifically for IT administration using the SynchroMed infusion system: Lioresal® and Gablofen® (baclofen injection) for treating severe spasticity and Infumorph® (morphine sulfate) and Prialt® (ziconotide) for treating intractable pain. Baclofen can be quite effective in the long-term management of severe spasticity with minimal development of tolerance [28]. For many chronic pain patients, monotherapy with either morphine or ziconotide can produce a clinically acceptable efficacy and side effect profile. A clinical practice that is common in pain management is to use analgesics and combinations of analgesics not specifically tested and approved for use in the IT infusion device. These practices represent off-label use and often are performed by physicians based on their medical judgment. In the case of IT pain management, off-label use of analgesics and analgesic mixtures are used to increase efficacy, reduce side effects and increase the analgesic durability of IT therapy [5]. In order to make drug admixtures or to make less expensive alternatives of the branded products such as Lioresal and Infumorph, IT formulations may be made by compounding pharmacists using pharmaceutical-grade drug powders and sterile saline. Potential issues related to these off-label practices will be discussed in the section on challenges.

Insulin is the most common drug infused into the SQ space and is indicated for the treatment of type 1 and insulin-requiring type 2 diabetes using ambulatory pump technologies. Other drugs administered into the SQ space using ambulatory infusion technology include apomorphine, a dopamine agonist, for treating advanced Parkinson's disease in Europe; the vasoactive prostanoid drug, treprostinil, for treating pulmonary arterial hypertension; and in China, gonadotropin-releasing hormone for the treatment for certain types of infertility and hypogonadism. In patients who cannot tolerate SQ infusion of treprostinil because of local site reactions or pain, it can also be infused using an external pump connected to an indwelling central venous catheter [29].

Another continuous infusion therapy that was first approved for clinical use in Europe and is also being developed for use in the United States is intraintestinal administration of a gel formulation of the combination levodopa and carbidopa (Duodopa®) (Figure 7.5). In some advanced Parkinson's patients, this therapy is an alternative to brain surgery such as deep brain electrical stimulation.

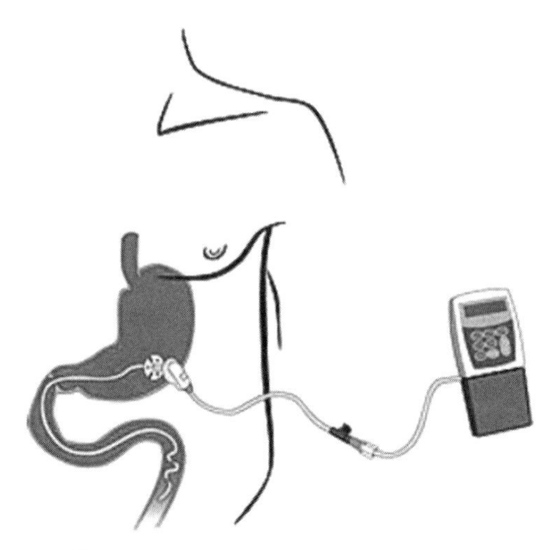

FIGURE 7.5 Duodopa® enteric infusion system for Parkinson's disease by AbbVie. (Reproduced with permission from Abbvie.)

7.4 CLINICAL APPLICATIONS

Implantable infusion systems have been used primarily to provide IT therapy for intractable pain and spasticity [5,30–32]. To a lesser degree, they have also been used for hepatic arterial infusion for treating liver cancer [6] and for intraperitoneal infusion of insulin for type 1 diabetes [33]. Some of the new indications being explored with implantable infusion systems will be described later in this chapter. In this section, a detailed look at a representative infusion therapy, IT baclofen for the treatment of severe spasticity, will be discussed in detail.

For many years, oral baclofen has been used to treat muscle spasticity arising from a variety of causes including injuries to the central nervous system. These types of injuries may limit the normal function of upper motor neurons in the brain to provide inhibitory input to lower motor neurons in the spinal cord; this input is needed to maintain normal motor output, muscle function, and tone. Skeletal muscle may become hypertonic leading to painful spasms and protracted muscle rigidity making normal ambulation and other motor tasks difficult or impossible. Baclofen is a centrally acting small molecule that is classified pharmacologically as a $GABA_B$ agonist, an agent that binds to and activates GABA receptors on neurons in the CNS. GABA, gamma-amino-butyric acid, is the most abundant inhibitory neurotransmitter in the central nervous system and in general inhibits or reduces neural activity by acting on two different types of GABA receptors. Activation of $GABA_B$ receptors in the dorsal horn of the spinal cord reduces motor neuron output and thus decreases muscle tone. Oral baclofen can produce muscle relaxation, which may be sufficient for some spasticity patients and should be assessed before progressing to IT baclofen. Oral and IT baclofen produce efficacy by activating the same receptors in the spinal cord, demonstrating that at least some baclofen is able to cross the BBB. Although baclofen is a small molecule, at physiologic pH it exists as a zwitterion with a single positive (amine group) and a single negative charge (carboxylic acid group), which limits transfer across the BBB. Higher oral doses can result in increases in spinal concentrations and greater efficacy but at the expense of higher brain levels and baclofen-induced side effects such as sedation and dizziness. Patients who display these dose-limiting side effects in response to oral baclofen may benefit from IT baclofen infusion.

IT baclofen infusion is used for severe spasticity of either spinal or cerebral origin, which is not adequately managed with oral medications [34]. Spinal origin spasticity includes diseases such as multiple sclerosis and spinal cord injury, while spasticity of cerebral origin includes diseases such as stroke, cerebral palsy, and traumatic brain injury. Before a pump implant is considered, a candidate patient is screened for a clinical response to IT baclofen using a bolus administered via a lumbar puncture; in Europe, an IT catheter may be placed and screening performed using infusion by way of an external pump to provide a longer term trial of baclofen response and tolerability.

A patient who has a successful trial response and elects to proceed to a pump implant must undergo surgery to implant the delivery system under local, regional, or general anesthesia. For baclofen pumps, this is often performed by neurosurgeons, anesthesiologists, or general surgeons and generally takes 2–3 h. The IT space is accessed using a spinal needle generally placed below the L2 vertebra to avoid injuring the spinal cord and the IT catheter is advanced through the needle so that the catheter tip is located at the desired location in the neuraxis. Placement of the catheter on the paramedian aspect of the intervertebral space with a shallow oblique trajectory into the subarachnoid space has been shown to produce an implant that results in the least amount of catheter complications such as breaks or dislodgments that may occur with midline placement [10,11]. The catheter body or tip is visible under fluoroscopy. The surgeon can advance the catheter tip to the desired location in the spinal column in order to achieve the desired clinical effect, for example, a patient with spastic legs will generally have the catheter tip placed in the lumbar or low thoracic space, whereas a patient with spastic arms and legs will have the catheter tip positioned at a mid-to-high thoracic location [31].

Once IT access has been successful and the catheter is positioned, the extradural catheter is placed below the skin using a blunt tunnelling surgical tool to create a path from the posterior midline around the flank and to the pump pocket. The pump pocket is generally made subcutaneously in the lower lateral abdominal wall just above the belt line; in small patients with less body mass, the pump may be implanted below the fascia of abdominal muscle to minimize the risk of erosion of the device through the skin. The pump is secured in the pocket with sutures placed around the metal loops provided on the perimeter of the pump body or in a mesh pouch made of polyester. Before the pump is implanted, it is prepared in the operating room in a warmed saline bath and filled with drug solution. The pump tubing is primed with drug solution. After the entire delivery system is implanted, a priming bolus may be programmed to rapidly advance drug solution to the distal tip of the IT catheter.

After implant, the daily dose of baclofen is usually set to a dose equal to twice the bolus dose that produced a successful screening response; the daily rate is initiated at

150 mcg/day using baclofen at a concentration of 500–2000 mcg/day if the patient had a successful screen with a 75 mcg bolus of 50 mcg/ml baclofen [34]. After surgery baclofen patients are often managed by physiatrists or neurologists. IT baclofen is continuously infused and the daily dose (infusion rate) is adjusted based on the signs and symptoms of the individual patient. To reach at an optimal dose for a patient, it is not unusual for simple continuous dose adjustments to be made over the course of 6–18 months after implant, depending on the underlying disease producing the spasticity [28]. If a patient does not respond sufficiently to simple continuous infusion, the SynchroMed pump can be used to delivery intermittent boluses of baclofen at a maximal rate of 1 ml/h. In some patients, this added kinetic energy and drug volume can help to distribute baclofen within the CSF of the spinal cord and improve therapeutic outcomes [35,36]. Intermittent boluses may be explored when outcomes produced by continuous infusion do not reach the level of efficacy achieved in the screening trial, when the daily baclofen dose reaches a predetermined daily maximum, or when side effects develop in response to escalating continuous doses. In Europe and Canada, patients may use a handheld programmer to give themselves extra boluses of baclofen. The amount and frequency of patient-administered boluses are limited by the physician who programs the patient programmer before dispensing. Patient-administered doses can be used when the patient's spasticity is increased in response to unpredictable events such as stress, concomitant illness, or cold weather. In the United States, the patient programmer is FDA approved only for use in pain patients for the treatment of breakthrough pain. In addition to bolus delivery, the SynchroMed system allows for flexible dosing in which the infusion rate may be changed depending on the individual needs of the patient including daily and weekly schedules. For example, some patients with multiple sclerosis benefit from higher infusion rates at night to control painful muscle spasms that can interfere with sleep; the daily infusion rate is lower so that the patient has sufficient muscle tone to ambulate or perform other activities of daily living [2,37].

Depending on the rate of infusion, an implantable infusion pump can last 4–7 years with power supplied by the integrated battery. Patients visit their clinics or hospitals for pump refills, which are required at 2–6 month intervals. Patient visits also allow for assessment of their underlying condition, the performance of the delivery system, and dose adjustments. With the telemetry of the programmer, the pump can be inspected to check the device performance and therapy parameters. It is very important to maintain IT baclofen administration once it has been initiated to prevent baclofen withdrawal, a potentially life-threatening clinical syndrome [38,39]. To minimize this risk in patients discontinuing therapy, IT baclofen should be gradually tapered, withdrawal symptoms assessed and IT baclofen reestablished if needed, and dose tapering continued until the delivery system can be safely explanted.

In addition to implantable systems, external delivery systems are used for the treatment of several chronic disease states. Long-term infusion of insulin is used for the treatment of type 1 and insulin-requiring type 2 diabetes but is also used to deliver dopamine agonists for Parkinson's disease [13,40], and vasoactive agents for treating pulmonary arterial hypertension [29,41]. Intraintestinal infusion of dopaminergic agents have also been developed for advanced Parkinson's disease (Figure 7.5) [42,43].

7.5 ADVANTAGES

Because chronic device-based infusion therapies involve a greater level of invasiveness, associated risks, and complications such as infection and upfront costs compared to oral or transdermal drug delivery, infusion therapies must provide significant advantages in order to be adopted by health care professionals, patients, and payers [44,45]. One of the major advantages of SQ, IV, or IT infusion over oral medication is that the gastrointestinal (GI) tract is bypassed. By circumventing the GI tract, greater concentrations of active drug are able to gain access to the body, that is, the bioavailability of the drug is increased. Many therapeutic agents or potential drug candidates are poorly absorbed from the GI tract, degraded by the harsh chemical environment of the GI tract (peptides and proteins exposed to stomach acid and proteases), and/or metabolized, that is, modified to inactive or less active chemical species, by enzymes located in the lining of the gut or in the liver, the so-called first-pass metabolism. Many new therapeutic agents are the result of advances in biotechnology and include compounds such as monoclonal humanized antibodies and recombinant proteins (e.g., Humalog, a recombinant human insulin analog) that have no significant bioavailability when administered orally. Other biological agents such as inhibitory ribonucleic acid (antisense oligonucleotides or small inhibitory RNA) and gene therapy are not candidates for either oral or transdermal administration.

In addition to maximizing bioavailability, avoiding the GI tract can also prevent or reduce GI-related irritation and side effects such as nausea and gastroesophageal reflux. Enteric infusion of levodopa for Parkinson's disease bypasses the stomach and delivers the drug directly to the small intestine; this provides a significant therapeutic advantage in many PD patients. Because PD affects the autonomic and the enteric nervous systems in addition to the CNS, gastric emptying is often compromised and delayed [46]. This can produce a phenomenon known as delayed "on time" after oral dopaminergic therapy because the medication becomes bioavailable only after absorption across the small bowel. Some patients with PD also experience difficulty swallowing (dysphagia), which also makes parenteral forms of drug administration more important.

IT infusion bypasses the GI tract and the BBB. One of the greatest challenges in treating diseases of the CNS with standard oral and parenteral routes of drug administration is achieving sufficient quantities of the therapeutic agent in the target tissue. The BBB limits or prevents all but small lipophilic molecules crossing from the vasculature space into the CNS primarily because of the tight junctions between capillary endothelial cells in the brain and spinal cord [47]. Peptides and proteins that have become a useful class of parenteral therapeutics do not readily cross the BBB. IV-administered antibody-based clinical strategies for treating Alzheimer's disease have so far been unsuccessful, which may reflect low levels of target engagement within the CNS [48,49]. A simple although invasive means to access the CNS is to

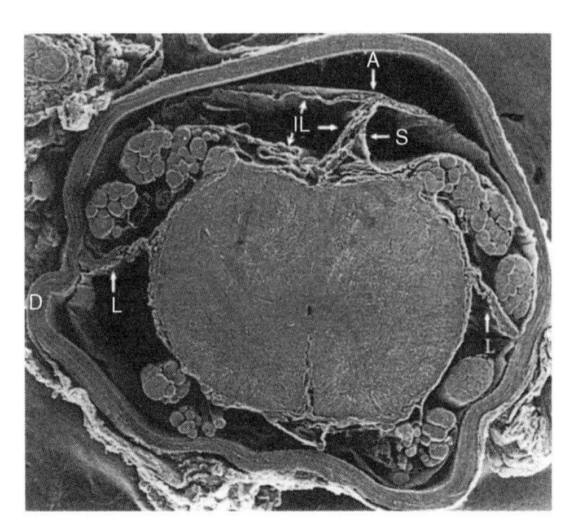

FIGURE 7.6 Anatomy of spinal cord and meninges. Dura mater (D), dentate ligaments (L), arachnoid membrane (A), dorsal septum (S). Cerebral spinal fluid (CSF) is located in the black areas within intrathecal between the arachnoid membrane and the spinal cord [98]. (Reproduced with permission from Alan Peters.)

mechanically breach this barrier with a catheter that allows agents to be directly administered into the CNS. This includes delivery to the CSF of the spinal subarachnoid space (the most common and least invasive route to the CNS), to CSF contained in the lateral ventricles of the brain [50], and delivery directly into specific targets in brain tissue (intraparenchymal delivery) [51–53]. The IT space is also referred to as the subarachnoid space because of its location directly below the arachnoid membrane of the meninges (Figure 7.6). The dura mater is located above the arachnoid membrane and CSF is located between the arachnoid and the pia mater, the membrane that directly opposes the spinal cord and brain tissue. Anesthesiologists frequently take advantage of the IT route of administration to deliver analgesics for acute surgical or other interventional procedures. Medication is also commonly administered acutely into the epidural space located above the dura mater, the outermost meningeal membrane. Local anesthetics such as bupivacaine used to relieve pain associated with childbirth are generally administered into the epidural space. Epidural delivery does not bypass the BBB but relies on lipophilic small molecules such as bupivacaine, clonidine, and fentanyl to permeate the dura and arachnoid membranes to gain access to the CSF [54]. Unlike IT infusion, chronic epidural drug administration has not proven feasible because of the difficulty of maintaining catheter patency due to epidural fibrosis [55]. Epidural and IT routes of administration are identified collectively as intraspinal or neuraxial administration.

Although many small molecules are able to penetrate the BBB to some extent and affect the CNS, dose-limiting side effects are often associated with oral delivery. Both morphine and baclofen can be effective in treating pain and spasticity, respectively, when given orally, but at much higher doses than required using the IT route. The IT route allows for lower dosing, fewer side effects, and greater efficacy by affecting targets in the dorsal horn of the spinal cord. For potential expensive biologic therapies, significant cost savings may be achieved with lower dosing requirements offered with IT infusion. By delivering on the CNS side of the BBB, systemic side effects such as renal or liver toxicity can be reduced or prevented by limiting systemic drug concentrations. By delivering to the spinal cord via subarachnoid infusion, supraspinal, that is, brain-related, dose-limiting side effects such as sedation or dizziness commonly seen with oral dosing of baclofen and morphine can be reduced substantially. Lipophilic drugs such as fentanyl, which more rapidly reach the target neurons in the dorsal horn than hydrophilic opioids such as morphine, are also susceptible to rapid clearance from the CNS since they readily permeate the meninges and the blood–spinal barrier. On the contrary, lipophilic analgesics may be preferable for patient-controlled spinal analgesia where the objective is rapid relief of breakthrough pain after a bolus of drug is administered.

IT infusion can provide regional targeting of the drug to the CNS, which is affected by the location at which the drug-infusion tip of the catheter is implanted. IT infusion can be used to more selectively affect the brain by positioning the catheter tip in the cervical subarachnoid space [11]. The catheter tip may also be positioned to target a particular dermatome associated with pain or to selectively target the arms, legs, or both pairs of extremities [31]. To more specifically target the brain as opposed to the spinal cord, drugs can be infused directly into the lateral ventricles of the brain, the same structures accessed by neurosurgeons to drain CSF to treat hydrocephalus with shunt devices. Substructures within the brain can be targeted by using stereotactic neurosurgery to place catheters and infusing drugs intraparenchymally [51].

Continuous drug infusion systems provide significant advantages over traditional methods of administration by providing a variety of dosing options for patient-tailored dosing not feasible with fixed increment dosage forms such as pills or transdermal patches. Infusion pumps allow precise dose adjustments based on the specific needs of an individual patient. Changes in dosing can be made by a variety of options: changes in infusion rates, drug concentration, and patterns of delivery such as intermittent bolusing, which may or may not be superimposed over a therapeutic continuous infusion rate. Infusion rates may be programmed to accommodate patterns in clinical symptoms related to day–night cycles or other predictable patterns of patient activity such as weekday verses weekend activities. Bolus dosing can be used to accommodate unpredictable patterns of behavior such eating, exercise (SQ insulin), stress (IT baclofen), and breakthrough pain using handheld patient programmers [17,18]. Bolus dosing adjustments include rate (amplitude), duration, and frequency of boluses in conjunction with the basal rate of continuous infusion.

A significant advantage of continuous infusion therapies is the ability to provide stable drug levels of the compound in the blood, biological fluid, or target tissue that are difficult to produce with oral administration, especially when drugs are rapidly eliminated from the body. Although extended-release oral formulations and transdermal formulations have provided pharmacokinetic advantages over traditional instant-release oral formulations, continuous parenteral administration affords the opportunity for maximal control and stability of drug levels. This advantage is perhaps best appreciated in the pharmacotherapies of PD. Continuous infusion of dopaminergic agents provides improved outcomes in PD patients by eliminating or reducing the highs and lows in drug concentrations in the blood associated with oral dopaminergic therapy [56,57]. Continuous infusion produces improved pharmacokinetics, which translates to improved pharmacodynamics (clinical outcomes) in this patient population [40,58]. High levels of levodopa and dopamine agonists may be associated with side effects such as "peak-dose dyskinesias" (uncontrolled muscle movements) and at low blood levels poor symptom control. These effects are generally more severe with (1) agents that have short half-lives, for example, levodopa, and (2) as the disease progresses. Apomorphine and enteric infusion of levodopa-carbidopa gel are designed to produce continuous dopamine stimulation by improving the stability of blood levels, thereby diminishing side effects and increasing efficacy relative to oral dopaminergic therapy [13,40,59]. Switching from oral to continuous infusion therapy in moderate-to-advanced stage Parkinson's disease can greatly improve patient outcomes.

Because long-term use of device-based infusion systems can be more expensive and invasive, patients may be screened for an acute response with a SQ or IT exposure before proceeding to pump implant [34,60]. A trial of IT morphine or baclofen is generally conducted with bolus lumbar punctures or with a short-term infusion using an IT catheter connected to an external syringe pump. Only if the patient has a successful trial procedure is the pump system implanted. In a similar manner, Parkinson's patients are screened for an acute response to apomorphine before considering continuous SQ infusion therapy. Screening procedures used today are useful for drugs that produce rapid responses that can be readily measured such as IT baclofen and SQ apomorphine. Unfortunately, drugs that need longer exposures to produce biological effects, for example, growth factor for a neurodegenerative disease or antiamyloid antibody for Alzheimer's disease, cannot be screened acutely before implant. In these cases, chronic animal and clinical safety and efficacy data will be critical to define the relevant patient populations and dosing strategies for optimal clinical outcomes.

7.6 INTERACTIONS

For therapy infusion systems, there exists at least the potential for a number of important interactions, including (1) reactions of the body to the device, (2) changes to the device in response to long-term implant in the body, and (3) interactions between the drug and the device. These interactions are critical since they can affect the safety and efficacy of the therapy system. Because implanted pumps are designed for long-term use, it is also important to understand how these interactions that may be subtle and slowly developing over time may affect the sustained performance of the delivery system.

The infusion system components that interact with the body must be biocompatible, that is, the ability of the device to perform its intended function, with the desired degree of incorporation in the body, without eliciting any undesirable local or systemic effects in that patient. For an implanted drug-delivery catheter, degree of incorporation in the body is critical since excessive incorporation or tissue reaction may lead to issues such as catheter occlusions [55] yet limited incorporation can produce issues such as catheter dislodgments, which may be addressed with improved surgical and device-enhanced fixation methods. With SQ infusion systems, a variety of designs and materials are used for the cannula portion of the system, which interacts over several days with the body. This is because different patients respond differently (some with local site reactions) to different adhesives and cannula materials placed on and under the skin. Overreaction to the cannula materials may also affect how the drug is absorbed from the delivery site. Acute inflammation with increases in vasodilation and permeability may increase, whereas fibrosis may decrease absorption of insulin over time.

Any medical device that is implanted in the body for a long term must be biostable, that is, the chemical and physical stabilities of the device materials must be sufficient for the device to perform its intended function during the implantation period within the body. Since the primary function of an infusion system is to accurately deliver drugs for sustained periods of time, a component that is not biostable may affect this critical function leading to underdosing, dosing cessation, overdosing, or delivery of the drug to unintentional targets. A catheter that fractures in response to repetitive body motion is an example of a need for improved biostability [61].

The other critical interactions that need to be assessed with chronic infusion systems are the interactions between the drug and the device, particular device components in direct contact with the drug. The drug product must not adversely affect the materials and function of the pump over time. Testing is often conducted with components of the device immersed in drug solution over time at 37 °C for implanted systems to assess changes in critical mechanical properties of the material (tensile strength, hardness, etc.) as well as testing of the drug product in intact complete delivery systems maintained in a controlled temperature and humidity environment that simulate clinical use [21,22,62]. Although *in vitro* bench testing can be prolonged and anticipated stressors can be exaggerated, for example, setting the pump to maximal flow rates or cycling between low and high flow rates, ultimate long-term performance can only be assessed *in vivo*.

Chronic implants in animals are an important step in testing new devices and new drugs administered with previously approved delivery systems. Animal testing and clinical development help to ensure safe and effective performance in order to achieve regulatory approval to market the product but are not of sufficient duration to ensure long-term *in vivo* performance since a single implanted infusion systems can last many years.

To supplement the regulatory submission data, it is critical that new infusion therapies be monitored in the field for long-term safety and efficacy. Maintaining systems such as analysis of returned products after explant (end of product life or device malfunction) in addition to careful vigilance of product performance that is currently being used in the field is critical to establishing long-term performance. Medical device reporting is a mandatory requirement of the FDA for device manufacturers, importers, and user facilities when significant adverse events are suspected to be device related. Many device companies maintain a device registry to prospectively track the performance of certain types of devices such as implantable infusion systems over time. These data can be used to improve product performance for therapies that are currently in use and for next-generation therapies.

The materials of the delivery system in contact with the device may also affect the drug in important ways. Stability of the drug is critical under the clinical use conditions throughout the duration the drug is intended to be stored in the implanted reservoir of the device. Drugs may degrade under the elevated implant temperatures, exposure to materials in the fluid path of the pump, physical stresses induced by pumping mechanisms of the device, and residual drug or drug-related material remaining in the reservoir between refills. Depending on the particular therapeutic agent, stability indicating analytical methods that assay for the drug as well as major degradants or impurities need to be developed and validated. For biological agents, assessment of bioactivity after device exposure is critical. Ideally, this includes time-dependent assessment of both target and off-target effects. For proteins, an assessment of physical stability may be critical since aggregation can lead to loss of potency and increases in immunogenicity that may lead to the production of neutralizing antibodies within the body, which decrease the biological effects of the infused protein.

A critical concern related to drug–device interactions is the potential for the drug formulation to interact with the device materials located in the fluid path, leach chemicals from these materials, and deposit these materials into the body. To understand worst-case scenarios, the fluid path is generally exposed to organic solvents at elevated temperatures and studies continued until the amount of extracted chemical reaches zero or steady state. Depending on the identity of the chemical and the amount of the chemical extracted from the device materials, toxicity assessments and potentially supplemental toxicological studies may need to be conducted. Qualifying the leachables under true *in vivo* biological conditions is a compelling reason to use the entire clinical version of the delivery system when conducting animal studies. For implantable systems this may necessitate using larger species like dogs or sheep [19]. Testing in this manner serves to assess any toxicity related to the device, the infused drug, leachables, and also subtle interactions that can occur between the two components (drug plus leachable) that may not be apparent with bench testing.

7.7 CHALLENGES

With SQ and IT drug infusion, a major challenge can be adverse biological reactions that occur at local sites of drug infusion. Although device materials with direct tissue interaction may contribute to these responses, in the SQ and IT spaces, they often are related to the drug, the drug formulation, infused volume, dose, and concentration of the agent infused. For SQ infusion, ideally the daily volume infused at a single site on the skin is no greater than 1 ml/day, and the infusate is isosmotic and has a physiologic pH. Significant excursions from these ideal conditions nonetheless may be well tolerated depending on the specific infusate and the duration the infusion site is used. For example, some agents (clonazepam, fentanyl, and ketamine) used in palliative care infused with external syringe drivers have pH values between 3 and 4 and are well tolerated via SQ infusion [1]. SQ infusions beyond 3 days are associated with increased risk of site reactions and are not recommended [63]. Not surprisingly, the drug infused is a major determinant of site reactions. For example, dopamine agonists can activate T lymphocytes in the skin, which in turn increase vascular permeability and release chemotactic factors that may attract additional inflammatory cells to the local site [64]. Infusion of SQ apomorphine, a dopamine agonist that predominantly activates D_1- and D_2-dopamine receptors, is well known for producing local site reactions, for example, SQ nodules, which may be painful and long lasting, in most patients, at most sites the drug is infused [65].

Because infusion sites are changed daily and apomorphine treatment is long term, many patients discontinue therapy. This is because of the accumulation of skin nodules despite often excellent control of Parkinson's symptoms [13]. Prostanoid vasoactive drugs (e.g., treprostinil) may be infused into the subcutis to treat pulmonary arterial hypertension [41]. Activation of prostaglandin receptors in the skin can induce pain and significant local reactions in some patients [29]. Although SQ administration is the preferred initial route of administration, an alternative and more invasive route, intravenous infusion using an externalized vascular catheter and pump, may be used in these patients [41].

Chronic infusion into the IT space presents unique challenges related to the dynamics of CSF flow. Unlike the vascular space, mixing and distribution of drugs in the IT space can be significantly limited especially at the low flow rates typically used with implantable pumps [66]. CSF does not flow significantly in one direction in the subarachnoid space but rather oscillates to-and-fro with the cardiac cycle as demonstrated by real-time MRI imaging studies [67–69]. As the vasculature of the brain is filled with blood during systole, CSF moves away from the brain; during diastole there is an approximate equal and opposite movement of CSF toward the brain. The fact that the CSF is not a well-mixed fluid compartment is suggested by the observations that certain endogenous CSF constituents are present in long-standing gradients along the neuraxis [70,71]. The combination of low CSF flow and low IT infusion rates can produce high drug concentrations near the outflow port(s) of the catheter typically located near the catheter tip.

In terms of long-term IT infusion, two types of local toxicity have been well described: inflammatory mass formation [27] and spinal cord necrosis. An inflammatory mass or "granuloma" (Figure 7.7) is a space-occupying potentially spinal-cord-compressing focal collection of a mixed population of inflammatory cells that can be observed in animals and humans with long-term morphine infusion [72]. Morphine and other opioids can produce inflammatory masses especially if they are infused at high concentrations and doses [73]. A second type of local toxicity seen near the tip of IT catheters is neuronal necrosis. This type of response has been seen with numerous types of NMDA antagonists such as ketamine and memantine [74,75].

Another challenge with both external and implantable infusion systems is minimizing the incidence of infections. Because external pumps connected to catheters or cannulas create a direct path of migration for bacteria from the external environment to the body, the risk of injection is greater with external infusion systems used chronically than in systems that are fully implanted. SQ infusion can lead to local skin

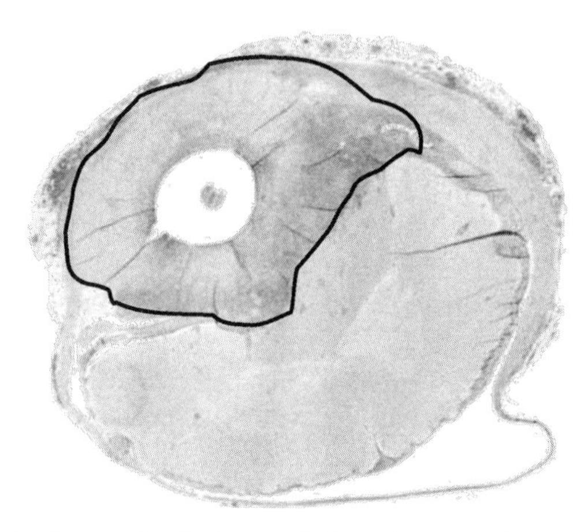

FIGURE 7.7 Photomicrograph of a large spinal-cord-compressing, catheter-tip granuloma (outlined) induced by continuous 28-day infusion of morphine (25 mg/ml, 12 mg/day) in a canine model. (Reproduced with permission from Medtronic, Inc.)

infections that can be minimized by careful preparation of the infusion site before use, rotating to a new infusion site and using a new infusion set at least every 3 days [63]. External pumps associated with enteric levodopa-carbidopa infusion are at risk for infection. Most commonly, local infection occurs where the infusion tube traverses the body wall and stomach (peristomal infection) and on rare occasion more severe peritonitis occurs [76]. Zibetti et al. reported an infection incidence of 13 in 25 patients followed for a mean of 3 years [77]. Intravenous infusion for treating pulmonary arterial hypertension (PAH) using external pumps increases the risk of potentially serious blood infections making SQ infusion the preferred initial route of administration [29]. Implantable IT infusion pumps have a relatively low infection risk (<5%) but this risk is greater in children implanted with baclofen pumps [78]. SynchroMed pump infections most often are associated with implant surgery rather than pump refills and generally affect the pump pocket. When an implanted pump system does become infected, it often, but not always, requires explant of the infusion system to treat the infection [78,79].

The most common infusion-system-related complication of implantable IT infusion systems are related to catheter problems including dislodgments, kinking, and integrity failure [80]. Likewise, for enteral levodopa–carbidopa infusion, catheter dislodgments from the small intestine and catheter kinking and obstruction are the most common complications; together they are the most common reason patients discontinue this therapy. In a 3-year follow-up trial of 25 patients on enteral levodopa–carbidopa, 58 catheter-related issues occurred [77]. Because of the significant efficacy, quality of life improvements, and the relative ease of fixing catheter-related complications in an accessible system, Duodopa therapy is an important option for some patients with advanced PD.

Perhaps the most serious challenge with implantable IT infusion systems are the medical complications produced from a sudden loss of therapy delivery. As noted earlier, this is primarily related to catheter issues but may be a result of an empty pump reservoir (missed refill) or a pump failure that is most common with off-label medications in SynchroMed. After 5 years, pumps used with on-label drugs have a 2.4% incidence of motor stall versus 4.5% when using off-label drugs [3]. Not only can infusion cessation result in return of pain or spasticity symptoms, but it can also produce serious withdrawal symptoms in pain patients, potentially life-threatening complications in baclofen patients and potentially serious rebound hypertension in patients being chronically infused with treprostinil [41] or IT clonidine [81].

7.8 DEVELOPMENT AND REGULATORY APPROVAL OF NEW INFUSION THERAPIES

The scope of this section is confined to the development of new agents for IT infusions that have already been approved for other routes of administration. An advantage of this type of development path is that often the primary systemic safety and toxicological data generated by the initial drug developer can be leveraged by regulatory bodies. If a new chemical entity is developed for long-term IT infusion, a complete toxicology and safety pharmacology package will need to be generated to assess systemic exposure of the new agent; this will include assessment of absorption, distribution, metabolism, and elimination from the body in animal and clinical studies. This type of advanced development program often is conducted as a partnership between the drug and the device companies.

An early part of development is procuring the GMP-quality drug substance, ideally from several vendors for comparative purposes in order to develop IT suitable parenteral formulations. Stability-indicating analytical methods need to be developed to assay for the drug substance as well as drug-related degradants and impurities. For small hydrophilic molecules, drug formulation is generally focused on aqueous-based vehicles with minimal excipients to adjust pH and osmolality in order to optimize tolerability in the IT space. Candidate formulations are often tested for stability when exposed to 37 °C in glass vials over time before proceeding to more expensive in-pump stability testing. Under simulated clinical use conditions, the delivery system is filled with the drug product, the specific drug formulation chosen to advance, and drug samples are collected over time from both the pump reservoir and the distal end of the catheter and analyzed for chemical and physical stabilities [22]. Pump function and delivery accuracy is assessed after long-term drug exposure. Pump components in contact with drug solution may be assessed following emersion in drug product [21]. Leachable and extractable studies are conducted to determine if the new drug product can release chemicals from the device that may enter the body. These chemicals are quantified, identified if levels reach a certain threshold, and assessed for toxicological potential.

Animal testing consists of several phases. Demonstration of improved efficacy (or decreased side effects) of the IT-administered drug relative to the oral equivalent is important to justify the need and advantages of a more invasive route of administration. For pain molecules, this is often done in a variety of rodent models with infusion conducted using osmotic mini pumps. Before proceeding to clinical assessment, the dose-dependent safety of the drug product must be clearly demonstrated in animals. Generally, this involves testing in two species of animals, a rodent and a nonrodent (large animal) species, for duration at least as long as the initially planned clinical study. Before proceeding to GLP (good laboratory practice) animal testing that will be used for regulatory submissions, dose-ranging studies are conducted to define tolerable doses, a toxic dose level, and a dose level that is not associated with any adverse events. Measurements of drug in CSF and plasma are conducted to demonstrate dose-related drug concentrations and to ensure exposure of the biological targets to the drug administered. After dose-ranging experiments in animals, it is common to meet with the regulatory body to establish agreement on a plan to advance the infusion therapy into humans. Ideally, in the large animal species, the exact same delivery system and drug product anticipated for human use are assessed in animals. This serves to qualify the entire drug–device system and any subtle interactions that may occur *in vivo*. In lieu of this, scaled versions of the device composed of the same materials may be used if needed to accommodate the smaller anatomy of the test animal (monkey versus human). Because IT toxicity is often manifested in spinal cord reactions at the tip of the catheter, it is important to evaluate the highest concentration of drug product that will be used clinically [73].

For new IT infusion therapies, clinical development generally consists of the same four phases of development as used for new drugs with phase 3 consisting of two pivotal trials to provide safety and efficacy data to support the regulatory filing, New Drug Application, for the specific drug, route of administration, and clinical indication. In conducting the clinical trials, patients are monitored for any adverse events that are classified as drug or device related. The device master file for the IT delivery system will be updated specifically with data related to use of the new drug in the device; the device–drug interaction data are a major part of this update. Phase 4 studies may be conducted to address specific additional requirements requested by the regulatory agency or to address additional questions the company or the institution developing the therapy may have.

7.9 NEW INDICATIONS IN DEVELOPMENT

Because of major advances in biotechnology producing large numbers of new molecules and potential therapies that are not candidates for standard oral administration, the importance of advancing long-term parenteral infusion technologies is critical in order for the full clinical potential of these therapies to be realized. On the other hand, infusion technology may also play a significant role in repurposing small molecules in order to improve targeting within the body, decrease dose-limiting side effects, and improve patient responses by producing improved pharmacokinetics. Therapies tailored to the individual patient by means of genetic testing and biomarker analysis has become a major goal for improving clinical outcomes; infusion therapy allows dosing to be customized to the individual patient and thus provides a device-enabled method of personalized medicine.

PD is a neurodegenerative disease involving loss of dopaminergic neurons in the substantia nigra of the brain; all therapies currently available, including long-term infusion therapies and deep brain stimulation, manage symptoms of the disease only (primarily motor symptoms). A strategy that continues to be investigated to not only treat PD but also potentially modify disease progression is delivery of GDNF, glial-derived neurotrophic factor, intraparenchymally to the striatum to prevent further loss of the remaining population of dopamine neurons, heal damaged neurons, and possibly regrow axons of neurons from the cell bodies in the substantia nigra to the synapses in the striatum [51,82,83]. Brain infusion of inhibitory RNA-based molecules is being assessed for Huntington's disease, in which the toxic protein, huntingtin, has been well described [53]. Targeted knockdown of huntingtin in the brain may improve clinical outcomes especially if administered early in the disease course or before the onset of disease, which can be diagnosed early in life with genetic testing. IT infusion of replacement enzymes created by recombinant technology for patients with a variety of lysosomal storage diseases that affect the CNS hold promise for rare but devastating diseases such as Niemann–Pick disease and Sanfilippo syndrome [24]. Although trials with systemically administered antibodies of Alzheimer's disease have so far been unsuccessful [49], delivery of antibodies directly into the CNS can increase antibody concentrations within the brain and potentially improve clinical efficacy [48].

Improving small-molecule delivery by using continuous infusion technology continues to be an area in which clinical therapies can be expanded, improved, and customized to meet the specific needs of individual patients. Drugs such as hydromorphone, bupivacaine, and clonidine, which are commonly used off-label by physicians to treat chronic pain, wait to be developed and approved as new IT infusion therapies so that thorough characterization of the interactions between these drugs and infusion systems can be conducted. Appropriate preclinical safety and toxicology studies also need to be performed and clinical trials completed, so that optimal clinical use of these commonly used off-label analgesics can be determined. Peptide molecules that target novel analgesic mechanisms in the spinal cord such as CGX-1066 (neurotensin receptor) and mambaglin-1 (central acid-sensitive ion channels) are promising new IT analgesics [84–86]. New IT therapies to target neurological diseases recalcitrant to standard medical approaches such as clonidine infusion for treatment-resistant neurogenic hypertension [87] and refractory major depression also hold promise for the future [88,89]. Prostanoid therapy for pulmonary arterial hypertension, which is infused SQ and IV with external pumps, is currently being assessed for continuous infusion using implantable pump technology [90,91]. New dopaminergic therapies are being assessed for SQ infusion such as levodopa/carbidopa and levodopa ethyl ester to overcome the shortcomings of SQ apomorphine infusion.

An important area that has not been systematically studied is the use of long-term infusion therapies in combination with medications administered by traditional routes. This subject has important safety and efficacy implications for treating patients, for example, understanding drug–drug interactions and optimizing the sustained efficacy of implantable infusion therapy systems. Patients treated with IT baclofen may experience changes in clinical response due to concomitant treatment with oral seizure medications or selective-serotonin reuptake inhibitors [31]. Morphine sulfate is only one of two drugs that are FDA approved for IT management of chronic pain and many patients may develop tolerance to its analgesic effects. Tolerance leads to dose escalation, which may be associated with increased side effects, the need for adjunctive IT medications, increased risk for catheter-tip inflammatory masses, and potentially opioid-induced hyperalgesia [92]. A clinical practice yet to be assessed in prospective controlled trials involves weaning patients off or minimizing systemic opioids before or shortly after initiating IT morphine infusion. Current evidence suggests that patients treated in such a manner can achieve pain relief with lower initial doses of IT morphine and that IT dose escalation is reduced over time compared to patients who remain on high systemic opioid therapy in combination with IT opioid infusion [93,94].

7.10 FUTURE POSSIBILITIES FOR INFUSION TECHNOLOGY

In addition to developing new infusion therapies by using currently available delivery devices with new drugs and for new clinical indications, future improvements in infusion technology are anticipated that will expand the clinical use of infusion therapies by making systems more reliable, user friendly, less expensive, and incorporate more sensing and diagnostic features. The Medtronic insulin infusion system can be used in combination with a glucose-sensing device component that continuously monitors insulin levels with a SQ sensor (Figure 7.8); the patient can use this real-time information in combination with intermittent finger-stick glucose measurements to adjust the amount of insulin infused. Ultimately, the goal is to develop a closed-loop delivery system often referred to as an "artificial pancreas" where the delivery system adjusts insulin therapy in a physiologic manner based on real-time blood glucose concentration. It is conceivable that sensing technology to allow closed-loop drug infusion may also be developed for diseases such as

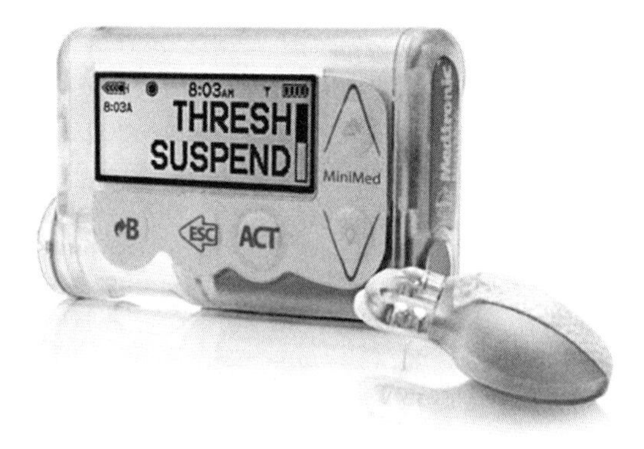

FIGURE 7.8 Minimed® insulin infusion pump with SQ glucose sensor manufactured by Medtronic. (Reproduced with permission from Medtronic, Inc.)

pulmonary arterial hypertension and essential hypertension where implanted pressure sensors communicate with the delivery system to adjust the amount of rapidly acting vasoactive therapy infused.

In terms of durability, it is anticipated that infusion systems will become more robust especially as catheter designs and fixation methods improve to reduce the number of catheter-related failures. New trilaminar catheter designs such as the Ascenda catheter is designed to reduce catheter breaks, kinks, and dislodgments but data from the clinical field is needed to substantiate these improvements. As battery technology continues to improve and mechanisms of fluid transfer become more efficient, the lifetime of implantable devices will increase to reduce the risks and costs of pump replacement surgeries. A technology that has already been adapted for implantable neural stimulators is the ability to recharge the device with an external charger so that the battery life does not dictate implant life [95,96]; this is technology that may be adapted to implantable pumps in the future.

As with other implantable devices such as pacemakers and defibrillators, advancing technology will also provide an opportunity for smaller devices that may not only be more acceptable to the patient but also reduce the risk of surgical complications and increase the regions in the body where these devices may be implanted, for example, above the neck for brain infusion. For implantable pumps, the drug reservoir serves as a major determinant of how small the overall device can be made. In general, smaller reservoirs require a trade-off in that more frequent refills are required. Increasing therapy potency, concentration, duration of effect, and formulation science may all provide means of reducing reservoir volume. Smaller implantable devices help reduce the risk of implant-related infections. Other technologies to limit infection may include pump components or accessories that elute antimicrobial agents or coatings that limit bacterial adherence and transfer from the skin to the pocket or inhibit the establishment of biofilms. Smaller external devices are less obtrusive for patients and have led to the development of SQ infusion systems sometimes referred to as patch pumps. The Omnipod insulin pump is an example of this type of small discrete device that is placed directly at the site of infusion [15,97]. This design eliminates tubing between the pump and the SQ cannula, which reduces the chances of unintended therapy interruptions.

Devices that are easier for physicians and patients to use will also increase the acceptance of this treatment modality. External pumps can be simplified and ergonomically designed with human factor analyses so that patients with limited motor skills, for example, PD and geriatric patients, can use these therapies. One way to simplify infusion therapy with external pumps is to package the drug in sterile, prefilled cartridges or syringes so the patient or caregiver is not burdened with transferring medication from a vial to the pump reservoir. Often, upon initiation of infusion therapy, dose titration for an individual patient can be a protracted process with numerous clinic visits and associated expenses incurred. Technologies such as selectable default drug titration schemes preloaded into the pump memory and ways for the physician or pharmacist to program the device remotely can make dose titration less burdensome for the health care system and patient.

Infusion devices of the future may also incorporate additional diagnostic features. Some of these features may be directly related to device function such as providing direct measurements and patient feedback on reservoir volume, flow path integrity, and drug stability. Other diagnostics may be incorporated that provide ancillary clinical information regarding the specific type of patients being treated such as activity levels or information on body position for pain and spasticity patients. Since infusion therapies involve the use of specific drug products with specific medical devices, safe gate measures such as a pump programmer, which reads codes from the drug vial label, may provide information to alert the physician regarding off-label use or prevent the use of specific off-label medications known to be problematic. Diagnostic information recorded by the device could be stored for retrieval at clinic visits using the programmer or the data could be transferred remotely either to the clinic or other intermediary health care service providers. By taking advantage of devices that patients already are using for long-term therapy infusion, integrated diagnostic features may provide enhanced long-term patient care.

REFERENCES

1. A. Dickman, J. Schneider, and J. Varga, *The Syringe Driver: Continuous Subcutaneous Infusions in Palliative Care*, Oxford University Press, Oxford, 2005, pp. 17–97.
2. N.G. Rainov and E. Buchser, Making a case for programmable pumps over fixed rate pumps for the management of fluctuations in chronic pain and spasticity: a literature review. *Neuromodulation* 5 (2) (2002) 89–99.
3. A. Rezai, D. Kloth, H. Hansen, et al., Physician response to Medtronic's position on the use of off-label medications in the synchromed pump. *Pain Phys.* 16 (2013) 415–417.
4. R.V. Duarte, J.H. Raphael, M.S. Haque, J.L. Southall, and R.L. Ashford, A predictive model for intrathecal opioid dose escalation for chronic non-cancer pain. *Pain Phys.* 15 (5) (2012) 363–369.
5. T.R. Deer, J. Prager, et al., Polyanalgesic Consensus Conference 2012: recommendations for the management of pain by intrathecal (intraspinal) drug delivery: report of an interdisciplinary expert panel. *Neuromodulation* 15 (5) (2012) 436–464.
6. M.K. Callahan, and N.E. Kemeny, Implantable hepatic arterial infusion pumps. *Cancer J.* 16 (2) (2010) 142–149.
7. M.M. Morris, T.G. Laske, K.T. Heruth, M.R. Ujhelyi, and J.W. Casas-Bejar, Slit valve catheters. US 2004/0176743 A1, Patent Application publication (to Medtronic, Inc.), September 9, 2004.
8. R.D. Penn, M.M. York, and J.A. Paice, Catheter systems for intrathecal drug delivery. *J. Neurosurg.* 85 (2) (1995) 215–217.
9. W.J. Dawes, J.M. Drake, and D. Fehlings, Microfracture of a baclofen pump catheter with intermittent under- and overdose. *Pediatr. Neurosurg.* 39 (2003) 144–148.
10. K.A. Follet, K. Burchiel, et al., Prevention of intrathecal drug delivery catheter-related complications. *Neuromodulation* 6 (1) (2003) 32–41.
11. A.L. Albright, M. Turner, et al. Best-practice surgical techniques for intrathecal baclofen therapy. *J. Neurosurg.* 104 (4 Suppl.) (2006) 233–239.
12. P.J. Patel, K. Benasi, et al., Randomized trial of infusion set function: steel versus teflon. *Diabetes Technol. Ther.* 16 (1) (2013) 15–19.

13. P. Hagell, and P. Odin, Apomorphine in the treatment of Parkinson's disease. *J. Neurosci. Nurs.* 33 (1) (2001) 21–38.
14. K.M. Ackland, A. Churchyard, et al. Panniculitis in association with apomorphine infusion. *Br. J. Dermatol.* 138 (1998) 480–482.
15. H. Zisser, The OmniPod insulin management system: the latest innovation in insulin pump therapy. *Diabetes Ther.* 1 (1) (2010) 10–24.
16. P.S. Davies, and B.S. Galer, Review of lidocaine patch 5% studies in the treatment of postherpetic neuralgia. *Drugs* 64 (9) (2004) 937–947.
17. J. Maeyaert, E. Buchser, J.P. Van Buyten, N.G. Rainov, and R. Becker, Patient-controlled analgesia in intrathecal therapy for chronic pain: safety and effective operation of the model 8831 personal therapy manager with a pre-implanted SynchroMed infusion system. *Neuromodulation* 6 (3) (2003) 133–141.
18. W. Ilias, B. le Polein, et al., Patient-controlled analgesia in chronic pain patients: experience with a new device designed to be used with implanted programmable pumps. *Pain Pract.* 8 (3) (2008) 164–170.
19. T.L. Gradert, W.B. Baze, W.C. Satterfield, K.R. Hildebrand, M.J. Johansen, and S.J. Hassenbusch, Safety of chronic intrathecal morphine infusion in a sheep model. *Anesthesiology* 99 (1) (2003) 188–198.
20. R. Rauck, R.J. Coffey, et al., Intrathecal gabapentin to treat chronic intractable noncancer pain. *Anesthesiology* 119 (3) (2013) 675–686.
21. K.R. Hildebrand, D.E. Elsberry, and V.C. Anderson, Stability and compatibility of hydromorphone hydrochloride in an implantable infusion system. *J. Pain Symptom Manage.* 22 (6) (2001) 1042–1047.
22. K.R. Hildebrand, D.D. Elsberry, and S.J. Hassenbusch, Stability and compatibility of morphine–clonidine admixtures in an implantable infusion system. *J. Pain Symptom Manage.* 25 (5) (2003) 464–471.
23. T.W. Van Haeften, Clinical significance of insulin antibodies in insulin-treated diabetic patients. *Diabetes Care* 12 (9) (1989) 641–648.
24. Z. Shahrokh, P. Callas, and L. Charnas, Intrathecl delivery of protein therapeutics to treat genetic diseaes involving the CNS. Frederick Furness Publishing, 2010, pp. 16–20. Available at www.ondrugdelivery.com.
25. G. Bennett, K. Burchiel, et al., Clinical guidelines for intraspinal infusion: report of an expert panel. PolyAnalgesic Consensus Conference 2000. *J. Pain Symptom Manage.* 20 (2) (2000) S37–S43.
26. H. Davson, M.B. Segal, The secretion of the cerebrospinal fluid, in Davson H. (Ed.), *Physiology of the CSF and Blood–Brain Barriers*. 1st ed., CRC Press, 1995.
27. T.R. Deer, J. Prager, et al., Polyanalgesic consensus conference 2012: consensus on diagnosis, detection, and treatment of catheter-tip granulomas (inflammatory masses). *Neuromodulation* 15 (5) (2012) 483–495.
28. N. Draulens, K. Vermeersch, B. Degraeuwe, et al., Intrathecal baclofen in multiple sclerosis and spinal cord injury: complications and long-term dosage evolution. *Clin. Rehabil.* 27 (12) (2013) 483–495.
29. Remodulin® (treprostanil) injection, United Therapeutics, Research Triangle Park, NC, May 2002.
30. E.F. Lawson, and M.S. Wallace, Advances in intrathecal drug delivery. *Curr. Opin. Anaesthesiol.* 25 (5) (2012) 572–576.

31. B. Ridley, and P.K. Rawlins, Intrathecal baclofen therapy: ten steps toward best practice. *J. Neurosci. Nurs.* 38 (2) (2006) 72–82.
32. P.M. Brennan, and I.R. Whittle, Intrathecal baclofen therapy for neurological disorders: a sound knowledge base but many challenges remain. *Bri. J. Neurosurg.* 22 (4) (2008) 508–519.
33. P.R. van Dijk, S.J. Logtenberg, et al., Complications of continuous intraperitoneal insulin infusion with an implantable pump. *World J. Diabetes* 3 (8) (2012) 142–148.
34. Lioresal® intrathecal (baclofen injection), Medtronic, Minneapolis, MN, March 2013.
35. L.E. Krach, R.L. Kriel, et al., Complex dosing schedules for continuous intrathecal baclofen infusion. *Pediatr. Neurol.* 37 (5) (2007) 354–359.
36. M.C. Scheiss, I.J. Oh, et al., Prospective 12-month study of intrathecal baclofen therapy for poststroke spastic upper and lower extremity motor control and functional improvement. *Neuromodulation* 14 (1) (2011) 38–45.
37. P.K. Rawlins, Intrathecal baclofen therapy over 10 years. *J. Neurosci. Nurs.* 36 (6) (2004) 322–327.
38. R.J. Coffey, T.S. Edgar, G.E. Francisco, et al., Abrupt withdrawal from intrathecal baclofen: recognition and management of a potentially life-threatening syndrome. *Arch. Phys. Med. Rehabil.* 83 (6) (2002) 735–741.
39. J.C. Ross, A.M. Cook, G.L. Stewart, B.G. Fahy, Acute intrathecal baclofen withdrawal: a brief review of treatment options. *Neurocrit. Care* 14 (1) (2011) 103–108.
40. A.J. Manson, K. Turner, et al., Apomorphine monotherapy in the treatment of refractory motor complications of Parkinson's disease: long-term follow-up study of 64 patients. *Mov. Disord.* 17 (6) (2002) 1235–1241.
41. F. Torres, and L.J. Rubin, Treprostanil for the treatment of pulmonary arterial hypertension. *Expert Rev. Cardiovasc. Ther.* 11 (1) (2013) 13–25.
42. D. Nyholm, Duodopa® treatment for advanced Parkinson's disease: a review of efficacy and safety. *Parkinsonism Relat. Disord.* 18 (8) (2012) 916–929.
43. M. Zibetti, A. Merola, V. Ricchi, et al., Long-term duodenal levodopa infusion in Parkinson's disease: a 3-year motor and cognitive follow-up study. *J. Neurol.* 260 (1) (2013) 105–114.
44. K. Kumar, S. Rizvi, and S. Bishop, Cost-effectiveness of intrathecal drug therapy in management of chronic nonmalignant pain. *Clin. J. Pain* 29 (2) (2013) 138145.
45. G. de Lissovov, L.S. Matza, H. Green, M. Werner, and T. Edgar, Cost-effectiveness of inthrathcal baclofen therapy for the treatment of severe spasticity associated with cerebral palsy. *J. Child. Neurol.* 22 (1) (2007) 49–59.
46. R. Hardoff, M. Sula, A. Tamir, A. Soil, A. Front, S. Badarna, S. Honigman, and N. Giladi, Gastric emptying time and gastric motility in patients with Parkinson's disease. *Mov. Disord.* 16 (6) (2001) 1041–1047.
47. M.M. Patel, B.R. Goyal, et al., Getting into the brain: approaches to enhance drug delivery. *CNS Drugs* 23 (1) (2009) 35–58.
48. D.R. Thakker, M.R. Weatherspoon, et al., Intraventricular amyloid-beta antibodies reduce cerebral amyloid angiopathy and associated microhemorrhages in aged Tg2576 mice. *Proc. Natl. Acad. Sci. USA* 106 (11) (2009) 4501–4506.
49. B. Vellas, M.C. Carillo, et al., Designing drug trials for Alzheimer's disease: what we have learned from release of the phase III antibody trials: a report from the EU/US/CTAD Task Force. *Alzheimers Dement.* 9 (4) (2013) 438–444.

50. B.G. Rocque, and A.L. Albright, Intraventricular versus intrathecal baclofen for secondary dystonia: a comparison of complications. *Neurosurgery* 70 (2013) 321–325.
51. A.E. Lang, S. Gill, et al., Randomized controlled trial of intraputamenal glial cell line-derived neurotrophic factor infusion in Parkinson disease. *Ann. Neurol.* 59 (3) (2006) 459–466.
52. N.K. Patel, and S.S. Gill, GDNF delivery for Parkinson's disease. *Acta Neurochir. Suppl.* 97 (2) (2007) 135–154.
53. D.K. Stiles, Z. Zhang, et al., Widespread suppression of huntingtin with convection-enhanced delivery of siRNA. *Exp. Neurol.* 233 (1) (2012) 463–471.
54. C.M. Bernards, and H.F. Hill, Physical and chemical properties of drug molecules governing their diffusion through the spinal meninges. *Anesthesiology* 77 (4) (1992) 750–756.
55. J.A. Aldrete, Epidural fibrosis after permanent catheter insertion and infusion. *J. Pain Symptom Manage.* 10 (8) (1995) 624–631.
56. C.W. Olanow, J.A. Obeso, et al., Continuous dopamine-receptor treatment of Parkinson's disease: scientific rationale and clinical implications. *Lancet Neurol.* 5 (2006) 677–687.
57. P. Jenner, A.C. McCreary, et al., Continous drug delivery in early- and late-stage Parkinson's disease as a strategy for avoiding dyskinesia induction and expression. *J. Neural. Transm.* 118 (2011) 1691–1702.
58. P. Jenner, Wearing off, dyskinesia and the use of continuous drug delivery in Parkinson's disease. *Neurol. Clin.* 31 (3 Suppl.) (2013) S17–S35.
59. G. Abbruzzese, P. Barone, U. Bonuccelli, L. Lopiano, and A. Antonini, Continuous intestinal infusion of levodopa/carbidopa in advanced Parkinson's disease: efficacy, safety and patient selection. *Funct. Neurol.* 27 (3) (2012) 147–154.
60. T. Deer, J. Prager, et al., Polyanalgesic Consensus Conference 2012: recommendations on trialing for intrathecal (intraspinal) drug delivery: report of an interdisciplinary expert panel. *Neuromodulation* 15 (5) (2012) 420–435.
61. R.D. Penn, M.M. York, et al., Catheter systems for intrathecal drug delivery. *J. Neurosurg.* 83 (2) (1995) 215–217.
62. K.R. Hildebrand, D.D. Elsberry, and T.R. Deer, Stability, compatibility, and safety of intrathecal bupivacaine administered chronically via an implantable delivery system. *Clin. J. Pain* 17 (3) (2001) 239–244.
63. Humalog® (insulin lispro injection, USP [rDNA origin]) injection, Lilly USA, Indianapolis, IN, 1996.
64. C. Sarkar, B. Basu, et al., The immunoregulatory role of dopamine: an update. *Brain Behav. Immun.* 24 (4) (2010) 525–528.
65. K.M. Ackland, A. Churchyard, et al., Panniculitis in association with apomorphine infusion. *Br. J. Dermatol.* 138 (1998) 480–482.
66. C.M. Bernards, Cerebrospinal fluid and spinal cord distribution of baclofen and bupivacaine during slow intrathecal infusion in pigs. *Anesthesiology* 105 (1) (2006) 169–178.
67. D.R. Enzmann, and N.J. Pelc, Normal flow patterns of intracranial and spinal cerebrospinal fluid defined with phase-contrast cine MR imaging. *Radiology* 178 (2) (1991) 467–474.
68. N. Alperin, E.M. Vikingstad, B. Gomez-Anson, and D.N. Levin, Hemodynamically independent analysis of cerebrospinal fluid and brain motion observed with dynamic phase contrast MRI. *Magn. Reson. Med.* 35 (5) (1996) 741–754.

69. R.A. Bhadelia, A.R. Bogdan, R.F. Kaplan, and S.M. Wolpert, Cerebrospinal fluid pulsation amplitude and its quantitative relationship to cerebral blood flow pulsations: a phase-contrast MR flow imaging study. *Neuroradiology* 39 (4) (1997) 258–264.

70. B. Weisner, and W. Bernhardt, Protein fractions of lumbar, cisternal, and ventricular cerebrospinal fluid: separate areas of reference. *J. Neurol. Sci.* 37 (3) (1978) 205–214.

71. I. Degrell, and E. Nagy, Concentration gradients for HVA, 5-HIAA, ascorbic acid, and uric acid in cerebrospinal fluid. *Biol. Psychiatry* 27 (8) (1990) 891–896.

72. T.L. Yaksh, S. Hassenbusch, et al., Inflammatory masses associated with intrathecal drug infusion: a review of preclinical evidence and human data. *Pain Med.* 3 (4) (2002) 300–312.

73. J.W. Allen, K.A. Horais, N.A. Tozier, K. Wegner, J.A. Corbeil, R.F. Mattrey, S.S. Rossi, and T.L. Yaksh, Time course and role of morphine dose and concentration in intrathecal granuloma formation in dogs: a combined magnetic resonance imaging and histopathology investigation. *Anesthesiology* 105 (3) (2006) 581–589.

74. T.L. Yaksh, N. Tozier, et al., Toxicology profile of *N*-methyl-D-aspartate antagonists delivered by intrathecal infusion in the canine model. *Anesthesiology* 108 (5) (2008) 938–949.

75. S.J. Hassenbusch, W.C. Satterfield, et al., Preclinical toxicity study of intrathecal administration of the pain relievers dextrophan, dextromethorphan and memantine in the sheep model. *Neuromodulation* 2 (4) (1999) 230–240.

76. D. Nyholm, Duodopa® treatment for advanced Parkinson's disease: a review of efficacy and safety. *Parkinsonism Relat. Disord.* 18 (8) (2012) 916–929.

77. M. Zibetti, A. Merola, et al., Long-term duodenal levodopa infusion in Parkinson's disease: a 3-year motor and cognitive follow-up study. *J. Neurol.* 260 (1) (2013) 105–114.

78. A.L. Albright, Y. Awaad, et al., Performance and complications associated with the synchromed 10-ml infusion pump for intrathecal balcofen administration in children. *J. Neurosurg.* 101 (1 Suppl.) (2004) 64–68.

79. S.M. Hester, J.F. Fisher, et al., Evaluation of salvage techniques for infected baclofen pumps in pediatric patients with cerebral palsy. *J. Neurosurg. Pediatr.* 10 (6) (2012) 548–554.

80. E.M. Dvorak, J.R. McGuire, et al., Incidence and identification of inthrathecal catheter malfunction. *PM R.* 2 (8) (2010) 751–756.

81. S.J. Hassenbusch, S. Gunes, S. Wachsman, and K.D. Willis, Intrathecal clonidine in the treatment of intractable pain: a phase I/II study. *Pain Med.* 3 (2) (2002) 85–91.

82. S.S. Gill, N.K. Patel, et al., Direct brain infusion of glial cell line-derived neurotrophic factor in Parkinson disease. *Nat. Med.* 9 (5) (2003) 589–595.

83. N.K. Patel, and S.S. Gill, GDNF delivery for Parkinson's disease. *Acta Neurochir. Suppl.* 97 (2) (2007) 135–154.

84. T. Deer, E.S. Krames, S. Hassenbusch, et al., Future directions for intrathecal pain management: a review and update from the interdisciplinary polyanalgesic consensus conference 2007. *Neuromoduation* 11 (2) (2008) 92–97.

85. S.E. Kern, J. Allen, J. Wastaff, et al., The pharmacokinetics of the conopeptide contulakin-G (CGX-1160) after intrathecal administration: an analysis of data from studies in beagles. *Anesth Analg.* 104 (6) (2007) 1514–1520.

86. S. Diochot, A. Baron, M. Salinas, et al., Black mamba venom peptides target acid-sensitive ion channels to abolish pain. *Nat. Lett.* 490 (2012) 552–557.

87. C.B. Komanski, R.L. Rauck, J.M. North, K.S. Hong, R. D'Angelo, and K.R. Hildebrand, Intrathecal clonidine via lumbar puncture decreases blood pressure in patients with poorly controlled hypertension. Neuromodulation (2015) E-pub ahead of print. DOI:10.1111/ner.12304.
88. L.B. Marangell, M.S. George, A.M. Callahan, et al., Effects of intrathecal thyrotropin-releasing hormone (protirelin) in refractory depressed patients. *Arch Gen. Psychiatry* 54 (3) (1997) 214–222.
89. A.M. Callahan, M.A. Frye, et al., Comparative antidepressant effects of intravenous and intrathecal thyrotropin-releasing hormone: confounding effects of tolerance and implications for therapeutics. *Biol. Psychiatry* 41 (3) (1997) 264–272.
90. S. Desole, C. Velik-Salchner, et al., Subcutaneous implantation of a new intravenous pump system for prostacyclin treatment in patients with pulmonary arterial hypertension. *Heart Lung* 41 (6) (2012) 599–605.
91. R. Ewert, M. Halank, L. Bruch, et al., A case series of patients with severe pulmonary hypertension receiving an implantable pump for intravenous prostanoid therapy. *Am. J. Respir. Crit. Care Med.* 186 (11) (2012) 1196–1198.
92. C.J. Woolf, Intrathecal high dose morphine produces hyperalgesia in the rat. *Brain Res.* 209 (1981) 491–495.
93. J.S. Grider, M.E. Harned, and M.A. Etscheidt, Patient selection and outcomes using low-dose intrathecal opioid trialing methods for chronic nonmalignant pain. *Pain Physician* 14 (2011) 343–351.
94. M. Hamza, D. Doleys, M. Wells, J. Weisbein, J. Hoff, M. Martin, C. Soteropoulos, J. Barreto, S. Deschner, and J. Ketchum, Prospective study of 3-year follow-up of low-dose intrathecal opioids in the management of chronic nonmalignant pain. *Pain Med.* 13 (10) (2012) 1304–1313.
95. M. Kaminska, D.E. Lumsden, et al., Rechargeable deep brain stimulators in the management of paediatric dystonia: well tolerated with a low complication rate. *Stereotact. Funct. Neurosurg.* 90 (4) (2012) 233–239.
96. L. Timmermann, M. Schupbach, F. Hertel, et al., A new rechargeable device for deep brain stimulation: a prospective patient satisfaction survey. *Eur. Neurol.* 69 (4) (2013) 193–199.
97. Y. Lebenthal, L. Lazar, et al., Patient perceptions of using the OmniPod system compared to conventional insulin pumps in young adults with type 1 diabetes. *Diabetes Technol. Ther.* 14 (5) (2012) 411–417.
98. A. Peters, S.L. Palay, and H. Webster, *The Fine Structure of the Nervous System: Neurons and Their Supporting Cells.* Oxford University Press, Oxford. 1991. Figure 13-1 p.395.

8

PROMUS ELEMENT™ PLUS®: A DRUG-ELUTING STENT

Yen-Lane Chen and Kimberly Robertson
Boston Scientific Inc., Maple Grove, MN 55311, USA

8.1 INTRODUCTION

Coronary artery disease (CAD) occurs when the arteries, which supply blood to the heart, become narrowed, or stenosed, by the deposition of atherosclerotic plaque. Severe stenosis of the vessel lumen can result in a dramatic decrease in the supply of oxygen and nutrients to the myocardium. This in turn can result in death. According to the American Heart Association, CAD was responsible for 400,000 deaths in the United States in 2008, which is 1 out of every 6 deaths [1].

The treatment of CAD was revolutionized in the late 1970s with the introduction of percutaneous transluminal coronary angioplasty (PTCA), as an alternative to coronary artery bypass graft (CABG) surgery. PTCA began as a minimally invasive procedure in which a balloon-tipped catheter is inserted into the vasculature through the femoral or radial artery. Guided by fluoroscopy, the balloon is threaded to the disease site in the artery, where it is inflated and thereby dilates the vessel with concomitant improvement in blood flow [2,3]. The simplicity combined with effectiveness of PTCA relative to CABG led to rapid acceptance by physicians. Nevertheless, the procedural complications of vessel recoil (acute, elastic reduction in diameter) and dissection (tear and damage) revealed the need for long-term mechanical support to maintain vessel dilation [4].

Bare metal stents (BMSs) were introduced as a means to minimize these complications, which further improved PTCA outcome. Although periprocedural

Drug–Device Combinations for Chronic Diseases, First Edition. Edited by SuPing Lyu and Ronald A. Siegel.

outcomes were markedly improved, a new phenomenon of in-stent restenosis appeared in patients. Here, the vessel is narrowed by robust tissue growth into the lumen of the stent, which was observed to affect 20–30% of patients within the first year following stent placement [5]. In-stent restenosis is believed to arise from a cascade of events stemming from endothelial denudation and/or other vessel wall injury, which occurs during PTCA or BMS placement [6,7]. The DES concept builds on the mechanical benefits of the BMS by incorporating a drug, which is released over time to prevent the excessive tissue growth associated with restenosis.

As a testament to their success, DESs are currently the treatment of choice among PTCA procedures. The clinical impact has been a reduction in reinterventions from about 25% with BMS to only 7% with DES [8]. DES usage peaked at nearly 90% of PTCA procedures in 2005 and had leveled out to approximately 70% by the end of 2008 [9]. Thus, DESs have significantly transformed treatment in the field of interventional cardiology, which had been previously dominated by BMS [10–13].

Introduction of a DES required the simultaneous development of a medical device as well as a drug delivery system; both are inherently complex and fraught with challenges in their own right. As an implant in critical blood vessels in the heart, attention to detail coupled with high quality was essential so that the treatment benefits could be provided while minimizing risk to the patients. More specifically, DES development required a thorough understanding of the drug-eluting coating, stent, and catheter as well as integration of these technologies through well-controlled manufacturing processes. Beyond providing a mechanical scaffold, the stent is a vehicle for drug delivery and, therefore, the design of the stent needs to achieve localized drug distribution. Likewise, the drug coating must be able to undergo expansion and compression during stent integration onto the delivery catheter, tracking through the tortuous vasculature, and eventually deployment in the target vessel.

Cypher™ and TAXUS™ were the first generation of DES combination products that were approved for use. In the subsequent decade, continuous advancements in the stent, catheter, and drug coating technology have been made that have resulted in further improvement in the clinical outcome. The PROMUS Element Plus everolimus-eluting stent system (Boston Scientific) represents a new-generation DES combination product. In the following, this product will be used as an example to illustrate the complex integration of science, engineering, and technology that was required to provide a superior drug–device combination product. The following sections are organized with an initial discussion of the major components of the system, including the stent, drug, polymer coating, and stent deployment system (catheter). Thereafter, the approach to the development of the formulation and associated characterization of the drug coating will be addressed. Finally, the preclinical and pharmacokinetic and pharmacodynamic evaluation of the drug in the arterial vessel of the porcine model will be provided.

8.2 STENT

The stent is critical for the initial vessel expansion as well as maintaining the dilated diameter of the blood vessel. This may appear deceptively simple as the subtle requirements of the mechanical properties of the stent are not obvious. Perhaps first, the stent must be threaded a rather longer distance through the vascular system, which necessitates flexibility to allow navigation along the tortuous route. Yet, once in place, the stent must have sufficient mechanical strength and durability to resist the repetitive compressional forces exerted on it during blood circulation. There is also a need for the stent to be radiopaque to facilitate correct placement of the stent. Finally, the surface must be amenable to the drug–polymer coating. In the following sections, the material, geometry, and properties pertaining to the stent are highlighted with respect to these required functions.

8.2.1 Stent Material

Both stainless steel (SS) (typically 316L) and cobalt–chromium alloys have a long history of use in coronary stents. These metals are notable for their excellent resistance to corrosion and inherent mechanical strength. More recently, advances in materials science and stent design have allowed the development of a novel platinum–chromium (PtCr) alloy, which has unique and desirable properties. A partial listing of the composition of three stents is given in Table 8.1. Incorporation of 33% platinum into chromium results in an alloy with higher density (9.9 g/cm^3) compared with 316L SS (8.0 g/cm^3) or L605 CoCr (9.1 g/cm^3) leading to enhanced radiopacity.

The radiopacity of several stents is displayed Figure 8.1, in which the Element stent is more visible in comparison to that of the CoCr Xience V™ stent and the TAXUS Liberté™ stainless stent.

Improved biocompatibility is also expected with PtCr due to the lower nickel content, since the latter metal has been associated with incidence of allergic reactions in patients. The lower nickel content may render the material more prone to corrosion; however, this is readily addressed by the application of a passive, corrosion resistant oxide layer that is comparable in durability to stainless steel [14]. Finally, the superior mechanical strength of the PtCr alloy combined with the stent architecture enabled a reduction in the stent strut width and thickness, which is discussed in the following section.

TABLE 8.1 Elemental Composition of Common Stent Materials

	Elemental Composition by Weight (%)		
Material	PROMUS Element PtCr Alloy	TAXUS Liberté 316L Stainless Steel	Xience V L605 Cobalt Chromium
Iron	37	64	3.0 max
Platinum	33	–	–
Cobalt	–	–	52
Chromium	18	18	20
Nickel	9	14	10
Tungsten	–	–	15
Molybdenum	2.6	2.6	–
Manganese	0.05 max	2.0 max	1.5
Density (g/cm^3)	9.9	8.0	9.1

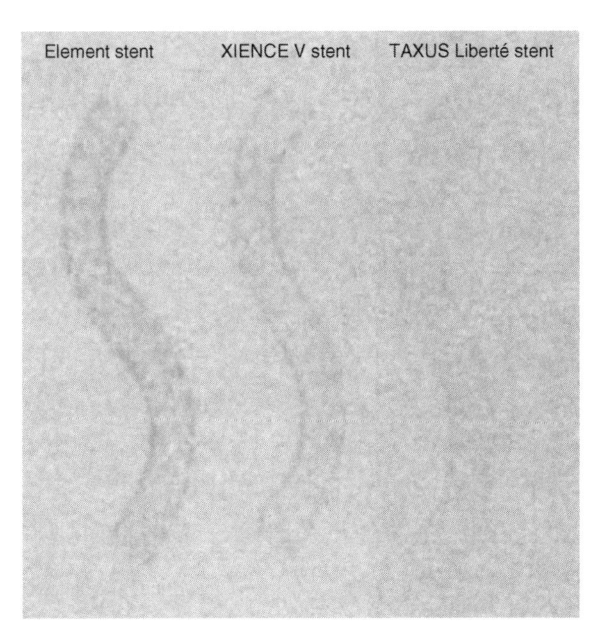

FIGURE 8.1 X-ray images of Element, Xience V, and Liberté stents (Simulation, 3.00 mm Diameter, 4 mm Copper Phantom.) (Reproduced with permission from Boston Scientific Inc.).

8.2.2 Geometric Pattern of Stents

The unique mechanical properties, flexibility in a collapsed state and resistance to recoil in an expanded state, are largely achieved through the elegant design of the geometric pattern of the struts, which refer to the individual armatures of the stent. Significant improvements have been incorporated into the advanced DES systems, which have resulted in thinner struts, enhanced conformable geometry, and more consistent surface-to-artery ratio (SAR) across the stent matrix. The outcome associated with the stent design has been studied in numerous clinical trials [15–20], which have provided substantial evidence confirming, what perhaps is intuitive, that thinner struts are associated with less restenosis. The introduction of a design with thinner struts has also reduced the projected area or as it is called, the stent profile. A low stent profile is associated with greater flexibility and enables deployment through more tortuous arteries, thereby allowing more complex lesions to be reached.

A schematic diagram depicting the geometry of the Element stent is given in Figure 8.2. This pattern provides the needed balance of seemingly opposing goals. That is, a greater mass favors inherent strength, radiopacity, and resistance to recoil, while a smaller mass favors flexibility and improved healing.

The Element stent design incorporates a dimensionally uniform pattern of serpentine segments, each with two offset connectors that reverse direction for alternate rows to maintain a balance of forces along the stent. This also allows each segment to operate almost independently of the other, improving deliverability and conformability. The peaks are widened at the crown to redirect the strain of expansion to the

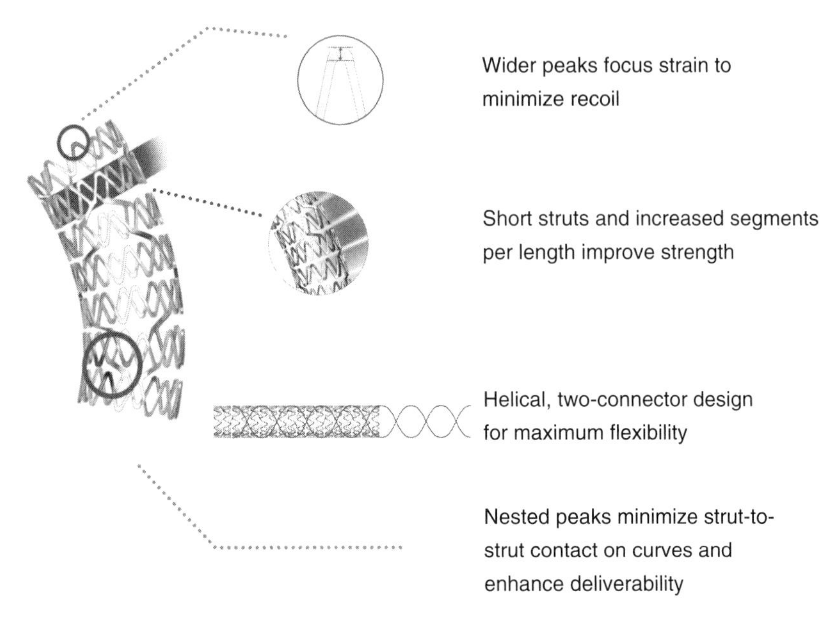

FIGURE 8.2 PROMUS Element workhorse stent. (Reproduced with permission from Boston Scientific Inc.) (See colour plate section.)

longitudinal portion, and this, along with the properties of the alloy, reduces the recoil of the expanded stent and helps to maintain luminal diameter. The offset and nested nature of the peaks also reduces potential strut-to-strut contact when the stent threads around a bend in the vasculature.

The Element Stent design incorporates four stent models across the range of diameters, whereas other stent platforms have only two or three stent models for the same range of diameters. For the Element stent, increasing the number of models allows the SAR to be optimized and provides more uniform drug distribution (discussed in detail below) and scaffolding over the range of diameters.

In Table 8.2, the geometric pattern and associated dimensions are given for PROMUS Element Plus, TAXUS Liberté, and Xience V. Among the three systems, there are clear differences in the geometric pattern, which has a direct impact on the stent to vessel contact area. The range of strut diameters and thicknesses is also provided, which fall in the micron-size range. Also provided in Table 8.2 are key bench test results, including radial strength, recoil, and conformability, which demonstrate the mechanical performance of the PROMUS Element Plus stent.

While resistance to compression, as measured by radial strength, relates to the ability of the stent scaffolding to maintain the vessel lumen, recoil is a measure of the ability of the stent to maintain its initial expansion diameter and minimize the risk of malapposition to the vessel. Conformability relates to the ability of a stent to support tortuous vessels without inducing undesirable vessel straightening. Increased stent rigidity limits the use of stents in both tortuous segments and at bends, which may result in a vessel hinge effect that has been associated with an increase in restenosis. It can be seen in Table 8.2 that the PtCr Element stent platform provides improved radial strength, compression resistance, and conformability compared to prior stent platforms.

TABLE 8.2 Geometric Characteristics of PROMUS Element Plus, TAXUS Liberté, and Xience V

	Specification	PROMUS Element Stent Platform	TAXUS Liberté Stent Platform	Xience V Stent Platform
Design Elements	Material	PtCr	316L SS	L605 CoCr
	Stent geometry			
	Number of model	4	3	2
	Strut width (μm)	91	76	91
	Strut thickness (μm)	81	97	81
	Surface-to-artery ratio	2.5 mm:17.6 3.0 mm:16.4	2.5 mm:15.5 3.0 mm:17.6	2.5 mm:16.8 3.0 mm:14.1
Bench testing	Radial strength (N/mm)	0.26	0.24	0.11
	Recoil (% OD)	3.0	3.2	5.0
	Conformability (N mm)	0.04	0.09	0.32

Definitions: Surface-to-artery ratio (SAR) = the percentage of the vessel surface area covered by stent material when deployed to a given diameter. Consistent SAR may result in more uniform dosing across vessel diameters. Too high of an SAR (>30%) may negatively impact vascular response.
Abbreviations: CoCr = cobalt chromium; NA = data not available; N/mm = Newtons/millimeter; PtCr = platinum chromium; SS = stainless steel; %OD = percent observed decrease. Also define conformability, N mm.
Bench data based on 2.5 mm diameter stents.

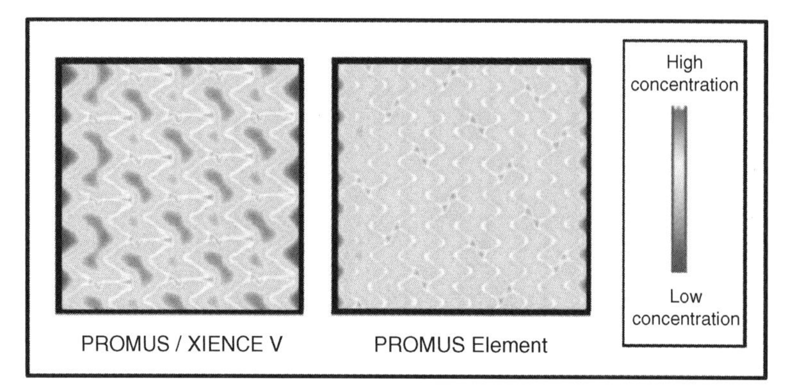

FIGURE 8.3 Computer simulation of drug distribution in vascular tissue. *Note:* Modeling results are for demonstration purposes only and may not necessarily be indicative of clinical performance. (Reproduced with permission from Boston Scientific Inc.) (See colour plate section.)

8.2.3 Stent Considerations as a Platform for Drug Delivery

The advances in stent geometry provided significant improvement but could not be implemented without a careful consideration of the ramifications for drug delivery. Here again, there are opposing design features with a thinner profile providing a better clinical outcome while posing a challenge in achieving uniform drug distribution in the treated artery [21]. Nevertheless, the geometric development of the Element stent was purposefully planned with simultaneous consideration of optimal mechanical performance and efficiency as a platform for drug delivery. Figure 8.3 depicts the drug distribution in the blood vessel based on computer modeling of two distinct stent patterns. Xience V has larger struts, which are distributed with larger intervening spaces, whereas PROMUS Element has smaller struts yet spaced closer together, which favors a more uniform distribution of drug in the blood vessel.

While the main mechanical properties pertaining to flexibility and strength have been addressed above, other stent performance factors were also evaluated. These included integrity, expansion uniformity, foreshortening, corrosion stability, side branch access, and MRI compatibility. These are clearly important, but are not discussed here.

8.3 DRUG COATING MATRIX

The second and equally important aspect in developing an advanced drug-eluting stent is consideration of the delivery of drug. The three main components are the drug, the primer layer, and the polymer, which controls the rate and extent of drug release.

Drug elution from PROMUS Element stent relies on two layers. The inner layer is a polymer coating of poly(*n*-butyl methacrylate) (PBMA), which is directly applied to the stent and serves as a primer for improving the adhesion of the outer layer. The outer layer is a drug–polymer matrix that contains a copolymer of vinylidene fluoride and hexafluoropropylene (PVDF-HFP), which is blended with the antiproliferative

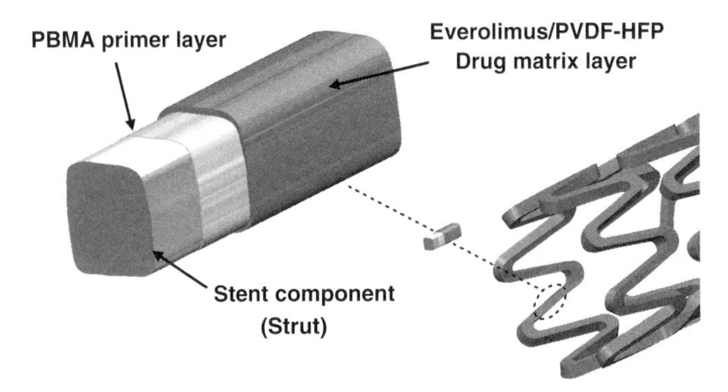

FIGURE 8.4 PROMUS Element Plus dual-layer coating design. (Reproduced with permission from Boston Scientific Inc.)

drug, everolimus. Figure 8.4 is a schematic diagram depicting the disposition of these components on the PROMUS Element stent.

8.3.1 Active Pharmaceutical Ingredient

Everolimus is the active pharmaceutical ingredient in the PROMUS Element drug-eluting stent system, and the chemical structure is given in Figure 8.5. It is a semisynthetic macrolide (Novartis, Basel, Switzerland) synthesized by addition of a 2-hydroxyethyl group in the 40th position of rapamycin (INN: Sirolimus) [22]. Rapamycin possesses antifungal, immunosuppressive and, most importantly, antiproliferative properties.

Both everolimus and rapamycin have similar mechanisms of action at the cellular and molecular level [23,24]. Everolimus forms a complex with the cytoplasmic protein FKBP-12. This complex in turn binds to and interferes with the function of

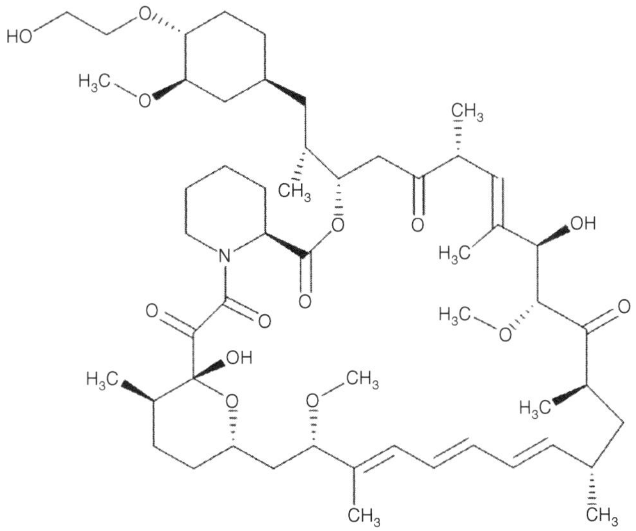

FIGURE 8.5 Chemical structure of everolimus. (Reproduced with permission from Boston Scientific Inc.)

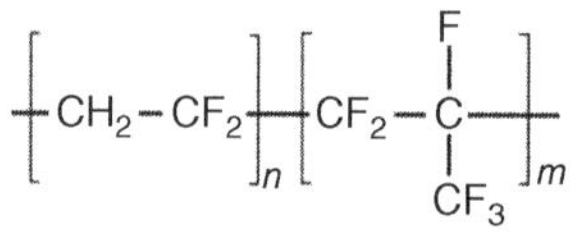

$$\left[CH_2 - CF_2 \right]_n \left[CF_2 - \underset{CF_3}{\overset{F}{C}} \right]_m$$

FIGURE 8.6 Chemical structure of drug-eluting polymer, PVDF–HFP–poly(vinylidene fluoride-*co*-hexafluoropropylene). (Reproduced with permission from Boston Scientific Inc.)

FRAP (FKBP-12 Rapamycin Associated Protein), also known as mTOR (mammalian target of rapamycin). In the presence of everolimus, the growth factor-stimulated phosphorylation of p70 S6 kinase and 4E-BP1, two key players in the initiation of protein synthesis, is inhibited [25,26]. At the cellular level, everolimus inhibits growth factor-stimulated cell proliferation in a reversible manner by blocking cell cycle progression at the G1 phase. This reversible blockage has implications for the delivery of drug.

Both *in vitro* and *in vivo* studies suggest that everolimus has similar pharmacological and toxicological profiles to sirolimus. However, due to its distinct physical and chemical properties, it differs with respect to the pharmacokinetics and intracellular distribution. Results from several *in vitro* cell assays and preclinical studies have provided critical evidence to support the effectiveness of everolimus in inhibiting smooth muscle cell (SMC) proliferation as a mechanism for preventing restenosis after stenting procedures.

8.3.2 Polymer

Stable polymers, such as poly(*n*-butyl methacrylate), polyisobutylene (Translute)™, and hexafluoropropylene monomers (PVDF-HFP), have been used in the DES coating to modulate the drug release. The polymer in PROMUS Element DES drug coating layer is a semicrystalline random copolymer of vinylidene fluoride and hexafluoropropylene (PVDF-HFP). The chemical structure of PVDF-HFP is shown in Figure 8.6. This polymer is highly resistant to hydrolytic, oxidative, and enzymatic degradation.

8.3.3 Mechanical Considerations for the Coating

In identifying the optimal copolymer composition, consideration of the mechanical properties as well as the drug release characteristics was needed. The ratio of VDF and HFP in the copolymer is used to achieve optimal elasticity in which greater than 600% elongation is achieved at the break point. In addition, appropriate toughness was obtained, as measured by the tensile strength, which is greater than 2000 psi. The hardness corresponded to a Shore D 60 valued at 2 mm thickness. Finally, the PVDF-HFP copolymers have a low level of crystallinity (20%). This provides the appropriate balance of mechanical strength and capability of elongation, such that the coating can then withstand stent crimping and other induced stresses that occur during the manufacturing, transcatheter delivery, and stent expansion.

Figure 8.7 shows a field-emission scanning electron micrograph of the coated Element stent initially processed (a), when integrated on the catheter (b), and after deployment (c). The areas of the coating, which undergo greater elastic and compressional deformation, are clearly revealed.

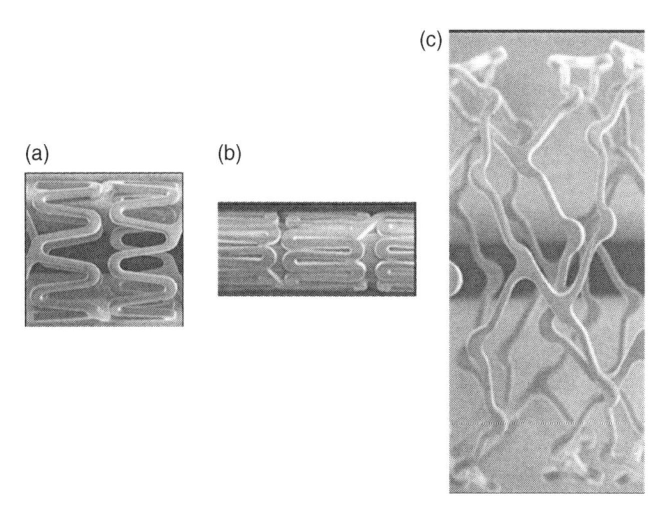

FIGURE 8.7 Field emission images of PROMUS Element Plus stent as processed, crimped, and deployed to demonstrate the compression and deformation experienced by the coating. (a) Coated Element stent initially processed. (b) Integrated on the catheter. (c) After deployment (Courtesy of Boston Scientific Inc.). (Reproduced with permission from Boston Scientific Inc.)

In addition to meeting the mechanical needs through deployment, the stent and coating must be sufficiently robust to endure the repetitive strain associated with the pulsatile nature of blood flow. The resiliency of the system is evident in the postelution image, where the geometry and coating were preserved for over 3 months *in vivo*.

8.3.4 Characterization of the Drug Coating Matrix

8.3.4.1 Drug Coating Thickness Coating thickness was measured trough SEM evaluation of a cross section of a resin-embedded PROMUS Element stent after focused ion beam (FIB) milling. A protective layer of platinum was first applied to preserve the surface of the coating and a high energy beam was then used to mill out a thin section of the coating (Figure 8.8a). The cross section of the coating was then examined by field emission scanning electron microscopy (FESEM) to characterize the thickness of the primer and active layers (Figure 8.8b).

The image of the cross section shows stratifications of the stent, PBMA-primer layer, and active everolimus/PVDF-HFP active layer. The primer layer was estimated to be 2 μm thick, and the drug matrix layer was 5 μm.

Additional characterization was provided by atomic force microscopy (AFM). In Figure 8.9, AFM images of a polymer coating on the PROMUS Element stent are shown. The control figure provides both measurements of the height (Figure 8.9a) and phase (Figure 8.9b) images in which a smooth and uniform appearance is evident. In contrast, the presence of drug reveals a distinct pattern appearing as darkened regions with phase contrast. This topology suggests a partial phase separation where both drug- and polymer-rich domains are present. This result was corroborated by

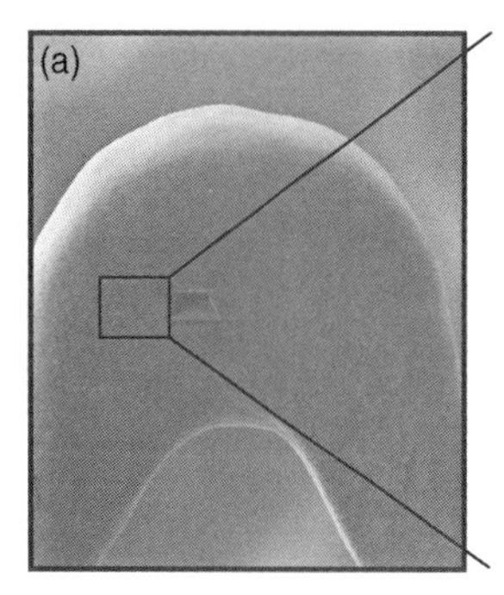

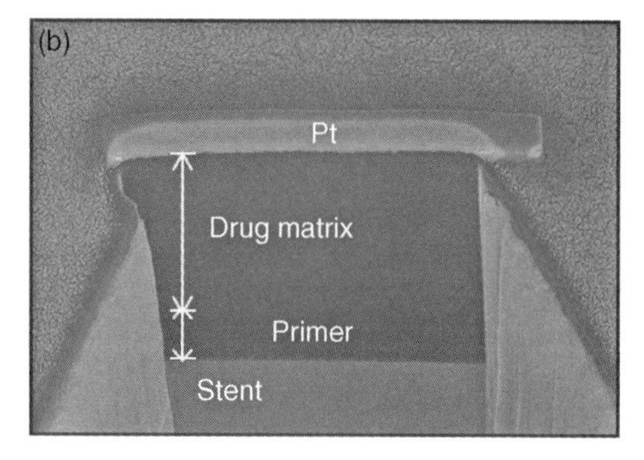

FIGURE 8.8 Characterization of the coating layers with FIB and FESEM. Part (a) shows the area of coating that was removed with the FIB. Part (b) shows exposed primer and drug matrix layer. (Reproduced with permission from Boston Scientific Inc.)

differential scanning calorimetry (DSC), where distinct glass transitions of everolimus and PVDF-HFP were observed.

In addition to SEM, AFM, and DSC, confocal Raman spectroscopy was also used to image the drug and polymer distribution in the PROMUS Element stent coating. The images below are vertical scans of a cross section of the PROMUS Element stent. As shown in Figure 8.10, the top of each image is above the coating, while the bottom of each image corresponds to a region within the stent strut. These results indicate that the drug-rich portion is uniformly distributed throughout the PVDF-HFP layer thickness, and little mixing if any, occurs between the drug/PVDF-HFP and PBMA primer layer.

Discrete active (everolimus and PVDF-HFP) and primer (PBMA) layers are evident with minimal mixing apparent between layers. Similar distributions are also displayed for the PVDF-HFP present in the active layer and PBMA primer layer.

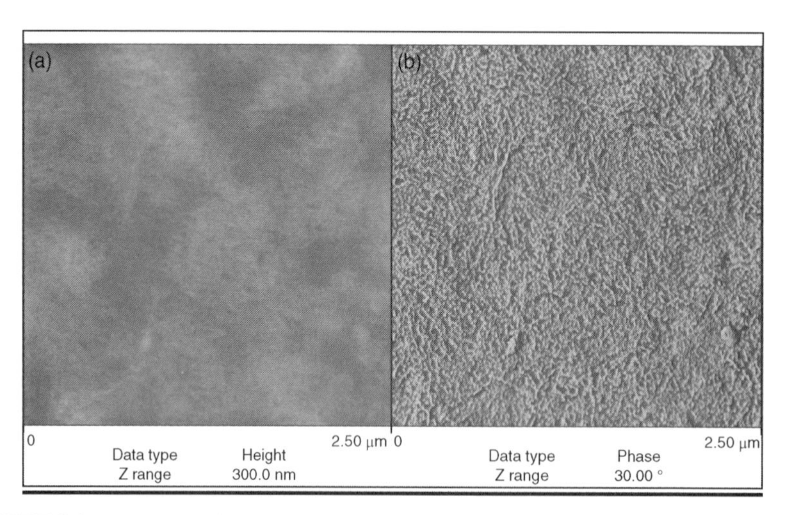

FIGURE 8.9 Representative AFM images of coating on PROMUS Element stent. (Reproduced with permission from Boston Scientific Inc.)

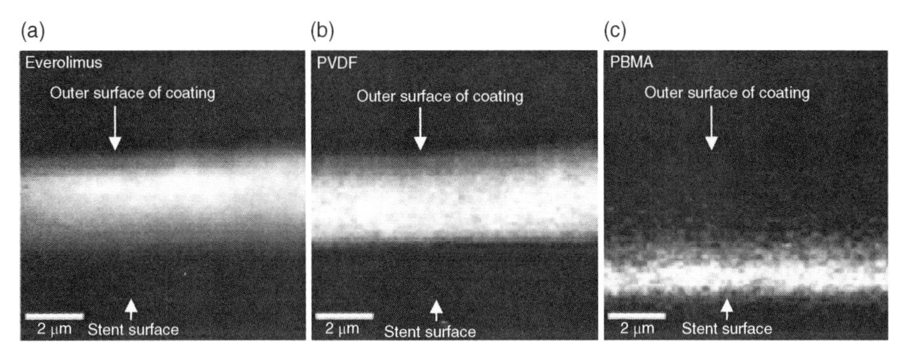

FIGURE 8.10 Confocal Raman spectroscopy of the PROMUS Element coating. Everolimus and PVDF–HFP from the active layer are shown in parts (a) and (b), respectively. The corresponding PBMA primer layer is shown in part (c). (Reproduced with permission from Boston Scientific Inc.)

Time-of-flight–secondary ion mass spectroscopy (TOF-SIMS) was used to map drug distribution in the surface layer of the coating. The everolimus molecular image was generated by summation of all the peaks in the positive everolimus ion spectrum. The everolimus ion image in Figure 8.11 shows that drug is evenly distributed across the stent surface that is consistent with AFM and SEM images, but here the technique identifies drug with chemical specificity.

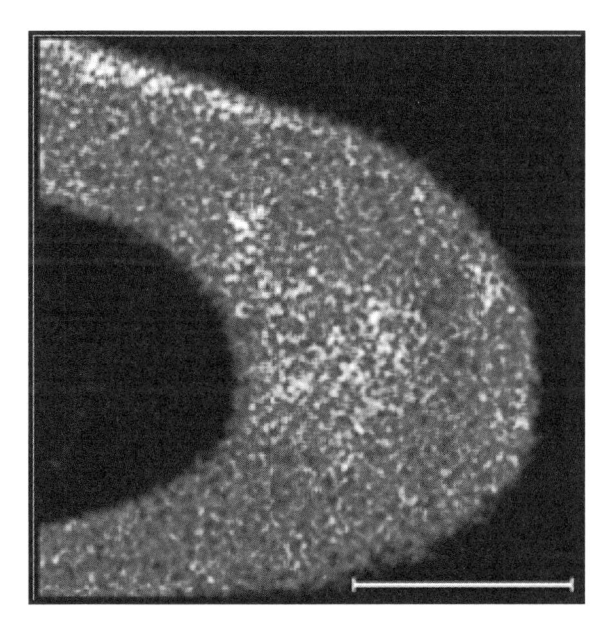

FIGURE 8.11 TOF-SIMS image of PROMUS Element drug coating. (Reproduced with permission from Boston Scientific Inc.)

8.3.5 Stent Delivery System

It finally remains to address the catheter, which is used to deliver the stent to the intended site. The general requirements for the catheter include pushability, trackability, stent securement, and stent deployment. Pushability refers to the ease with which the stent can be moved from the site of insertion in the femoral artery to the site of deployment, and trackability denotes the ability of the system to follow the guide wire through the tortuous anatomy. Stent securement is critical as separation of the stent from the catheter when threading through the vasculature can be fatal. Finally, stent deployment involves the process in which the stent is properly positioned within the blood vessel, expanded, and the catheter is retracted in a safe, facile manner.

The stent is centered on the balloon between two radiopaque marker bands, which in conjunction with fluoroscopy, aid in the positioning of the stent during the procedure and system for postdeployment dilation. The balloon extends beyond the stent ends, but only to the degree that ensures complete expansion of the stent during deployment. Balloon compliance (the increase in balloon diameter for a given increase in inflation pressure) is key to being able to size the stent to fit the target vessel.

The PROMUS Element Plus stent is available on three different types of delivery systems: monorail (MR), rapid-exchange delivery system, and over-the-wire (OTW) delivery system catheter. Only the monorail design will be discussed here. The major components of the monorail delivery system for the PROMUS Element Plus are manifold/strain relief, proximal shaft, midshaft, distal shaft, shaft hydrophilic coating, markerbands, balloon, and bumper tip. A schematic diagram of the monorail delivery catheter is provided in Figure 8.12.

Similar to other BSC monorail coronary stent systems and coronary dilatation catheters, the PROMUS delivery system has a coaxial lumen at the distal section of the catheter. The outer lumen is used for inflation of the balloon and the inner lumen permits the use of guide wires to facilitate advancement of the stent delivery system to and through the stenosis. Shown in Figure 8.13 is a PROMUS Element Plus stent crimped on a balloon catheter.

The balloon is designed to deploy the stent to a known diameter and length at a recommended pressure. A mandrel is placed into the inner lumen to protect the patency of the catheter. It is worth noting that the balloon for the PROMUS Element

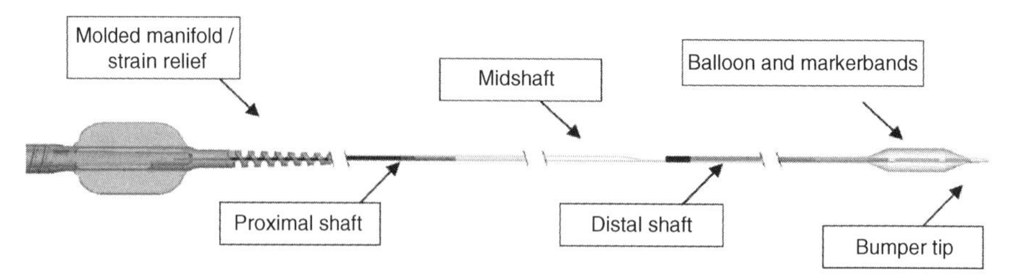

FIGURE 8.12 Monorail delivery catheter. (Reproduced with permission from Boston Scientific Inc.)

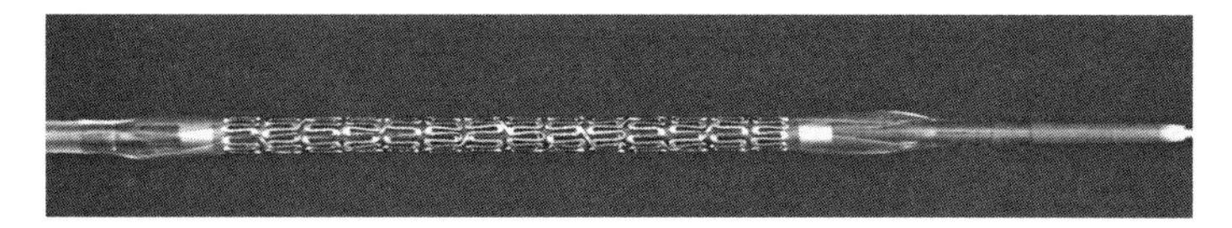

FIGURE 8.13 PROMUS Element Plus on stent delivery system. (Reproduced with permission from Boston Scientific Inc.)

Plus system is composed of coextruded two different thermoplastic materials (PEBAX®); this dual-layer design is to provide a more flexible balloon while maintaining the necessary strength and balloon compliance profile. The catheter's tip is tapered to facilitate advancement of the catheter to and through the stenosis and is designed to enable the device to follow the guidewire. Finally, a hydrophilic coating is applied to the catheter shaft to improve trackability of the device.

8.3.6 Additional Considerations for Delivery and Deployment of DES

As with the stent, there are a number of other factors that were examined before clinical introduction but are not covered in detail. These include: the stent–balloon interface, which must not have strong interactions that might cause loss of coating to the balloon surface, size of the crimped state, which ideally has a minimal profile without pinching/damaging the balloon, securement of the stent over the balloon, which must be maintained during delivery but must also be readily detached following stent expansion.

Once the integration of the drug-coated stent to the catheter system is complete, the finished units are placed in packaging and undergo an ethylene oxide sterilization cycle. The packaging must protect the device from physical damage during transport and maintain a sterile barrier throughout the shelf life of the device. The sterilization cycle is carefully selected to ensure sterility while minimizing impact on device performance.

Before considering the detailed *in vivo* performance of the DES, it is of value to reflect on the profound ramifications of this drug–device combination. The motivation of drug design is to achieve the aspiration of Paul Ehrlich, who originally described the concept of a "magic bullet," which entailed chemical properties that specifically targeted bacteria while leaving human cells unscathed. The DES system perhaps better achieves the aspiration of ideal drug delivery, by exploiting the fundamental physical chemistry of the drug with engineering principles pertaining to mass transport. That is, placement and release of drug at the site of action must necessarily provide a higher concentration at the location where the drug is needed. At this point, the reader has hopefully gained an appreciation of the detailed complexity involved in the development of the DES. It will be demonstrated below that the DES is both safe and effective for coronary artery disease.

In the following, the ultimate goal is to provide a connection among the drug coating characteristics, *in vivo* release and local distribution to the overall clinical outcomes.

8.4 DRUG RELEASE KINETICS

8.4.1 *In Vivo* and *In Vitro* Drug Release

The *in vivo* release of drug was determined from the stent explanted from the coronary artery of domestic swine at various time points following stent implantation. The release of everolimus from the polymer matrix into the surrounding arterial tissue is shown in Figure 8.14. The percentage of everolimus released during the first 24 h was approximately 25–30%, rising to about 50% at 1 week, and reaching 90% at 90 days postimplantation.

Initially, everolimus at or near the surface of the stent coating is rapidly dissolved and burst released into the surrounding media; thereafter, the release of everolimus transitions to a slower sustained release.

For the purpose of product quality assurance, an *in vitro* drug release method was developed to serve as a predictive measure of the *in vivo* release profile. The *in vitro* everolimus release profile in the *in vitro* release media is shown in Figure 8.15. As intended, the release of everolimus is much faster than that obtained *in vivo*, as an *in vitro* test cannot be prolonged to over 3 months and still serve as a predictive measure of product quality. However, the critical feature is that the measured release profiles *in vitro* can be related to the measured release profiles *in vivo* with proper scaling of the time axis. This is important in obtaining the *in vitro–in vivo* correlation (IVIVC).

The microstructure of the stent coating postdrug elution was evaluated using SEM, following the *in vitro* release in the release media for 3 days as displayed in Figure 8.16a and b. Prior to the elution of everolimus, the coating surface appeared to be relatively smooth prior to the everolimus elution as seen in 8.16a. At day 3, more than 85% of the drug was released and the coating matrix assumed a porous appearance as shown in 8.16b.

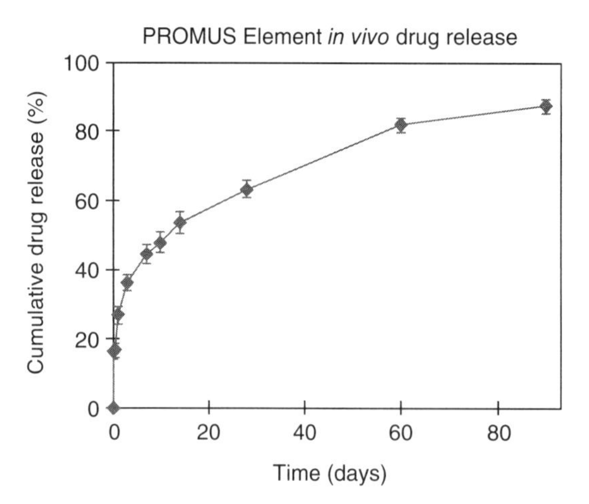

FIGURE 8.14 *In vivo* everolimus cumulative percentage release along with Weibull dissolution model-fit curves PROMUS Element (mean ± standard deviation, $n = 9–12$). (Reproduced with permission from Boston Scientific Inc.)

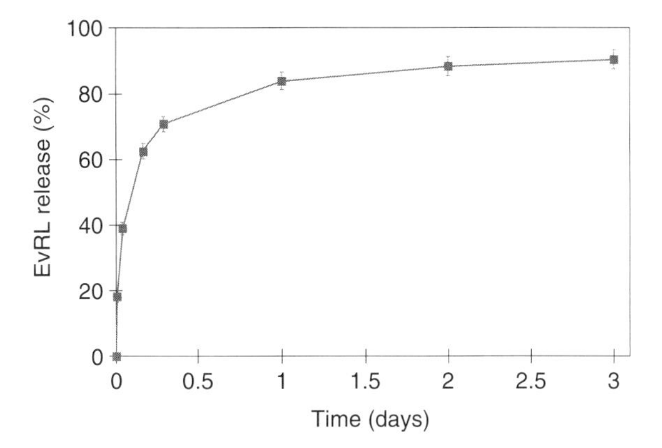

FIGURE 8.15 *In vitro* everolimus cumulative percentage release for PROMUS Element and Xience V (PROMUS) (mean ± standard deviation, $n = 12$) in release media. (Reproduced with permission from Boston Scientific Inc.)

8.4.2 Quantitative Assessment of the Mechanism of Drug Release

The second major point of consideration is the rate, extent, and mechanism of drug release. The images above, substantiating the existence of partial phase separation, are critical in providing an explanation for the measured release rates *in vitro* and *in vivo*. A mechanistic pore–shell dissolution–diffusion model was developed based on the microstructure of the coating matrix and mechanism of release. This model incorporates the findings from the coating characterization to construct a physically relevant model of the drug mechanism release. As developed, the model consists of an idealized cylindrical pore–shell microstructure, as shown in Figure 8.17, and incorporates both dissolution of solid drug and diffusion of dissolved drug within the polymer as the two mechanisms of drug release.

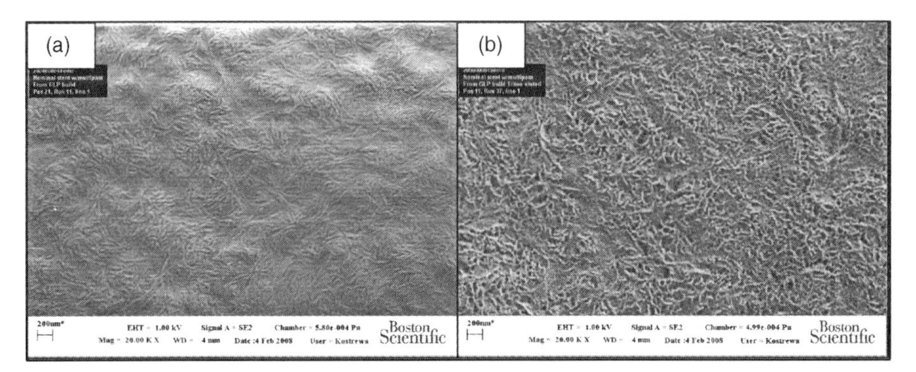

FIGURE 8.16 Field-emission scanning electron microscopy images of the PROMUS Element stent coating microstructure at 20,000× magnification. (a) Pre-elution. (b) Three-day everolimus elution in Triton X-405 *in vitro* release media. (Reproduced with permission from Boston Scientific Inc.)

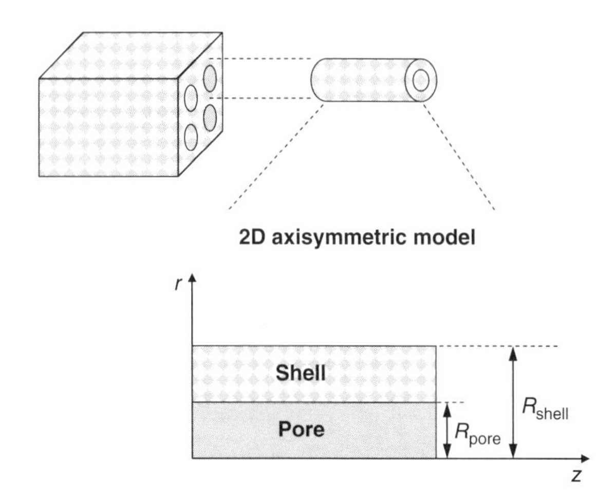

FIGURE 8.17 Idealized pore–shell microstructure used in the alternative computational mode. (Reproduced with permission from Boston Scientific Inc.)

Although the pore–shell geometry is a simplified representation of the coating microstructure that was observed in the characterization studies, it can capture the essential features for describing the release of drug. The pore is initially composed of solid drug directly connected to the surface, while the shell contains solid drug particles encapsulated by polymer along with drug, which is miscible in polymer. By accounting for the coating microstructure, the factors controlling release from a phase-separated system can be modeled.

Drug release in the pore can be characterized as initial dissolution of solid drug at the surface and subsequent diffusion of the dissolved drug through the media-filled pores beyond the dissolution front [27]. In the limit of fast dissolution compared to diffusion and at drug concentrations far above the solubility limit in the release media, the square root of time release approaches a square root of time described by Higuchi.

Drug release in the shell can be characterized as a dissolution–diffusion process through a heterogeneous media using effective-medium approximations for diffusivity and solubility [28]. Again, in the limit of fast dissolution and at concentrations far above the solubility limit, the release approaches that described by Higuchi using properties of the effective medium.

The total release through the pore–shell system depends, *inter alia*, on the volume fraction of drug contained within the pores, the permeability of drug in the release medium and the polymer, and the radii of the pore and shell. For example, if the permeability in the pore is much higher than that in the shell, the release will appear biphasic, with fast drug release from the pore occurring first followed by a slow release of drug from the shell.

In addition to the characterization studies, many of the parameters needed for the model such as drug diffusivity and solubility in both the release media and the polymer matrix were experimentally measured. The remaining model parameters were then fit to the *in vivo* drug release. Results of these fits are shown in Figure 8.18.

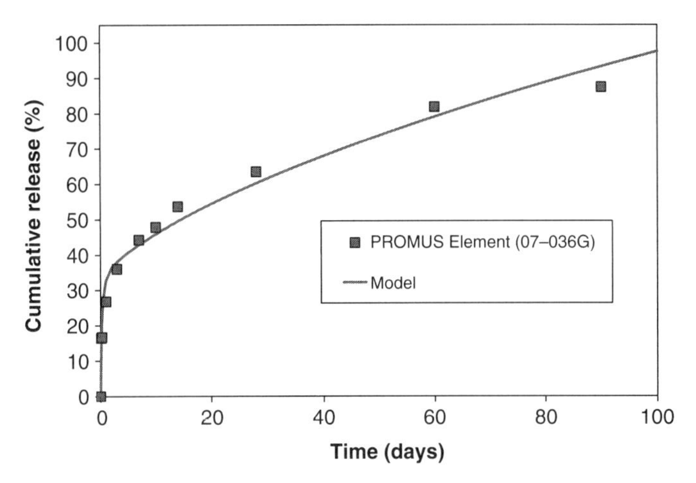

FIGURE 8.18 Pore–shell MOR model of *in vivo* drug release profile for PROMUS Element. (Reproduced with permission from Boston Scientific Inc.)

The model can explain both *in vitro* and *in vivo* drug release profiles. The agreement between the model results and experimental drug release data supports the hypothesis that the coating consists of a phase-separated system composed of a surface-connected, drug-rich phase (pores) and a polymer-encapsulated drug phase (shell).

8.5 LOCAL AND VASCULAR PHARMACOKINETICS

8.5.1 Drug Concentration Time Profiles

Of perhaps greatest importance to the therapeutic effectiveness of the DES as well as potential side effects is the arterial tissue concentration near the site of deployment and the plasma concentrations. The concentration time course of everolimus in coronary arterial tissue following deployment of the PROMUS Element is given in Figure 8.19. As can be seen, the peak everolimus concentration in arterial tissue occurs at about 4 h after implantation. This is followed by a gradual decline in concentration over several weeks. At day 90, the arterial tissue concentrations approached zero, indicating the vessel is exposed to everolimus for approximately 3 months with implantation of PROMUS Element. Peak levels of everolimus in the unstented proximal and distal arterial vessel segments were 50- and 10-fold lower, respectively, in comparison to that in stented segments. Relative to arterial flow, this concentration data reflects a slight downstream effect with distal concentrations trending higher in comparison to proximal concentrations. However, these levels are minimal and substantiate achievement of local delivery of drug.

To determine the regional and systemic tissue exposure profiles of everolimus following stent implantation, the myocardium immediately adjacent to the stented vessel as well as the lungs, liver, kidney, and spleen were collected and analyzed for everolimus. The concentrations of everolimus were at least 50-fold lower than that

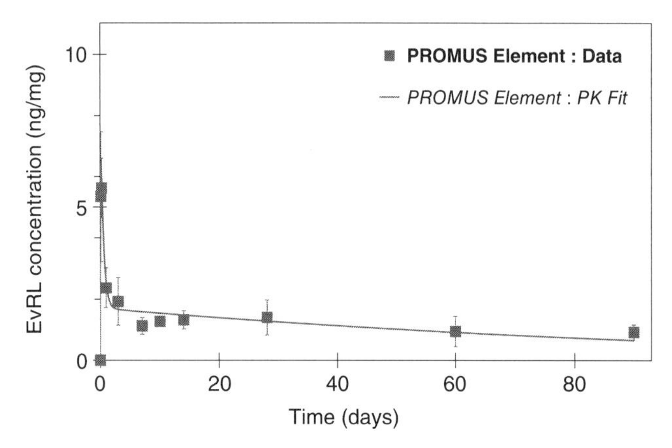

FIGURE 8.19 Stented coronary arterial tissue everolimus (EvRL) concentrations along with two-compartment pharmacokinetic model-fit curves for PROMUS Element (mean ± standard deviation, $n = 9$–12). (Reproduced with permission from Boston Scientific Inc.)

measured in the artery at 3 and 6 h where C_{max} was observed, with the duration of measurable levels restricted to 28 days (Figure 8.20). These limited levels of everolimus may be due to transport from the stent through the vessel and into the subjacent myocardium, along with some resulting from distal blood flow. The results obviously demonstrate the intrinsic value of local drug delivery achieved by the DES system.

In blood, the PROMUS Element peak concentration of 0.8 ng/ml occurs within the first 30 min following implant and approaches zero by 3 days, as displayed in Figure 8.21; no everolimus was detected at 72 h. Importantly, the peak blood levels were less than the minimum systemic level of everolimus recommended for the therapeutic range for preventing organ transplant rejection (3 to 8 ng/ml) when delivered orally [29–33].

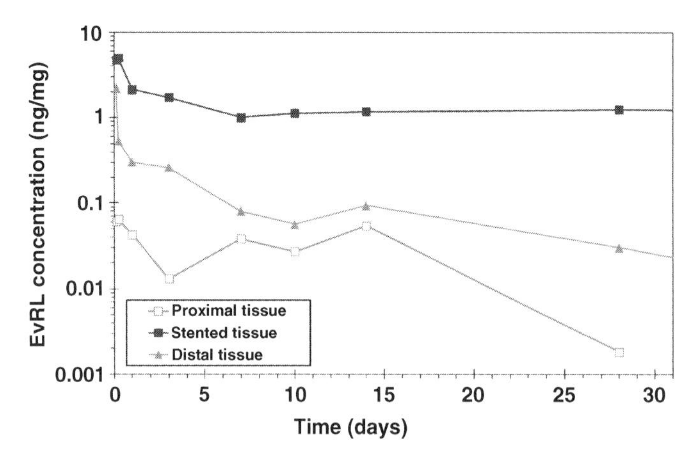

FIGURE 8.20 Stented, proximal unstented, and distal unstented coronary arterial tissue everolimus concentrations for PROMUS Element (mean, $n = 9$–12). (Reproduced with permission from Boston Scientific Inc.)

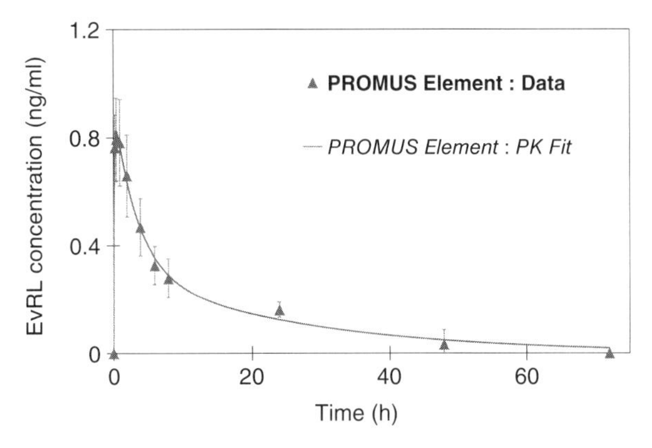

FIGURE 8.21 Blood everolimus concentrations along with two-compartment pharmacokinetic model-fit curves following implant of three 8 mm PROMUS Element stents (mean ± standard deviation, $n = 6$–8). (Reproduced with permission from Boston Scientific Inc.)

The arterial tissue and blood concentration profiles correlate with expectations based on the *in vivo* release profiles. Due to rapid dissolution of surface-associated everolimus, an equally rapid increase in arterial tissue concentrations is measured. However, not all everolimus that is released by the stent is taken up by the arterial tissue and available for the suppression of smooth muscle cell proliferation. Rather, a fraction of this burst release is lost to the bloodstream, leading to quantifiable systemic levels of everolimus for a period of time. The term local bioavailability is used to represent that fraction of drug distributed into the site of action relative to the amount of drug released. As release of everolimus from the polymer matrix slows and/or the vessel begins to encapsulate the stent struts, the local bioavailability increases. As a result, a pseudo-steady-state everolimus concentration is reached in the arterial tissue at the same time blood concentrations are decreasing to nondetectable levels.

The everolimus concentrations in the specific peripheral organs were at least 100-fold lower in relation to the stented arterial tissue levels at the same time point (Figure 8.22). Average peak everolimus concentrations, reached in the first 3–6 h, were all below 6 pg/mg. By day 3, everolimus concentrations in the peripheral organs were below the lower limit of quantification (0.8 pg/mg). These everolimus concentrations in the tested peripheral organs are minimal, which suggests that the peripheral organ exposure to everolimus after stenting is very limited.

8.5.2 Pharmacokinetic Modeling

The *in vivo* cumulative percentage release as a function of time was fit using the Weibull dissolution model. The Weibull dissolution model is expressed by the following equation:

$$\%\text{Release} = F_{\text{inf}}\left(1 - \mathrm{e}^{-(\text{time}/\beta_{vitro})^{\gamma_{vitro}}}\right) \tag{8.1}$$

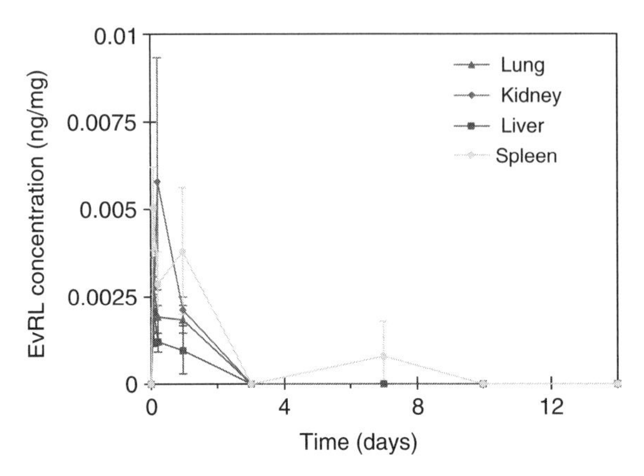

FIGURE 8.22 Peripheral organ everolimus concentrations following implant of three 8 mm PROMUS Element stents (mean ± standard deviation, $n = 2$–3). (Reproduced with permission from Boston Scientific Inc.)

where F_{inf} is the amount of drug release occurring at infinite time (no units), β_{vitro} is the time to achieve 63.2% of infinite release (days), and γ_{vitro} is a unitless parameter accounting for the shape of the release profile.

Using the above expression for the release of everolimus, the concentrations as a function of time in both the arterial tissue and blood were fit using a two-compartment model with linear absorption. While the arterial tissue concentrations are modeled on a per stent basis, the systemic everolimus blood concentrations were measured in animals, which had three implanted stents. The two-compartment pharmacokinetic model is expressed by the following equation, in which the transport steps were modeled as first-order processes.

$$\text{Conc}(t) = Ae^{-\alpha t} + Be^{-\beta t} + Ce^{-K01t} \tag{8.2}$$

where

- $\alpha = K_{\text{elim}}\, K_{21}/\beta$; $\beta = 1/2$ $\{(K_{12} + K_{21} + K_{\text{elim}}) - [(K_{12} + K_{21} + K_{\text{elim}})^2 - 4\,K_{21}\,K_{\text{elim}}]^{1/2}\}$
- $A = \dfrac{K_{\text{abs}}D(K_{21} - \alpha)}{V(K_{\text{abs}} - \alpha)(\beta - \alpha)}$; $B = \dfrac{K_{\text{abs}}D(K_{21} - \beta)}{V(K_{\text{abs}} - \beta)(\alpha - \beta)}$; $C = \dfrac{K_{\text{abs}}D(K_{21} - K_{\text{abs}})}{V(\alpha - K_{\text{abs}})(\beta - K_{\text{abs}})}$
- Conc(t): concentration in arterial tissue (ng/mg) at time t (days)
- D: amount of drug that is delivered to the artery (μg)
- V: apparent volume of distribution of drug in tissue (μg/(ng/mg))
- K_{abs}: apparent first-order absorption rate constant (1/day)
- K_{elim}: apparent first-order elimination rate constant (1/day)
- K_{12} and K_{21}: apparent first-order partition rate constants (1/day)

The model-estimated everolimus dissolution parameters plus the everolimus arterial tissue and blood pharmacokinetic parameters are listed in Table 8.3. The

TABLE 8.3 Pharmacokinetic Parameters Estimated from Two-Compartment Models for Arterial Tissue and Blood Everolimus Concentration Profiles

Parameter	PROMUS Element
Everolimus arterial pharmacokinetics	5.62
C_{max} (ng/mg)	
t_{max} (days)	0.25
$AUC_{0\text{–}90\ day}$ ((ng/mg) day)	107.60
CL_{term} (g/day)	0.39
Everolimus blood pharmacokinetics	0.79
C_{max} (ng/ml)	
t_{max} (h)	0.50
$AUC_{0\text{–}72\ h}$ ((ng/ml) h)	10.1
CL_{term} (l/h)	12.5

C_{max} = maximum drug concentration.

resulting everolimus arterial tissue and blood concentration profiles are displayed in Figures 8.19 and 8.21, respectively.

The clearance of everolimus from both the arterial tissue and blood was biphasic where there was an initial rapid decline of everolimus concentration after C_{max} followed by a slower elimination phase. The first phase of clearance may be due to the immediate efflux of free everolimus followed by a slower clearance due to the concentration-dependent binding of everolimus in plasma. Therefore, a two-compartment model is most appropriate to describe the everolimus pharmacokinetics in arterial tissue and blood.

8.6 PHARMACODYNAMICS: VASCULAR RESPONSE

The vascular response of the local arterial tissue to the implantation of a stent and subsequent release of everolimus was determined for PROMUS Element overlapping stents at 7, 30, 90, 180, and 270 days. Bare metal stents and stents coated with polymer-only (PVDF-HFP) were control groups in the study. The data support the notion that the vascular tissue response to PROMUS Element in the noninjured porcine coronary artery model out to 270 days of follow-up is safe. Comparison of key safety parameters showed no differences in cardiac mortality, stent thrombosis, and myocardial infarction (MI). Strut coverage by neointima was essentially complete by 30 days and remained stable out to 270 days. Scanning electron microscopy showed nearly complete endothelial cell strut coverage of single and overlap stent segments by 7 days. Representative histological photomicrographs (cross sections of explanted stented vessels) of PROMUS Element overlap regions are shown in Figure 8.23 at 30, 90, and 270 days. The histological images at the same time points for BMS and polymer-only groups are also shown for comparison.

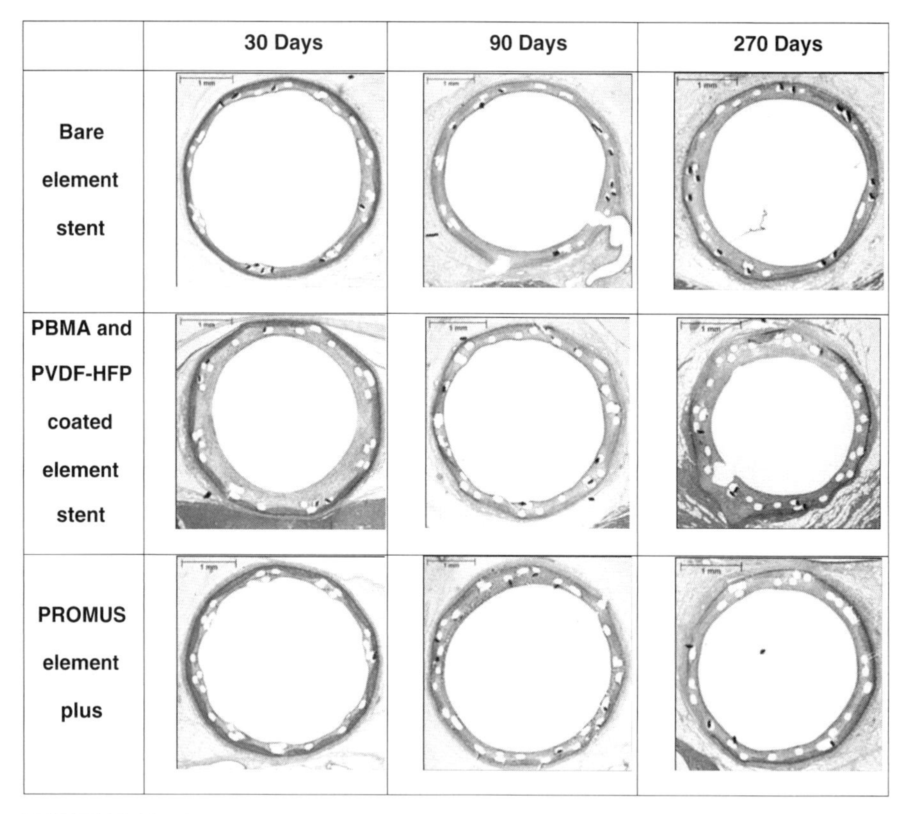

FIGURE 8.23 Representative photomicrographs of PROMUS Element, bare Element stent, and PBMA/PVDF-HFP-coated Element stent elastic–trichrome-stained sections of porcine coronary arteries implanted in an overlap configuration for 30, 90, and 270 days. (Reproduced with permission from Boston Scientific Inc.) (See colour plate section.)

8.6.1 Biocompatibility of the Stent and Coating

Since PROMUS Element is categorized as a permanent implantable device with circulation blood contact greater than 30 days, both short- and long-term biocompatibility testing are required to demonstrate that the device is safe for the intended clinical use. Biocompatibility testing should be conducted under GLP conditions in accordance with the U.S. FDA Recognized Consensus Standard ISO-10993-Part 1 (2003) for this category of devices. Acceptance criteria specific to each test were as outlined in the relevant standard.

In general, the biocompatibility testing of the stent and delivery system were conducted separately after sterilization with ethylene oxide (ETO). Wherever required, the devices were extracted in polar and nonpolar solvents for testing of the "extractables" and "leachables." For permanent implantable devices, assessments of chronic toxicity, carcinogenicity, and teratogenicity were also necessary.

8.7 CLINICAL

The safety and efficacy of the PROMUS Element were demonstrated in the PLATINUM clinical trials in which the pivotal trial was a prospective, multicenter, randomized, single-blind, noninferiority trial with 1532 patients for workhorse (WH); included in the PLATINUM trial were also several subtrials that covered small vessel (SV) and long lesion (LL) conditions, pharmacokinetics, and quantitative coronary angiography (QCA). The primary endpoint of PLATINUM WH was the 12-month target lesion failure (TLF) rate, defined as any ischemia-driven target lesion revascularization (TLR), myocardial infarction (MI, Q-wave, and non-Q-wave) related to the target vessel, or cardiac death related to the target vessel. This study met its primary endpoint (Figure 8.24), demonstrating that the PROMUS Element stent is noninferior to the predicate PROMUS (Xience V) stent in patients undergoing PCI of *de novo* coronary artery lesions. The 12-month rate of TLF was 2.9% in the Xience V group and 3.4% in the PROMUS Element group ($P_{\text{noninfereiority}} = 0.001$). The 12-month rate of TLF was not significantly different between the two stent groups. In addition, there was no statistical difference between the PROMUS Element and Xience V groups for other clinical safety endpoints such as myocardial infarction, q-wave-elevated myocardial infarction (QMI), non-q-wave elevated myocardial infarction (NQMI), definite and probable stent thrombosis, or ARC ST as defined by the Academic Research Consortium, as shown in Figure 8.24. However, TLR for the PROMUS Element group trended lower than that of the Xience V group at the

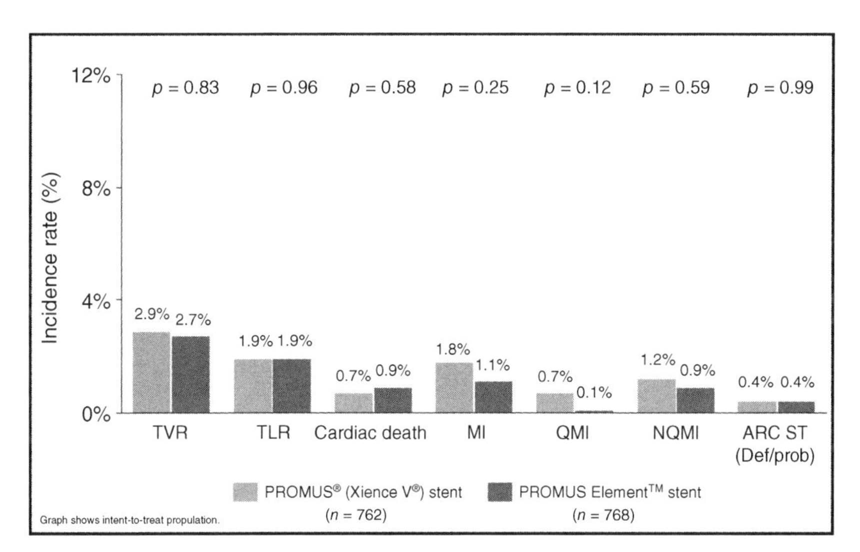

FIGURE 8.24 PROMUS Element™ Stent PLATINUM clinical trial 12 months data. TVR: target vessel revascularization; TLR: target lesion revascularization; MI: myocardial infarction; QMI: q-wave-elevated myocardial infarction; NQMI: non-q-wave-elevated myocardial infarction; ARC ST: definite and probable stent thrombosis as defined by the Academic Research Consortium. Presented by Gregg W. Stone, MD, ACC 2011. Xience V is a trademark of Abbott Laboratories group of companies. (Reproduced with permission from Boston Scientific Inc.) (See colour plate section.)

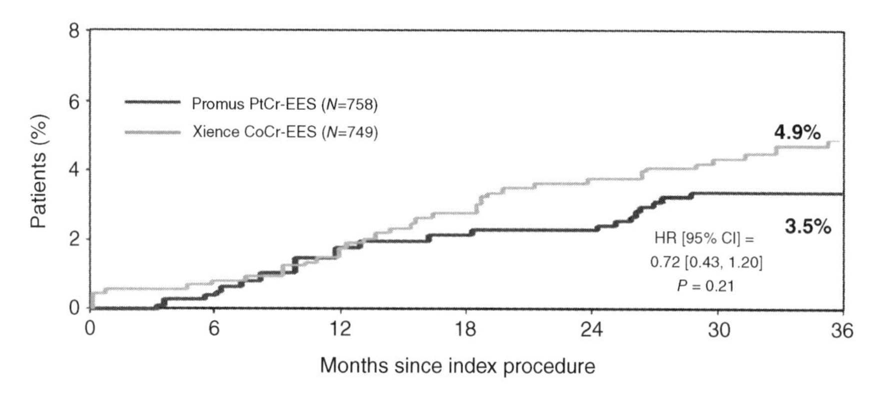

FIGURE 8.25 PROMUS Element Stent PLATINUM clinical trial (workhorse) ischemia-driven target lesion revascularization (ID-TLR) at 3 years. (Reproduced with permission from Boston Scientific Inc.)

3-year follow-up (Figure 8.25), although the difference was not statistically significant. The better long-term clinical outcomes of the PROMUS Element group could be attributed to the superior architectural design of the Element stent.

8.8 CONCLUSION

The development of a DES required the simultaneous considerations of a medical device as well as a drug delivery system; both of which are complex and challenging. A thorough characterization of the drug and polymer carrier was needed and integration of several technologies that pertain to stent, catheter, and manufacturing processing was achieved. Finally, the DES underwent the validation and assurance from the preclinical, biocompatibility studies, and quality control.

The success of the PROMUS Element Plus Everolimus Drug Eluting Coronary stent is a culmination of scientific understanding, engineering ingenuity, and thoughtful clinical design. The advancement of DES continues to evolve by embracing newer technology with the optimal goal of providing minimally invasive combination products to improve the quality of patients' lives.

In closing, given the complexity of the DES system, it is not possible to provide an exhaustive delineation of the product. Nevertheless, the description of the salient features hopefully demonstrates that a fully integrated approach is required in dealing with drug–device combination products exemplified in the DES. There remain nearly limitless possibilities for innovative products by combining multiple technologies across wide-ranging scientific disciplines, ultimately to improve the quality of patient care.

ACKNOWLEDGMENTS

We would like to acknowledge the entire PROMUS Element cross-functional team from R&D, PD, Preclinical, Clinical, Manufacturing, and Regulatory groups in Interventional Cardiovascular Division for making this product possible.

REFERENCES

1. V.L. Roger, A.S. Go, D. Lloyd-Jones, E. Benjamin, et al., Heart Disease and Stroke Statistics—2012 Update: a report from the American Heart Association. *Circulation* 125 (2012) e2–e220.
2. A. Gruntzig, Transluminal dilatation of coronary–artery stenosis. *Lancet* 1 (1978) 263.
3. A.R. Gruntzig, A. Senning, and W.E. Siegenthaler, Nonoperative dilatation of coronary–artery stenosis: percutaneous transluminal coronary angioplasty. *N. Engl. J. Med.* 301 (1979) 61–68.
4. D.S. Baim, (2005) [1958]. Percutaneous coronary revascularization, in D.L. Kasper, A.S. Fauci, D.L. Longo, E. Braunwald, S.L. Hauser, and J.L. Jameson (Eds.), *Harrison's Principles of Internal Medicine*, 16th ed., McGraw-Hill, New York, pp. 1459–1462.
5. J. Kunz and M. Turco, The DES landscape in 2011. *Cardiovasc. Interv. Today* (2012) 48–52.
6. M.W. Liu, G.S. Roubin, and S.B King, III, Restenosis following coronary angioplasty potential biological determinants and role of intimal hyperplasia. *Circulation* 79 (1989) 1374–1387.
7. R.S. Schwartz, and T.D. Henry, Pathophysiology of coronary artery restenosis. *Rev. Cardiovasc. Med.* 3 (Suppl. 5) (2003) S4–S9.
8. T.H. Lee, Slow rehabilitation of drug-coasted stents. *Harv. Heart Lett.* 18 (10) (2008) 1–2.
9. A.J. Epstein, et al., Coronary trends in the United States: 2001–2008. *JAMA* 305 (17) (2011) 1769–1776.
10. H.M. Garcia-Garcia, et al., Drug-eluting stents. *Arch. Cardiol. Mex.* 76 (3) 297–319.
11. D.R. Holmes, Jr., and D.O. Williams, Catheter-based treatment of coronary artery disease: past, present, and future. *Circ. Cardiovasc. Interv.* 1 (2008) 60–73.
12. J.W. Moses, M.B. Leon, et al., Sirolimus-eluting stents versus standard stents in patients with stenosis in a native coronary artery. *N. Engl. J. Med.* 349 (2003) 1315–1323.
13. G.W. Stone, et al., A polymer-based, paclitaxel-eluting stent in patients with coronary artery disease. *N. Engl. J. Med.* 350 (2004) 221–231.
14. M. Haidopoulos, Development of an optimized electrochemical process for subsequent coating of 316 stainless steel for stent applications. *J. Mater. Sci. Mater. Med.* 17 (2006) 647–657.
15. A. Kastrati, et al., Intracoronary stenting and angiographic results: strut thickness effect on restenosis outcome (ISAR-STEREO) trial. *Circulation* 103 (2001) 2816–2821.
16. R. Hoffmann, et al., Relation of stent design and stent surface material to subsequent in-stent intimal hyperplasia in coronary arteries determined by intravascular ultrasound. *Am. J. Cardiol.* 89 (2002) 1360–1364.
17. J. Hausleiter, et al., Impact of lesion complexity on the capacity of a trial to detect differences in stent performance: results from the ISAR-STEREO trial. *Am. Heart J.* 146 (2003) 882–886.
18. S.Z. Rittersma, et al., Impact of strut thickness on late luminal loss after coronary artery stent placement. *Am. J. Cardiol.* 93 (2004) 477–480.
19. J. Pache, et al., Intracoronary stenting and angiographic results: strut thickness effect on restenosis outcome (ISAR-STEREO-2) trial. *J. Am. Coll. Cardiol.* 41 (2003) 1283–1288.

20. M. Turco, et al., Reduced risk of restenosis in small vessels and reduced risk of myocardial infarction in long lesions with the new thin-strut TAXUS™ Liberté stent: one-year results from the TAXUS™ ATLAS Program. *J. Am. Coll. Cardiol. Cardiovasc. Interv.* 1 (2008) 699–709.
21. H. Takebayashi, G.S. Mintz, S.G. Carlier, Y. Kobayashi, T. Yasuda, et al., Nonuniform strut distribution correlates with more neointimal hyperplasia after sirolimus-eluting stent implantation. *Circulation* 110 (2004) 3430–3434.
22. B. Nashan, Review of the proliferation inhibitor everolimus. *Expert Opin. Investig. Drugs* 11 (12) (2002) 1845–1857.
23. J.J Augustine, and D.E. Hricik, Experience with everolimus. *Transplant Proc.* 36 (Suppl. 1) (2004) S500–S503.
24. B. Nashan, Review of the proliferation inhibitor everolimus. *Expert Opin. Investig. Drugs* 11 (12) (2002) 1845–1857.
25. G. Nakazawa, A.V. Finn, M.C. John, et al., The significance of preclinical evaluation of sirolimus-, paclitaxel-, and zotarolimus eluting stents. *Am. J. Cardiol.* 100 (2007) 36–44.
26. H.J. Eisen, E.M. Tuzcu, R. Dorent, et al., Everolimus for the prevention of allograft rejection and vasculopathy in cardiac transplant recipients. *N. Engl. J. Med.* 349 (2003) 847–858.
27. G. Frenning, Theoretical investigation of drug release from planar matrix systems: effects of a finite dissolution rate. *J. Control. Release* 92 (2003) 331–339.
28. V. Barocas, A dissolution–diffusion model for the TAXUS drug-eluting stent with surface burst estimated from continuum percolation. *J. Biomed. Mater. Res. B* (2009) 267–274.
29. H.J. Eisen, E.M. Tuzcu, R. Dorent, et al., Everolimus for the prevention of allograft rejection and vasculopathy in cardiac transplant recipients. *N. Engl. J. Med.* 349 (2003) 847–858.
30. B. Kaplan, H. Tedesco-Silva, and R. Mendez, North/South American, double-blind, parallel group study of the safety and efficacy of Certican. *Am. J. Transplant.* 1 (2001) 475.
31. S. Vitko, R. Margreiter, and W. Weimar, International, double-blind, parallel group study of the safety and efficacy of Certican (RAD) versus mycophenolate mofetil in combination with Neoral and steroids. *Am. J. Transplant.* 1 (2001) 474.
32. J.M. Kovarik, B. Kaplan, H.T. Silva, et al., Pharmacokinetics of an everolimus-cyclosporine immunosuppressive regimen over the first 6 months after kidney transplantation. *Am. J. Transplant.* 3 (2003) 606–613.
33. J.M. Kovarik, B. Kaplan, H. Tedesco Silva, et al., Exposure response relationships for everolimus in *de novo* kidney transplantation: defining a therapeutic range. *Transplantation* 73 (2002) 920–925.

9

INFUSE® BONE GRAFT

Steve Peckham, John M. Zanella, and William F. McKay
Spinal Division, Medtronic plc, 2600 Sofamor Danek Drive Memphis, TN 38132 USA

9.1 INTRODUCTION

Once properly stabilized, broken bones generally heal on their own. However, severe fractures are at risk of nonunion, and some procedures, such as spinal fusion or dental implant cases, need new bone to form in areas where it did not exist previously. INFUSE® Bone Graft was developed to stimulate and enhance bone healing and to act as a replacement for harvesting and transplanting a patient's own bone to a site where bone formation is needed. The INFUSE Bone Graft kit (Figure 9.1) consists of several components: recombinant human bone morphogenetic protein-2 (rhBMP-2), also known as dibotermin alfa, absorbable collagen sponge (ACS), sterile water for injection, needles, and syringes. INFUSE Bone Graft is available in six different kit sizes (Figure 9.2). The kit sizes correlate to different total graft volumes with varying sizes of ACS (Table 9.1).

rhBMP-2, the active agent in INFUSE Bone Graft, is a manufactured version of a naturally occurring protein, BMP-2, that is important for bone formation and healing. The rhBMP-2 component in INFUSE Bone Graft is provided as a lyophilized powder in vials designed to deliver 1.05, 4.2, or 12 mg of protein. Using the syringe and needle, the sterile water for injection, provided in 5 or 10 ml vials, is injected into the rhBMP-2 vial to reconstitute the lyophilized protein at a concentration of 1.5 mg/ml. The rhBMP-2 solution has a pH of 4.5, is clear, colorless to slightly yellow in color, and essentially free of plainly visible particulate matter. The rhBMP-2 solution is buffered by 2.5% Glycine, 25.3 mM L-glutamic acid, 1.875 mM NaCl, 0.5% D-sucrose, and 0.01% polysorbate 80. The reconstituted rhBMP-2 is drawn up

Drug–Device Combinations for Chronic Diseases, First Edition. Edited by SuPing Lyu and Ronald A. Siegel.

INFUSE® Bone Graft with LT-CAGE® Device incorporates technology developed by Gary K. Michelson, MD

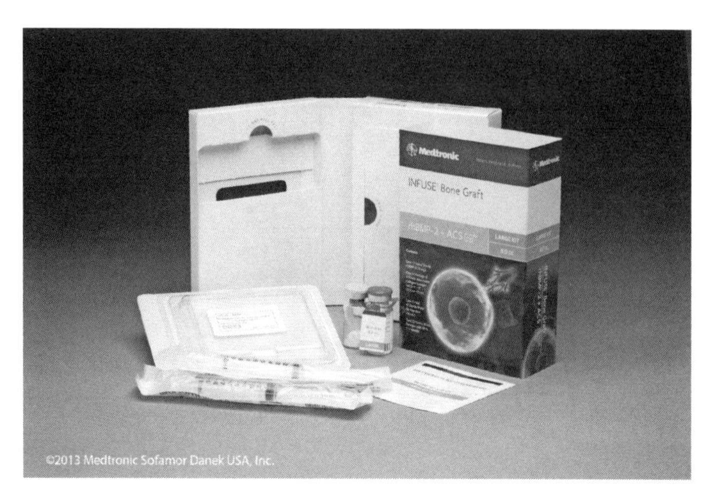

FIGURE 9.1 The contents of an INFUSE Bone Graft large kit. (Images provided by Medtronic. INFUSE® Bone Graft with LT-CAGE® Device incorporates technology developed by Gary K. Michelson, MD.)

FIGURE 9.2 The six INFUSE Bone Graft kit sizes. The kit sizes correlate with a different total graft volume. (Images provided by Medtronic. INFUSE® Bone Graft with LT-CAGE® Device incorporates technology developed by Gary K. Michelson, MD.)

into another syringe and is applied to the ACS (Figure 9.3), a soft, white, pliable, absorbent implantable matrix made from bovine type I collagen obtained from the deep flexor (Achilles) tendon. The ACS acts as a carrier for delivery and retention of the rhBMP-2 at the site of implantation and as a scaffold for new bone formation.

The application of rhBMP-2 to the ACS occurs at the time of surgery. Once combined to form INFUSE Bone Graft, the rhBMP-2/ACS implant is allowed to stand for a prescribed amount of time (no less than 15 min and no more than 2 h) before placement at the surgical site. The implant site is prepared utilizing standard

TABLE 9.1 The Commercially Available INFUSE Bone Graft Kit Sizes

INFUSE® Bone Graft Kits	Total Graft Volume	Sterile Absorbable Collagen Sponge (ACS)	mg rhBMP-2	Concentration rhBMP-2
7510050 XX SMALL KIT	0.7 cc	(1) ACS ½" x 2" (1.25 cm x 5.08 cm)	1.05 mg	1.5 mg/cc
7510100 X SMALL KIT	1.4 cc	(1) ACS 1" x 2" (2.5 cm x 5.08 cm)	2.10 mg	1.5 mg/cc
7510200 SMALL KIT	2.8 cc	(2) ACS 1" x 2" (2.54 cm x 5.08 cm)	4.2 mg	1.5 mg/cc
7510400 MEDIUM KIT	5.6 cc	(4) ACS 1" x 2" (2.54 cm x 5.08 cm)	8.4 mg	1.5 mg/cc
7510600 LARGE KIT	8.0 cc	(6) ACS 1" x 2" (2.54 cm x 5.08 cm)	12.0 mg	1.5 mg/cc
7510800 LARGE II KIT	8.0 cc	(1) ACS 3" x 4" (7.62 cm x 10.16 cm)	12.0 mg	1.5 mg/cc

surgical techniques, after which INFUSE Bone Graft is placed where bone formation is desired according to the approved indications.

Today, INFUSE Bone Graft is regulated in the United States as a combination product. The rhBMP-2 on the ACS carrier at 1.5 mg/ml has received three separate Premarket Approval (PMA) approvals:

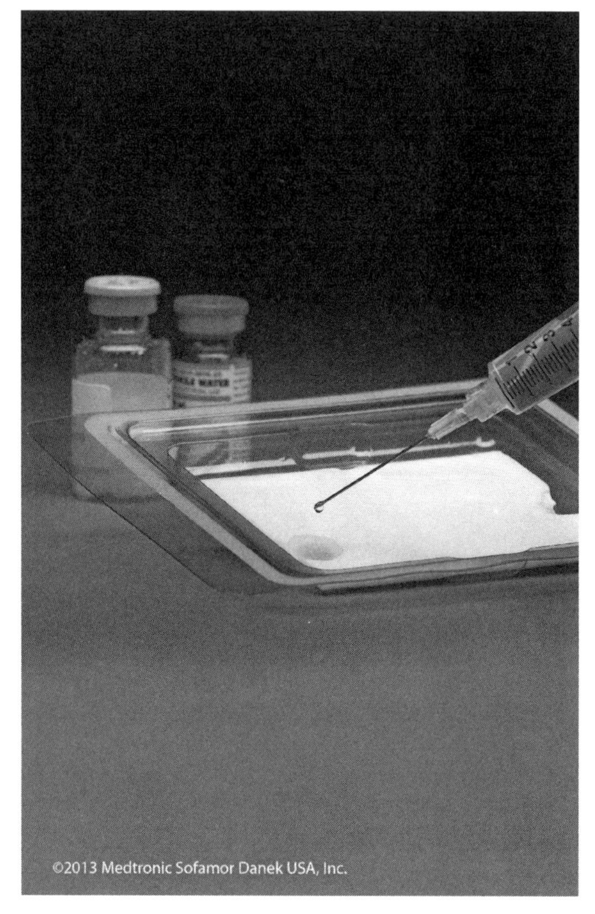

FIGURE 9.3 Using the syringe and needle, sterile water for injection is taken from the vial and injected into the rhBMP-2 vial to reconstitute the lyophilized protein. The reconstituted rhBMP-2 is drawn up into another syringe and is applied to the ACS. (Images provided by Medtronic. INFUSE® Bone Graft with LT-CAGE® Device incorporates technology developed by Gary K. Michelson, MD.)

1. Anterior lumbar interbody fusion with the LT-CAGE® Lumbar Tapered Fusion Devices, INTER FIX™ Threaded Fusion Devices, or INTER FIX RP Threaded Fusion Devices.
2. Acute open tibial fracture with intramedullary nail fixation within 14 days of injury.
3. Sinus augmentations, and for localized alveolar ridge augmentations for defects associated with extraction sockets.

9.1.1 Patient Needs

There are many instances when bone formation or regeneration is needed. This would include healing of traumatic fractures with or without segmental bone loss, fusion of degenerated spinal segments to relieve pain or address instability, and growth of bone in the maxilla to support dental implant placement. In some cases, the addition of a bone growth factor may accelerate healing or reduce the number of secondary interventions required to achieve healing and return to function. For other indications, such as spinal fusion, the goal is to form bone where it would not normally grow. In these instances, bone grafting is used to stimulate bone formation. Historically, the most common grafting technique has been to transplant bone from one part of a patient's body to the site of desired bone formation (autograft). Autograft will grow and remodel over time, filling in defects or stabilizing a spinal segment. Autograft is most often taken from the iliac crest, jaw, or tibia.

While autograft is effective, there are significant limitations that make the concept of an autograft replacement attractive to surgeons and patients. One is the limited volume of the autograft. Harvest takes time, often requires a second surgical site, and is associated with the risk of both major and minor complications. Major complications are rare and include herniation [1], vascular injury [2–4], deep infection [5–7], neurologic injury [8], hematoma [9–11], fracture [12–14], and pelvic instability [15,16]. Minor complications are reported more often with rates in the range of 9.5–24% of patients [17,18]. The most common complaint with autograft harvest is persistent pain at the harvest site. Some studies [18,19] claim harvest site pain can be present for years after the original harvest procedure, while other studies [20,21] suggest that harvest site pain is often overstated. Despite the controversy over the rate and severity of graft site harvest pain, it is clear that many surgeons and patients are looking for alternatives to autograft and an autograft replacement has significant value.

9.1.2 Background

The introduction of rhBMPs for clinical bone healing applications was the culmination of decades of research and development. The journey from discovery to conceptualization, to eventual introduction of a commercial product was dependent on individual and corporate vision, close interaction with regulatory authorities in an evolving landscape, and most importantly a recognition and understanding of patient and surgeon needs.

Any history of INFUSE Bone Graft must begin with a description of the pioneering work of Marshall Urist, M.D. Urist was an orthopedic surgeon and Director of the Bone Research Laboratory at the University of California, Los Angeles. With his interest in bone regeneration, Urist was recruited by the Atomic Energy Commission in the 1950s to study the incorporation of radioactive materials into bone, looking for ways to minimize the effects and risk of radiation exposure. As part of that research, Urist performed many experiments involving the implantation of demineralized bone matrix (DBM), bone with the inorganic mineral removed to leave behind the organic collagen matrix (Figure 9.4), into the soft tissue of rodents. Urist discovered that DBM would stimulate the formation of new bone tissue in muscle where bone would not normally form. He recognized the significance of the discovery and published his paper "Bone: Formation by Autoinduction" in the journal *Science* in 1965 [22].

As Urist's research continued and his understanding of bone formation evolved, he postulated that specific proteins located in the bone matrix were responsible for the osteoinductive potential of DBM, which is defined as the ability to stimulate *de novo* bone formation in a site where bone does not normally occur. It was at this point that

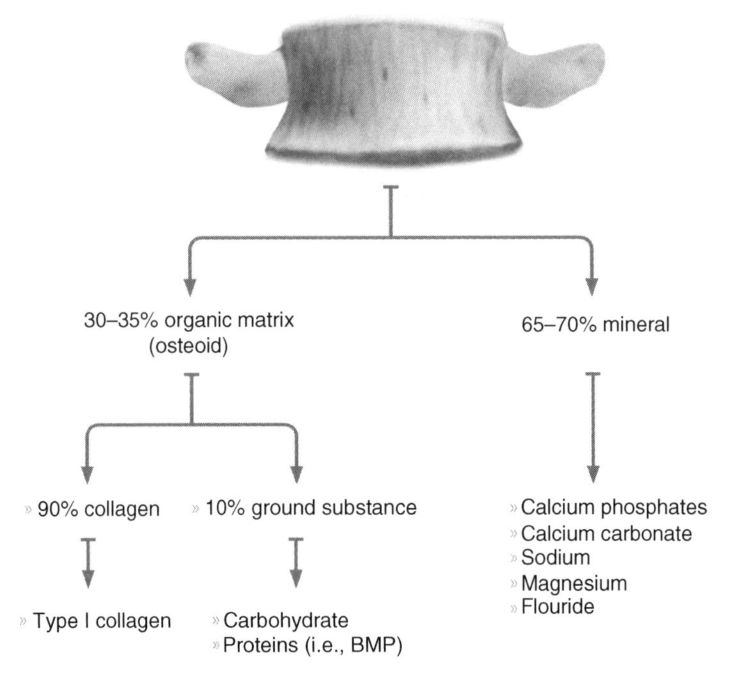

FIGURE 9.4 The composition of bone. Bone consists of both mineral and organic components. (Images provided by Medtronic.)

Urist designated the bone-inducing factors as bone morphogenetic proteins (BMP) [23,24]. This work also led to basic science studies investigating the mechanism of action of BMPs in inducing new bone formation. Over the years, the role of BMPs in the bone formation cascade has been elucidated.

Osteoinductive BMPs exert their influence over bone formation in multiple ways. First, there is chemotaxis, in which the mesenchymal stem cells capable of differentiating into bone forming osteoblasts are recruited to the site of BMP implantation [25]. BMP can also stimulate proliferation of mesenchymal stem cells, further increasing the number of potential osteoblasts [26,27]. Finally, BMP binds to receptors on the surface of stem cells and preosteoblasts and induces those cells to become osteoblasts through a series of intracellular signals [28,29]. In addition, BMP can indirectly enhance the development of new blood vessels, which is important in supplying nutrients to the developing bone [30]. These steps represent the generally accepted method by which BMPs stimulate and induce bone formation clinically. Despite not having all of the information around the mechanism of BMP-induced bone formation, by building on his early research, Urist was able to demonstrate in nonclinical animal models and in clinical case series that BMP extracted from bone was capable of enhancing and stimulating bone healing [31–34].

While Urist's BMP formulations appeared to be effective, the nature of the product was not conducive to widespread commercial application. Extracting proteins from human cadaveric bone would constrain the availability of the product to the limited supply of donated tissue. Further, isolating the natural protein would result in a product that was suboptimal in consistency, purity, and biological activity. Recognizing the importance of BMPs, and the application of these proteins to an established patient need, scientists at Genetics Institute in Cambridge, MA (now Pfizer, Inc.), began the work to identify and characterize the proteins extracted from bone matrix [35–37]. This work identified a family of BMPs that could be manufactured in large quantities by well-characterized recombinant production methods. At this point, commercial application of BMP became a real possibility.

A number of different BMPs were identified and characterized. rhBMP-2 was the first of the BMPs to be extensively studied by Genetics Institute. Early work clearly demonstrated the osteoinductive potential of rhBMP-2 in the same rat implantation assay that Urist used to measure bone induction. Later work would confirm that rhBMP-2 acts upon both mesenchymal stem cells capable of differentiating into osteoblasts and on preosteoblasts that are partially committed to osteoblast differentiation in order to stimulate bone formation. With the osteoinductive protein identified and selected, the next phase of development consisted of nonclinical safety and efficacy evaluations in anticipation and support of future clinical investigations.

9.1.3 Manufacturing

The INFUSE Bone Graft kit contains several components, all manufactured by a separate entity and ultimately assembled by a Medtronic plc facility.

Pfizer, Inc. manufactures rhBMP-2 for Medtronic using well-established molecular biology techniques. There are two phases in the manufacturing flow for rhBMP-2. The first phase is the creation of a working cell bank. This phase requires the recombination of the human gene for BMP-2 into the DNA of Chinese hamster ovary (CHO) cells for replication. The replication process allows the cells to grow and multiply into a homogeneous population of cells capable of producing rhBMP-2. A single batch of rhBMP-2 production cells is distributed into small vials, creating a cell bank for future production of rhBMP-2.

The second phase (Figure 9.5) includes producing, purifying, sterilizing, and validating the protein. To produce rhBMP-2, cells from the cell bank are transferred to a bioreactor, where large-scale production of the protein occurs. After a growth period, the rhBMP-2 is filtered away from the cells and further purified through a series of chromatography columns. The protein is filter-sterilized. This production method results in extremely pure (>98%) solutions of a single BMP. The final product step for the purified rhBMP-2 is filling it into vials and freeze-drying.

The ACS is manufactured by Integra LifeSciences Corporation from bovine type I collagen obtained from the deep flexor tendon. The collagen is purified, freeze-dried, and cross-linked to produce a moldable absorbent sponge. Finally, the sponges are cut to size, packaged, and sterilized using ethylene oxide (EtO).

9.1.4 Regulatory

In the United States, INFUSE Bone Graft is a combination product regulated as a medical device and marketed under a PMA. In the 1990s, the development of an

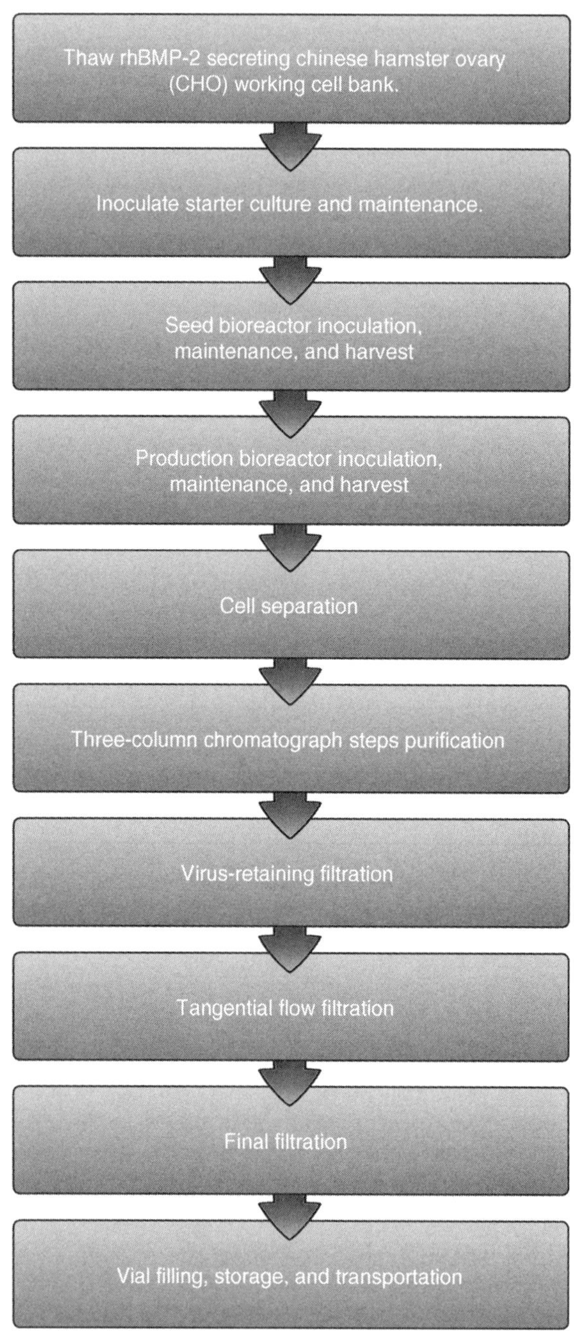

FIGURE 9.5 Flow diagram of the recombinant human bone morphogenetic protein-2 manufacturing process. (Images provided by Medtronic.)

orthopedic PMA combination product represented uncharted territory for both the company doing the development and the regulatory agency performing the review. Early in the development, it was unclear which division of the U.S. FDA would be responsible for the regulatory review and what requirements the product would need to meet in order to enter into clinical investigations.

While it was ultimately determined that the primary review responsibility in the United States falls under the Center for Devices and Radiological Health (CDRH), any information related to the rhBMP-2 portion of the device was also reviewed by the Center for Drug Evaluation and Research (CDER). Satisfying one branch of the FDA is not always simple, so meeting the needs of two branches that view the products from very different perspectives offered a unique set of challenges.

9.1.5 Preclinical Safety Testing

Genetics Institute, a biopharmaceutical company, approached the nonclinical development from the perspective that, regardless of how the final product was regulated, the recombinant protein would be evaluated as a drug. Therefore, a very extensive pharmacologic and toxicology data package was prepared. The first component of this data package was an extensive series of toxicology studies to establish the safety of the protein alone. Though rhBMP-2 is a locally acting osteoinductive agent, the potential for systemic drug effects needs to be addressed. Acute and chronic toxicity studies in two species, as well as a series of reproductive toxicology studies, were performed [38]. There were no toxic effects noted in rats or canines with intravenously (IV) administered rhBMP-2 of up to 5.3 mg/kg [38]. There were no pharmacologic effects, such as bone formation, observed at any site. To put this in context, the dose evaluated was equivalent to a 70 kg human receiving the combined amount of rhBMP-2 from 30 large INFUSE Bone Graft kits directly into the bloodstream. Two 28-day chronic IV administration studies at doses of up to 0.16 mg/kg/day in rats and canines also failed to demonstrate any systemic toxicity or bone formation as a result of these large doses of protein [38].

Fertility and teratology studies were performed to assess the potential for reproductive toxicity [38]. IV dosing of rhBMP-2 had no effect on maternal or paternal mating performance and reproductive parameters. Administration of rhBMP-2 IV during pregnancy did not lead to maternal toxicity, embryolethality, or gross fetal abnormalities. These systemic toxicity studies would not normally be considered for an implantable medical device, but they are viewed as important data in the safety analysis of combination drug products. The results of these studies demonstrate the systemic safety profile of the rhBMP-2 protein.

In addition to the basic questions about systemic toxicology and reproductive toxicology that would be of interest for the drug component of a combination product, there is the potential for specific safety questions based on the known pharmacologic action of the drug or therapeutic protein. The primary mode of action of rhBMP-2 as a morphogen is characterized by cell differentiation rather than proliferation; nevertheless, the idea that BMP is a potential growth factor led to questions about the possibility for rhBMP-2 to influence the growth rate of cancer cells. In order to address this issue early on, Genetics Institute performed *in vitro* studies to examine the growth potentiating activity of rhBMP-2 on both human tumor cell lines and primary tumor cell isolates. In a study of human tumor cell lines [38], 14 cell lines, including osteosarcoma, breast, prostate, and lung cancers, were either unaffected or growth-inhibited by the addition of rhBMP-2 at several concentrations up to 1000 ng/ml. In a second study looking at primary tumor isolates [39], 65 samples were collected, including breast, ovarian, and lung cancers. Of the 65 samples tested, none showed growth stimulation and 16 demonstrated growth inhibition with exposure to rhBMP-2 at 10, 100, or 1000 ng/ml. While these data do not include an assessment of every potential cancer cell line, and they were *in vitro* assessment studies, the results did not signal a cause for concern with respect to rhBMP-2 on tumor growth.

While systemic studies are of interest in assessing the overall safety profile of a pharmacologic agent, they do not necessarily show how the agent will be delivered in this combination product. In the case of INFUSE Bone Graft, the rhBMP-2 is delivered locally at the site of desired bone formation on a carrier matrix. Early work clearly demonstrated that even in lower order animals, a carrier matrix is required to achieve consistent bone formation [40–42]. This is due to the fact that rhBMP-2 independent of the carrier is very rapidly catabolized and cleared. Pharmacokinetic studies of the protein in circulation have yielded half-life on the order of minutes in both rats and nonhuman primates [38]. In the case of a combination product for bone healing, the carrier matrix is required to be more than a simple drug delivery device. Bone healing requires three components: (1) a scaffold on which new bone can form, (2) cells to produce the bone matrix, and (3) bone growth factors to stimulate and direct the cells to form bone. The rhBMP-2 provides the signal, the local tissue provides the cells, and the carrier matrix provides the scaffold compatible with bone formation. Thus, pharmacokinetics, local retention, and release rate from the carrier are important parameters for a combination device, but the form and composition of the carrier are just as important.

A number of materials had the potential to serve as the rhBMP-2 carrier matrix. The requirements of the carrier were that it retain the rhBMP-2 at the site of implantation, that the material itself be cohesive and moldable to facilitate implantation and minimize the risk of migration, and that it provide an environment compatible with new bone formation. The candidate materials were synthetic or natural polymers, calcium-containing ceramics or cements, or combinations of different materials from these categories.

In addition to the requirements for appropriate handling properties and bone compatibility, there was an inherent advantage to choosing a material with which regulatory agencies were already familiar and one that was already approved as a marketed medical device. This stipulation did not limit the options significantly and had the potential to ease the regulatory data requirements by focusing the review on the new information being evaluated for safety and efficacy. The material that was chosen as the carrier matrix for INFUSE Bone Graft is a bovine type I collagen sponge. The cross-linked sponge is cohesive and moldable. Bovine collagen has a long history as an implantable material used to achieve hemostasis during surgery. Finally, as the predominant component of bone matrix, collagen is very compatible with cell attachment and bone formation. These hemostatic materials are regulated by FDA as medical devices.

After the carrier matrix was chosen, toxicology of implanted product and biocompatibility studies were performed to supplement the systemic toxicology data. Standard biocompatibility tests were carried out on the combination product as would be performed for any implantable medical device in contact with blood without any significant findings [38]. Long-term implantation studies in rabbits [38] did not yield any effects other than the local effects expected with the known pharmacologic action of the protein (i.e., bone formation at the site of implant).

The bulk of the safety data required to support clinical investigation and, ultimately, product approval was on the active ingredient and the biocompatibility of the device or the combination product. However, the ultimate application of the product had to be considered to determine whether there were any indication-specific safety questions that needed to be addressed. In the case of rhBMP-2/ACS, a spine-specific safety study was initiated in canines [43] to address concerns about the potential effect of rhBMP-2 on neural tissue. In this study, access to the spinal cord was achieved by removing the bony lamina from a segment from the back of the spine. rhBMP-2/ACS was laid directly on top of the dura of the spinal cord. During the following 12 weeks, clinical neurological and radiographic evaluations were performed, followed by histological analysis of the neural tissue at the end of the study. The results showed neither any evidence of mineralization of the neural tissue nor any clinical or neurological abnormalities in the animals. This indication-specific study was important for providing data regarding the safety of the product with use near the spinal cord and nerve roots.

Another area of investigation that could be easily overlooked is compatibility of the protein with the preparation materials and stability of the protein during product preparation. A series of studies were completed to examine these issues and ensure that a consistent and active product would be delivered to the patient. Compatibility studies were performed with the rhBMP-2 solution and the syringe and needle to ensure that the rhBMP-2 does not bind to the syringe or needle materials and that exposure to the materials does not inactivate or otherwise alter the protein. The same type of study was performed with the hydrated ACS to ensure that the protein does not preferentially bind to the ACS packaging tray during the soak time and that the rhBMP-2 extracted from the ACS was the same as it was before being added to the carrier. While stability studies and shelf-life testing for the packaged product were expected and routine, additional stability studies were performed on the reconstituted rhBMP-2 solution in the vial to demonstrate that there were no effects if the protein was reconstituted and the operating room staff waited several hours before adding the solution to the ACS.

Anyone with orthopedic medical device experience would quickly recognize that the level of detail and the consideration and quantity of testing that goes into drug development is much different than that typically seen for medical devices. With the assessment of the local and systemic nonclinical safety of the rhBMP-2 and the combination product completed, the development focus shifted to pharmacology studies based on the intended clinical and commercial use of the product in order to provide nonclinical efficacy data to support clinical investigations.

9.1.6 Pharmacology Testing

Before designing a nonclinical efficacy study plan, it is vital to understand the ultimate intended clinical indication. For medical device approvals, the commercial indication should be almost identical to what is studied in the clinical investigation of the combination product. The nonclinical efficacy studies should mimic the clinical indication as closely as possible. The early work by Genetics Institute on rhBMP-2/ACS focused on orthopedic trauma and oral maxillofacial indications, so the first nonclinical efficacy studies initiated were a series of long bone segmental defect studies performed in lower order animal models. These studies included rat femora [44], rabbit radii and ulnae [45], canine radii [38], and nonhuman primate radii [38]. In these studies, rhBMP-2/ACS was acting as an autograft replacement. The size of the excised segment in each model was chosen such that the defect is considered critical size (i.e., the defect is large enough that it will not heal spontaneously if left unfilled). Depending on the size of the animal, different types of hardware fixation may be used to provide initial stability during bone healing.

In another series of studies [46–48], the ability of rhBMP-2/ACS to affect fracture healing was assessed. In these studies, bone osteotomies were performed to remove a small segment of bone, such that the defect will spontaneously heal over time. The primary outcome in these cases was time to healing as a model for acceleration of fracture healing. In rabbit [47,48] and goat [46] long bone studies, rhBMP-2/ACS accelerated fracture healing by increasing the size of the fracture callus and accelerating the callus development and maturation. The outcomes from these studies showed that rhBMP-2-induced bone integrates with existing bone and remodels in a manner consistent with normal bone. In addition, the bone forming process in response to rhBMP-2 was self-limiting. The presence of natural BMP inhibitors in the local tissue combined with the gradual loss of BMP from the implant site ensures that only a defined volume of bone will form [38]. The orthopedic trauma studies formed the basis of support for the clinical program examining the use of rhBMP-2/ACS in the treatment of open tibia fractures.

The first entry into the oral and maxillofacial market centered on filling extraction sockets or sinus lift, procedures to support dental implant placement. Treatment in these cases is based on the fact that, in the absence of dentition, the bone that will normally support the teeth will become weaker and recede if there is no stimulus (i.e., loading) for the body to maintain the bone mass; this is Wolff's law [50]. In many cases, there is not enough bone in the superior jaw to ensure stable dental implant placement. The surgical procedure includes making a window into the sinus above the area where the implants will be placed, and filling that area of the sinus with bone grafting material. The newly formed bone will then support the dental implant. In the case of extraction sockets, the bone graft material can be placed at the time of tooth extraction to prevent bone loss and fill in the defect with new bone to support the dental implant.

Due to issues with animal size, rats were not used for sinus lift procedures, but were used in cranial defect models to show proof of concept for rhBMP-2-induced bone formation above the neck. The bulk of the oral and maxillofacial work was performed in canines and goats, with final confirmatory and dosing work done in nonhuman primates [50–53]. These nonclinical investigations showed that rhBMP-2/ACS, when protected from compression, will form normal, vital bone in oral maxillofacial applications. The bone formation is capable of supporting dental implant placement, and is equivalent to or superior to the use of autograft in these models. These data were used to support initiation of a series of clinical investigations focused on sinus lift procedures and filling extraction socket defects.

The development of INFUSE Bone Graft for spinal applications was similar to the orthopedic trauma and oral maxillofacial indications in the sense that the data package built upon all of the general drug toxicology and safety data, and a series of nonclinical efficacy studies, was used to support initiation of the clinical investigations.

The intent in spinal fusion is to grow a bone bridge between vertebral bodies in order to stabilize the spinal segment. Every spinal fusion procedure needs some type of grafting material to stimulate bone formation where bone would not normally form. Spinal fusion of the lower, or lumbar, spine is the most common application for bone grafting. There are two general means of achieving spinal fusion. Bone graft can be placed along the back of the spine between prepared transverse processes in a posterolateral fusion, or the bone graft can be placed into the prepared disk space between two vertebral bodies in what is known as an interbody fusion. While the surgical approach for an interbody fusion may vary, every case requires the use of one or two interbody devices. These devices are made of metal, plastic, or bone, and generally have space in the center for the placement of bone graft.

The purpose of the interbody fusion device is to hold the disk space open and to increase stability until bone grows across the space and the fusion is complete. Thus, one difference between spinal and trauma or oral maxillofacial applications is the addition of a third component to the combination device for spinal fusion. The presence of the additional device somewhat complicates the nonclinical development as animal models must mimic the surgical procedure as much as possible. In spinal fusion, the interbody space of rats and rabbits is too small for interbody fusion implants, so posterolateral fusion models are used for product screening and proof of concept in lower order animals. The bulk of the nonclinical efficacy testing was performed in sheep due to the larger disk spaces available and a range of motion similar to human spines [54]. However, sheep form bone relatively easily, so nonhuman primates were used for final product configuration and confirmatory testing. In the sheep and nonhuman primate studies, the rhBMP-2/ACS was implanted into the disk space inside of threaded, titanium interbody fusion devices similar to those intended for use in the clinical investigations.

In the sheep study, all of the rhBMP-2/ACS treated interbody spaces fused compared to 37% of those treated with iliac crest autograft inside the fusion device [55]. The rhBMP-2/ACS treated spaces also had more bone and less fibrous tissue. In the nonhuman primate [56], two concentrations of rhBMP-2/ACS were tested: 0.75 versus 1.5 mg/ml. Both achieved consistent fusion, but the higher concentration formed denser, more mature bone, leading to the choice of evaluating the higher concentration in the clinical evaluation of INFUSE Bone Graft. Across a range of studies from rats and rabbits to nonhuman primates, rhBMP-2/ACS was effective in inducing spinal fusion, and was equivalent to or better than autograft whenever the two were compared.

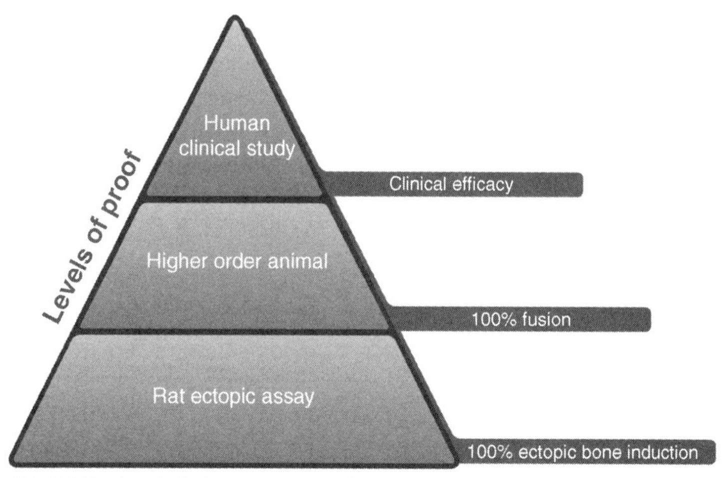

FIGURE 9.6 The burden of proof pyramid for the nonclinical assessment of bone graft substitutes. (Images provided by Medtronic.)

One of the observations regarding the nonclinical study outcomes was that both the length of time and the concentration of rhBMP-2 required to achieve bone healing increased when moving from lower order animals to higher order animals. Thus, the concentration required to form bone in a rat may be orders of magnitude lower than that required for a nonhuman primate. The importance of this observation cannot be overstated as the implications are that rat and rabbit studies are acceptable for initial screening and proof of concept; however, those models are not adequate to provide clinical dosing estimates, and they have no value in predicting eventual clinical performance.

Out of these observations came what is often referred to as "the burden of proof" (Figure 9.6). Higher order animal studies are necessary to obtain clinical dosing estimates to minimize the risk of ineffective clinical concentrations. In order to minimize the use of higher order animals, the nonclinical efficacy story should be built by moving from lower order animals to larger animals for increased difficulty in stimulating bone formation and anatomic models more applicable to the human condition before finally moving to higher order animals for concentration and dosing.

9.1.7 Post-Market Studies

Nonclinical research on the safety of rhBMP-2 did not end with the initiation of clinical investigations or the eventual PMA approval. During discussions with the FDA on the condition of product approval, there were two areas identified that warranted further investigation: (1) the effects of anti-BMP-2 antibodies on fetal development and (2) expansion of the data on the effects of rhBMP-2 on the growth rate of cancer cells both *in vitro* and *in vivo*. Patients in both the investigational and control groups from multiple clinical investigations were tested for anti-BMP-2 antibodies. The positive antibody response rate was low (<5%) and transient in all of the INFUSE Bone Graft studies [57]. Early studies did not assess the neutralizing capacity or the ability of the anti-BMP-2 antibodies to inactivate BMP-2; however, more recent testing of samples from other rhBMP-2 clinical investigations [58] has not identified any neutralizing antibodies.

The theoretical issue is the possibility that a mother's anti-BMP antibodies could be transferred to a growing fetus and the antibodies could inactivate BMP-2 in the fetus. BMP-2 is highly conserved across species and is vital to development. Mouse embryos in which the BMP-2 gene is knocked out do not survive until birth [59]. In order to gain information on reproductive toxicology for BMP antibodies, a rabbit study[1] was designed in which does were immunized with rhBMP-2 prior to mating. Fetuses were collected at day 28 of gestation and assessed for abnormalities in dozens of different tissues including skeletal. In the initial assessment, the only finding in the investigational group that warranted further review was delayed ossification in the frontal and parietal bone in 3 of 108 fetuses. This finding was eventually determined not to be significant based on the fact that the rate was within historical control data, delayed ossification is reversible, and the finding did not correlate with either antibody titer or antibody neutralizing capacity. This reproductive toxicity study is another example of the additional studies that may be considered for a combination product that would not normally be relevant to implantable medical devices.

The second area of post-approval nonclinical research was related to the effects of rhBMP-2 on cancer cell growth. While the early studies on tumor cell lines and tumor cell isolates did not show any increase in growth rate, and, in one case, showed growth inhibition, those studies did not assess the presence of BMP receptors or the potential for *in vivo* effects. In the post-approval series of studies[1], the theoretical concern was that rhBMP-2 implanted distant from a tumor could migrate to the tumor site and influence the growth rate. The investigation was broken into three phases. First, a series of cell lines were tested for the presence of BMP receptors RNA. This would suggest that the cells were at least capable of responding to rhBMP-2 through the receptors. In the second phase, the 10 cell lines identified from phase one as potentially being capable of a BMP receptor-mediated rhBMP-2 response were exposed to rhBMP-2 *in vitro* to look for effects on growth rate. Except for one prostate cancer cell line that was growth inhibited, rhBMP-2 did not affect cancer cell growth rates *in vitro*. In the third phase, tumor cells from five cell lines that had BMP receptors and two that did not were implanted into nude mice. Once the tumors reached a predetermined size, rhBMP-2 on ACS was implanted on the opposite flank of the mouse. The concentration of rhBMP-2 was 4–40 times higher than the concentration that has been shown to form bone in mice and rats (0.1 mg/ml). The growth rate of the tumors was compared with that in mice implanted with buffer on the ACS carrier. There was no increase in tumor growth rate or metastasis in the group implanted with rhBMP-2/ACS compared to the group implanted with buffer/ACS. Independent of these results, there is a contraindication against implanting INFUSE Bone Graft in the vicinity of a resected or extant tumor, in patients with an active malignancy or patients undergoing treatment for a malignancy [57].

[1]Medtronic Internal Study Documents.

In addition to post-approval commitments mandated by a regulatory agency, there is a very high likelihood that nonclinical research projects post-commercialization will be dictated by concerns, questions, or adverse events that are not observed during a clinical investigation but occur after widespread commercial use of the product. A clinical investigation is generally performed under very well-controlled circumstances with specific inclusion and exclusion criteria for patients and explicit instructions in terms of surgical technique and product application and dosing. The commercial application does not necessarily mimic these well-controlled circumstances.

One question that arose after commercial release of INFUSE Bone Graft related to the possibility of cancellous bone remodeling or resorption. A nonclinical research project was initiated in an attempt to elucidate potential contributing factors to bone remodeling in response to rhBMP-2 [60]. In this study, drill hole defects of controlled and consistent volume were created in the distal femur of sheep, which is easily accessible and provides ready access to cancellous bone. The drill hole defects were filled with (1) Buffer/ACS at a volume appropriate to fill the defect size, (2) the appropriate rhBMP-2 concentration for a sheep (0.43 mg/ml) on ACS at a volume appropriate to fill the defect size, (3) 0.43 mg/ml rhBMP-2/ACS at twice the defect volume, (4) an rhBMP-2/ACS implant of the right volume but 3.5 times higher in concentration (1.5 mg/ml), and (5) an rhBMP-2/ACS implant that had both higher concentration and twice the defect volume. Defects that were implanted with hyperconcentrated product or overstuffed with the appropriate concentration showed moderate bone resorption relative to the controls. Defects that were overstuffed with hyperconcentrated implants showed extensive osteoclastic activity and large zones of bone resorption. In all cases, the bone resorption was transient with evidence of new bone filling the space over time.

Since the ACS is easily compressible, it becomes clear that care should be taken to avoid local hyperconcentration by overfilling a grafting location, particularly if the INFUSE Bone Graft is implanted in direct access to cancellous bone. Additional information related to the potential for increased risk based on overstuffing or hyperconcentrating was added as a consequence of the nonclinical data [57]. This change represents an example of how ongoing nonclinical research work postcommercialization is an important component of combination product management. As new research topics are identified, having research and development resources available to address those topics should be recognized as an ongoing need.

9.2 CONCLUSION

The process for development and commercialization of a combination product is very different from either drug- or device-only development. Aspects of each must be incorporated, but there is really no standard approach that can be applied to a particular area. There is no "one size fits all" plan for combination product commercialization. There are, however, broader concepts that should be considered that can help guide the development plan. First, the drug must initially be evaluated independently. Even if the administration and pharmacologic action is localized, drug safety will need to be examined from a systemic perspective; however, concepts such as pharmacokinetics, adsorption, distribution, metabolism, and excretion will be less relevant to a locally administered and locally active agent. Next, the importance of the delivery system cannot be overlooked. It is important to understand the role and requirements of the delivery system. Often the optimal release or retention profile is not known, making it impossible to design a delivery system from the ground up. It is more likely that release will end up being a characterization study rather than a design input.

In the case of bone growth factors, decisions on product composition and formulation would be based on nonclinical efficacy rather than a demonstration of nonclinical pharmacokinetics. It is vital to construct an adequate nonclinical data package prior to entering into clinical investigations. Otherwise, there is an increased risk of reaching the end of a clinical investigation and finding that the wrong carrier or protein concentration was chosen.

Medical device companies typically do not have the same scale of clinical expenditure and tolerance for failed clinical investigations that large pharmaceutical companies have come to expect, so additional time and effort spent in the nonclinical phase is well spent. While, it is impossible to anticipate every use of the eventual commercial product and generate data to address and answer all of the potential questions, a well-designed nonclinical program can provide a sufficient framework for building a successful clinical program and, ultimately, regulatory approval. Because nonclinical studies can take a long time to design and execute, waiting until the questions are officially asked can lead to long delays and frustration for the eventual users of the approved combination product.

REFERENCES

1. S.P. Cowley and L.D. Anderson, Hernias through donor sites for iliac-bone grafts. *J. Bone Joint Surg. Am.* 54 (1972) 83–101.
2. F.P. Carinella, G.A. De Laria, and R.L. DeWald, False aneurysm of the superior gluteal artery: a complication of iliac crest bone grafting. *Spine* 15 (1990) 1360–1362.
3. F. Escalas and R.L. DeWald, Combined traumatic arterioveneous fistula and ureteral injury: a complication of iliac bone-grafting. *J. Bone Joint Surg. Am.* 59 (1977) 270–271.
4. B. Kahn, Superior gluteral artery laceration: a complication of iliac bone graft surgery. *Clin. Orthop.* 140 (1979) 204–207.
5. A. Zoma, R.D. Sturrock, W.D. Fisher, et al., Surgical stabilization of the rheumatoid cervical spine: a review of indications and results. *J. Bone Joint Surg. Br.* 69 (1987) 8–12.
6. D.J. Weiland and P.C. McAfee, Posterior cervical fusion with triple-wire strut graft technique: one hundred consecutive patients. *J. Spinal Disord.* 4 (1991) 15–21.
7. D.P.K. Chan, K.S. Nhian, and L. Cohen, Posterior upper cervical fusion in rheumatoid arthritis. *Spine* 17 (1992) 268–272.
8. J.G. Edelson and H. Nathan, Meralgia paresthetica: an anatomical interpretation. *Clin. Orthop.* B122 (1977) 255–262.
9. A. DePalma, R. Rothman, G. Lewinnek, et al., Anterior interbody fusion for severe cervical disc degeneration. *Surg. Gynecol.* 134 (1972) 755–758.

10. S. Sacks, Anterior interbody fusion of the lumbar spine. *J. Bone Joint Surg.* 47B (1965) 211–223.
11. R.N. Stauffer and M.B. Coventy, Posterolateral lumbar spine fusion. *J. Bone Joint Surg.* 54A (1972) 1195–1204.
12. S.C. Guha and M.D. Poole, Stress fracture of the iliac bone with subfascial fermoral neuropathy: unusual complications at a bone graft donor site: case report. *Br. J. Plast. Surg.* 36 (1983) 305–306.
13. F. Reale, D. Gambacorta, and G. Mencattini, Iliac crest fracture after removal of two bone plugs for anterior cervical fusion. *J. Neurosurg.* 51 (4) (1979) 560–561.
14. C.S. Ubhi and D.L. Morris, Fracture and herniation of bowel at bone graft donor site in the iliac crest. *Injury* 16 (1984) 202–203.
15. M.B. Coventry and E.M. Tapper, Pelvic instability: a consequence of removing iliac bone for grafting. *J. Bone Joint Surg. Am.* 54 (1972) 83–101.
16. S. Lichtblau, Dislocation of the sacro–iliac joint. *J. Bone Joint Surg. Am.* 44 (1962) 193–198.
17. E.E. Keller and W.W. Triplett, Iliac bone graft: review of 160 consecutive cases. *J. Oral Maxillofac. Surg.* 45 (1987) 11–14.
18. B.N. Summers and S.M. Eisenstein, Donor site pain from the ilium: a complication of lumbar spine fusion. *J. Bone Joint Surg. Br.* 71 (1989) 677–680.
19. S.W.S. Laurie, L.B. Kaban, J.B. Mulliken, et al., Donor-site morbidity after harvesting rib and iliac bone. *Plast. Reconstr. Surg.* 73 (1984) 933–938.
20. J.M. Howard, S.d. Glassman, and L.Y. Carreon, Posterior iliac crest pain after posterolateral fusion with or without iliac crest graft harvest. *Spine J.* 11 (2011) 534–537.
21. E.J. Carragee, C.M. Bono, and G.J. Scuderi, Pseudomorbidity in iliac crest bone graft harvesting: the rise of rhBMP-2 in short-segment posterior lumbar fusion. *Spine J.* 9 (2009) 873–879.
22. M.R. Urist, Bone: formation by autoinduction. *Science* 150 (3698) (1965) 893–899.
23. M.R. Urist, A morphogenetic matrix for differentiation of bone tissue. *Calcif. Tissue Res.* (1970) 98–101.
24. M.R. Urist and B.S. Strates, Bone morphogenetic protein. *J. Dental Res.* 50 (6) (1971) 1392–1406.
25. J. Fiedler, G. Roderer, K.P. Gunther, and R.E. Brenner, BMP-2, BMP-4, and PDGF-bb stimulate chemotactic migration of primary human mesenchymal progenitor cells. *J. Cell Biochem.* 87 (3) (2002) 305–312.
26. K. Akino, T. Mineta, M. Fukui, T. Fujii, and S. Akita, Bone morphogenetic protein-2 regulates proliferation of human mesenchymal stem cells. *Wound Repair Regen.* 11 (5) (2003) 354–360.
27. A. Wilke, F. Traub, H. Kienapfel, and P. Griss, Cell differentiation under the influence of rh-BMP-2. *Biochem. Biophys. Res. Commun.* 284 (5) (2002) 1093–1097.
28. H. Cheng, W. Jiang, F.M. Phillips, et al., Osteogenic activity of the fourteen types of human bone morphogenetic proteins (BMPs). *J. Bone Joint Surg. Am.* 85A (8) (2003) 1544–1552.
29. A. Yamaguchi, T. Katagiri, T. Ikeda, et al., Recombinant human bone morphogenetic protein-2 stimulates osteoblastic maturation and inhibits myogenic differentiation *in vitro*. *J. Cell Biol.* 113 (3) (1991) 681–687.

30. M.M. Deckers, R.L. van Bezooijen, G. van der Horst, et al., Bone morphogenetic proteins stimulate angiogenesis through osteoblast-derived vascular endothelial growth factor A. *Endocrinology* 143 (4) (2002) 1545–1553.
31. K. Takagi and M.R. Urist, The role of bone marrow in bone morphogenetic protein-induced repair of femoral massive diaphyseal defects. *Clin. Orthop. Rel. Res.* 171 (1982) 224–231.
32. K. Sato and M.R. Urist, Induced regeneration of calvaria by bone morphogenetic protein (BMP) in dogs. *Clin. Orthop. Rel. Res.* 197 (1985) 301–311.
33. O.S. Nilsson, M.R. Urist, E.G. Dawson, T.P. Schmalzried, and G.A. Finerman, Bone repair induced by bone morphogenetic protein in ulnar defects in dogs. *J. Bone Joint Surg. Br.* 68 (4) (1986) 635–642.
34. E.E. Johnson, M.R. Urist, and G.A. Finerman, Repair of segmental defects of the tibia with cancellous bone grafts augmented with human bone morphogenetic protein: a preliminary report. *Clin. Orthop. Rel. Res.* 236 (1988) 249–257.
35. E.A. Wang, V. Rosen, P. Cordes, et al., Purification and characterization of other distinct bone-inducing factors. *Proc. Natl. Acad. Sci. USA* 85 (24) (1988) 9484–9488.
36. V. Rosen, J.M. Wozney, E.A. Wang, et al., Purification and molecular cloning of a novel group of BMPs and localization of BMP mRNA in developing bone. *Connect. Tissue Res.* 20 (1–4) (1989) 313–319.
37. A.J. Celeste, J.A. Iannazzi, R.C. Taylor, et al., Identification of transforming growth factor beta family members present in bone-inductive protein purified from bovine bone. *Proc. Natl. Acad. Sci. USA* 87 (24) (1990) 9843–9847.
38. Medtronic Summary of Safety and Effectiveness Data for InFUSE™ Bone Graft. Available at http://www.accessdata.fda.gov/cdrh_docs/pdf/P000058b.pdf. (accessed December 18, 2013).
39. H. Soda, E. Raymond, S. Sharma, et al., Antiproliferation effects of recombinant human bone morphogenetic protein-2 on human tumor colony-forming units. *Anticancer Drugs* 9 (1998) 327–331.
40. H.-W. Liu, C.-H. Chen, C.-L. Tsai, and G.-H. Hsiue, Targeted delivery system for juxtacrine signaling growth factor based on rhBMP-2-mediated carrier–protein conjugation. *Bone* 39 (2006) 825–836.
41. J.I. Dawson and R.O.C. Oreffo, Bridging the regeneration gap: stem cells, biomaterials and clinical translation in bone tissue engineering. *Arch. Biochem. Biophys.* 473 (2008) 124–131.
42. H. Schliephake, Application of bone growth factors: the potential of different carrier systems. *Oral Maxillofac. Surg.* 14 (2010) 17–22.
43. R.A. Meyer, Jr., H.E. Gruber, B.A. Howard, et al., Safety of recombinant human bone morphogenetic protein-2 after spinal laminectomy in the dog. *Spine* 21 (1996) 2689–2697.
44. A.W. Yasko, J.M. Lane, E.J. Fellinger, et al., The healing of segmental bone defects, induced by recombinant human bone morphogenetic protein (rhBMP-2): a radiographic, histological, and biomechanical study in rats. *J. Bone Joint Surg. Am.* 74 (5) (1992) 659–670.
45. M. Bostrom, J.M. Lane, E. Tomin, et al., Use of bone morphogenetic protein-2 in the rabbit ulnar nonunion model. *Clin. Ortho. Relat. Res.* 327 (1996) 272–282.

46. R.D. Welch, A.L. Jones, R.W. Bucholz, et al., Effect of recombinant human bone morphogenetic protein-2 on fracture healing in a goat tibial fracture model. *J. Bone Miner. Res.* 13 (9) (1998) 1483–1490.
47. B.E. Bax, J.M. Wozney, and D.E. Ashhurst, Bone morphogenetic protein-2 increases the rate of callus formation after fracture of the rabbit tibia. *Calcif. Tissue Int.* 65 (1) (1999) 83–89.
48. M.L. Bouxsein, T.J. Turek, C.A. Blake, et al., Recombinant human bone morphogenetic protein-2 accelerates healing in a rabbit ulnar osteotomy model. *J. Bone Joint Surg. Am.* 83A (8) (2001) 1219–1230.
49. J. Wolff, *The Law of Bone Remodeling*, Springer, Berlin, 1986 (translation of the German 1892 edition).
50. T.J. Sigurdsson, M.B. Lee, K. Kubota, et al., Periodontal repair in dogs: recombinant human bone morphogenetic protein-2 significantly enhances periodontal regeneration. *J. Periodontol.* 66 (2) (1995) 131–138.
51. M. Nevins, C. Kirker-Head, M. Nevins, et al., Bone formation in the goat maxillary sinus induced by absorbable collagen sponge implants impregnated with recombinant human bone morphogenetic protein-2. *Int. J. Periodontics Restorative Dent.* 16 (1) (1996) 8–19.
52. D.L. Cochran, R. Schenk, D. Buser, et al., Recombinant human bone morphogenetic protein-2 stimulation of bone formation around endosseous dental implants. *J Periodontol.* 70 (2) (1999) 139–150.
53. S.H. Choi, C.K. Kim, K.S. Cho, et al., Effect of recombinant human bone morphogenetic protein-2/absorbable collagen sponge (rhBMP-2/ACS) on healing in 3-wall intrabony defects in dogs. *J. Periodontol.* 73 (1) (2002) 63–72.
54. H.J. Wilke, A. Kettler, and L.E. Claes, Are sheep spines a valid biomechanical model for human spines? *Spine* 22 (20) (1997) 2365–2374.
55. H.S. Sandu, J.M. Kabo, A.S. Turner, et al., rhBMP-2 augmentation of titanium fusion cages for experimental anterior lumbar fusion. Presented at the 11th Annual Meeting of the North American Spine Society, Vancouver, BC, October 1996.
56. S.d. Boden, G.J. Martin, Jr., W.C. Horton, et al., Laparoscopic spinal arthrodesis with rhBMP-2 in a titanium interbody threaded cage. *J. Spinal Disord.* 11 (1998) 95–101.
57. Just the Facts: Addressing INFUSE® Bone Graft Safety and the Potential for Cancer. Available at http://facts.infusebonegraft.com/hcp/cancer-safety/index.htm. (accessed January 9, 2014).
58. J.K. Burkus, M.F. Gornet, S.d. Glassman, et al., Blood serum antibody analysis and long-term follow-up of patients treated with recombinant human bone morphogenetic protein-2 in the lumbar spine. *Spine* 36 (25) (2011) 2158–2167.
59. H. Zhang and A. Bradley, Mice deficient for BMP2 are nonviable and have defects in amnion/chorion and cardiac development. *Development* 122 (1996) 2977–2986.
60. J.M. Toth, S.d. Boden, J.K. Burkus, J.M. Badura, S.M. Peckham, and W.F. McKay, Short-term osteoclastic activity induced by locally high concentrations of recombinant human bone morphogenetic protein-2 in a cancellous bone environment. *Spine* 34 (6) (2009) 539–550.

INDEX

Drug–Device Combinations for Chronic Diseases, First Edition. Edited by SuPing Lyu and Ronald A. Siegel.